TAKING SIDES

Clashing Views on Controversial
African Issues

D1081923

TAKING SIDES

Clashing Views on Controversial

African Issues

Selected, Edited, and with Introductions by

William G. Moseley
Macalester College

McGraw-Hill/Dushkin
A Division of The McGraw-Hill Companies

For Julia, Ben, and Sophie

Photo Acknowledgment
Cover image: © 2004 by PhotoDisc, Inc.

Cover Art Acknowledgment
Charles Vitelli

Library of Congress Cataloging-in-Publication Data
Main entry under title:
Taking sides: clashing views on controversial African issues/selected, edited, and with introductions by William G. Moseley.—1st ed.
Includes bibliographical references and index.
I. Africa—Politics and government—1989– II. Africa—Social conditions—1989–
III. Africa—Economic conditions—1989– IV. Africa—Civilization—21st century.
1. Moseley, William G. ed. 2. Series.
916
0-07-284517-1
ISSN: 1545-5327

Printed on Recycled Paper

Preface

I began my encounter with the African continent in the 1980s as an agricultural volunteer for the U.S. Peace Corps. I was sent to a rural village of 200 people in southern Mali. Quite unexpectedly, I found myself so intellectually captivated by the people and their use of the land in this community that I went on to work for roughly 10 years as a development professional. While employed by a variety of nonprofit and governmental organizations in Africa (in Mali, Zimbabwe, Malawi, Niger, and Lesotho) and in Washington, D.C., I was forced to grapple with many of the issues presented in this volume. Since becoming a college professor, I have been fortunate to share my fascination for the African continent with, and also learn from, some incredibly bright, insightful, and engaged students. Like other academics who have been bitten by the Africa bug, it is during my Africa course each spring that I find my own and my students' enthusiasm for the subject to be most infectious. It is hoped that this volume will serve as a useful platform for discussions in courses offered in a variety of departments dealing with contemporary African issues, including anthropology, African studies, development studies, economics, geography, history, international studies, political science, and sociology.

Prior to the publication of this volume, I essentially found myself assembling sets of articles that presented contrasting views on key topics in contemporary African studies. This was because I found that many students new to the study of Africa tended to enter into a course on the region with a number of preconceptions. I found that the best way to encourage students to grapple with these preconceptions was to offer readings that both supported and contradicted their initial leanings. Through the process of reading different arguments and discussing them with their classmates, individual students continually surprise me by the degree to which their positions may change during the course of a semester (far more than if I had tried to convince them of the validity of a certain perspective). Even if the perspective of a student does not change, I find there is real pedagogical value in being forced to grapple with both sides of an argument. In fact, I believe I would be doing my students a disservice if I did not let them know when an issue is controversial and expose them to different perspectives on the subject. Of course there is no obvious right or wrong to answer many of these issues, and some of the most rewarding class discussions occur when there is no clear majority of students for one position or the other.

In the introduction to each issue, I try to set the debate in its proper context. Sometimes I do this by providing historical background, while in other instances this is accomplished by contextualizing the arguments within broader intellectual debates or traditions. The readings selected for each of the 20 issues in this volume have been chosen for their relevance to the topic as well as the ease with which they may be understood. The selections hail from a variety of sources, including academic publications, the popular press, and the research reports of nonprofit think tanks. In the postscripts I generally attempt to clarify the arguments made, to raise further questions, to mention other contemporary events that have been similarly debated, and to make some suggestions for further reading.

A word to the instructor An *Instructor's Manual With Test Questions* (multiple-choice and essay) is available through the publisher, and a general guidebook, *Using Taking Sides in the Classroom*, which discusses methods and techniques for using the pro-con approach in any classroom setting, is also available. An on-line version of *Using Taking Sides in the Classroom* and a correspondence service for Taking S*ides* adopters can be found at http://www.dushkin.com/usingts/.

Taking Sides: Clashing Views on Controversial African Issues is only one title in the Taking Sides series. If you are interested in seeing the table of contents for any of the other titles, please visit the Taking Sides Web site at http://www.dushkin.com/takingsides/.

Acknowledgements I am grateful to three students from Macalester College, Cole Akeson, Roland McKay, and Jen Wichmann, who helped me collect articles and pull together the author biographies for this text. I wish to acknowledge Ron Ward and James Burns who provided me with invaluable feedback when I was initially developing a list of controversial African issues. I also must pay tribute to the excellent students in my classes and seminars who continually push me to think anew about many of the issues presented in this volume. I finally thank the editorial staff at McGraw-Hill/Dushkin, particularly Juliana Gribbins and Theodore Knight, for their copyediting, patience, insight, and encouragement.

William G. Moseley
Macalester College

Contents In Brief

Contents

Gavin Kitching, professor of political science at the University of New South Wales, left the field of African studies because he "found it depressing." According to Kitching, Africanist scholars have failed to see Africa's own ruling elites as the principal culprits for the continent's dire predicament. He suggests that we "have to ask what it is about the history and culture of sub-Saharan Africa that has led too its disastrous present." Jeff Popke, a professor of geography at East Carolina University, challenges the notion that "Africa's 'failure' is due to its backward and uncivilized culture." He suggests that Afro-pessimists are prone to ethnocentric assessments of Africa that judge the continent on Western rather than African terms. While African realities may not reflect our own assumptions about modernity, he argues that Africans are pursuing their own vision of development with great skill.

Marcus Colchester, director of the Forest Peoples Programme of the World Rainforest Movement, argues that rural communities in equatorial Africa are today on the point of collapse because they have been weakened by centuries of outside intervention. In Gabon, the Congo, and the Central African Republic, an enduring colonial legacy of the French are lands and forests controlled by state institutions that operate as patron-client networks to enrich indigenous elite and outside commercial interests. Robin M. Grier, assistant professor of economics at the University of Oklahoma, contends that African colonies that were held for longer periods of time tend to have performed better, on average, after independence.

Gerald Scott, an economist at Florida State University, argues that structural adjustment programs are the most promising option for promoting economic growth in Africa and asserts that mismanagement and corruption are responsible for prohibiting economic growth. Macleans A. Geo-Jaja, associate professor of economics and education at Brigham Young University, and Garth Mangum, professor emeritus of economics at the University of Utah, argue that structural adjustment programs and stabilization policies rarely have been effective. Rather, they contend that the available evidence indicates that these policies have "accentuated the deterioration in the human condition and further compounded the already poor economic conditions in many African countries."

Maria Julia, professor of social work at Ohio State University, argues that the work of non-governmental organizations (NGOs) is critical for facilitating the empowerment and development of poor women in Zimbabwe. She makes this point by examining the role of a micro-credit NGO that provides financial assistance, as well as educational and emotional support to female entrepreneurs. Giles Mohan, a lecturer in development studies at the Open University, presents a case study of non-governmental organization (NGO) intervention in northern Ghana. His examination reveals that tensions exist between the northern NGO and its partners, that local NGOs create their own mini-empires of client villages, and that some NGO officers use their organizations for personal promotion.

Dorothy Logie, a general practitioner and active member of Medact, and Michael Rowson, assistant director of Medact, argue that debt is a human-rights issue because debt and related structural adjustment policies reduce the state's ability to address discrimination, vulnerability, and inequality. Debt relief, if channeled in the right direction, could help reduce poverty and promote health. Robert Snyder, an associate professor of biology at Greenville College, counters that debt cancellation will only work if the factors that created debt in the first place are addressed. He uses a case study of Rwanda to demonstrate why political and social change must occur for debt forgiveness to work.

PART 2 AGRICULTURE, FOOD, AND THE ENVIRONMENT 89

Florence Wambugu, CEO of Harvest Biotech Foundation International, argues for the development and use of agricultural biotechnology in Africa to help address food shortages, environmental degradation, and poverty. She asserts that only wealthy nations have the luxury of refusing this technology. In a case study examining attempts to control the parasitic Striga weed, Brian Halweil, a research associate at the Worldwatch Institute, questions whether producing maize that is bio-engineered for herbicide resistance is really the best approach in the African context. He suggests that improved soil fertility management practices and mixed cropping are more appropriate and accessible strategies.

W. Thomas Conelly and Miriam S. Chaiken, both professors of anthropology at Indiana University of Pennsylvania, examine an area in Western Kenya that has very high population densities. Despite the wide variety of sophisticated practices that maintain a high level of agro-diversity, they conclude that intense population pressure has led to smaller land holdings, poorer diet quality, and declining food security. Michael Mortimore, a geographer, and Mary Tiffen, a historian and socio-economist, both with Drylands Research, investigate population and food production trajectories in Machakos, Kenya. They determine that increasing population density has had a positive influence on environmental management and crop production. Furthermore, they found that food production kept up with population growth from 1930 to 1987.

Oliver Maponga, chair of the Institute of Mining Research at the University of Zimbabwe, and Philip Maxwell, professor at the Western Australian School of Mines at Curtin University of Technology, describe a resurgence in the African mining industry in the 1990s after several lackluster decades. They assert that mineral and energy mining can make a positive contribution to economic development in Africa. Sunday Dare, a Nigerian journalist, describes how "much sorrow has flowed" from Africa's resource blessing. While Dare blames African leaders for corruption and resource management, he also implicates transnational corporations (TNCs) as key contributors to this problem. He states that TNCs have acted as economic predators that support repressive African leaders in order to garner uninterrupted access to resources. The result, Dare suggests, is that Africa's "raw materials are still being depleted without general development."

William D. Newmark, research curator at the Utah Museum of Natural History, University of Utah, and John L. Hough, global environment facility coordinator for biodiversity and international waters for the United Nations Development Programme, acknowledge the limited success of integrated conservation and development programs to date in Africa, but see great promise for success in the future. They call for more adaptive management in which activities are monitored, evaluated, and reformulated in an interactive fashion. Roderick P. Neumann, associate professor and director of graduate studies in the Department of International Relations at Florida International University, argues that protected area buffer zone programs have not lived up to their initial intent of greater participation and benefit sharing. Rather, these programs duplicate more coercive forms of conservation practice associated with parks and facilitate the expansion of state authority into remote rural areas.

World Bank economists Kevin M. Cleaver and Götz A. Schreiber argue
that Africa is engaged in a downward spiral of population growth, poor
agricultural performance, and environmental degradation. Academic geographers Thomas J. Bassett and Koli Bi Zuéli, counter that it is dominant
perceptions of environmental change, rather than concrete evidence, that
lie behind the widely held belief that Africa is engaged in an "environmental crisis of staggering proportions."

PART 3   SOCIAL ISSUES   189

Richard A. Shweder, professor of human development at the University of
Chicago, acknowledges the adverse reaction that most Westerners have to
female genital cutting (FGC), but he also notes that women from certain
African countries are repulsed by the idea of unmodified female genitals.
He suggests, "We should be slow to judge the unfamiliar practice of female
genital alterations, in part because the horrifying assertions by... activists
concerning the consequences of the practice... are not well supported with
credible scientific evidence." Liz Creel, senior policy analyst at the Population Reference Bureau, and her colleagues, argue that female genital cutting (FGC), while it must be dealt with in a culturally sensitive manner, is a
practice that is detrimental to the health of girls and women, as well as a violation of human rights in most instances. Creel et al. recommend that
African governments pass anti-FGC laws, and that programs be expanded
to educate communities about FGC and human rights.

Akin Jimoh, program director of Development Communications, a non-
governmental organization (NGO) based in Lagos, Nigeria, argues that
the AIDS epidemic in Africa is linked to a number of factors, including the

high cost of drugs. He describes how some of the big drug companies, in the face of international protests, begrudgingly agreed to lower the price of anti-HIV medications in Africa. "The companies, however, remain steadfast about keeping their patent rights, which would leave ultimate control over prices and availability in their hands." Siddhartha Mukherjee, a resident in internal medicine at Massachusetts General Hospital and a clinical fellow in medicine at Harvard Medical School, asserts that the availability of cheap anti-HIV drugs in Africa, without adequate health care networks to monitor their distribution and use, is dangerous. If such medications are not taken consistently and over the prescribed length of time, new strains of HIV are likely to develop more quickly that are resistant to these drugs. He states that investment in health care infrastructure must accompany any distribution of cheap anti-HIV medications.

Issue 13.   Is "Overpopulation" a Major Cause of Poverty in Africa?   220

Partha S. Dasgupta, a professor of economics at the University of Cambridge, contends that many African families have too many children because the benefits of having an additional child outweigh the costs (especially since the expense of childrearing is subsidized by the greater community). According to Dasgupta, there is a vicious cycle in operation in which overpopulation leads to the depletion of community resources, creating greater poverty and leading individual families to have even more children in an attempt to reverse this poverty. Bernard I. Logan, a professor of geography at the University of Georgia, argues that Africa is not overpopulated when absolute population numbers are compared to the resource base. The problem is that the terms of exchange between Africa and its European trading partners are unfairly set, diminishing the returns that Africans see on their own resources. This leads to a situation where Africans are subsidizing European consumption at the expense of their own livelihoods.

Issue 14.   Is Sexual Promiscuity a Major Reason for the HIV/AIDS Epidemic in Africa?   244

William A. Rushing, late professor of sociology at Vanderbilt University, explains the high prevalence of HIV/AIDS in Africa in terms of how Africans express and give social meaning to sex. He argues that the confluence of a set of sex-related behavioral patterns and gender stratification explains the HIV/AIDS infection rate. According to Rushing, these behavioral patterns include polygamous marriage practices, weak conjugal bonds, the transactional nature of sexual relations, the centrality of sexual conquest to male identity, and sex-positive cultures. Joseph R. Oppong, associate professor of geography at the University of North Texas, and Ezekiel Kalipeni, associate professor of geography at the University of Illinois at Urbana-Champaign, take issue with Rushing's conclusions. They contend that his analysis is Americentric, suffers from overgeneralizations, and problematically depicts Africans as sex-positive (and by implication, promiscuous and immoral). They assert that Rushing's cultural stereotypes are far too general to provide any meaningful insight into the AIDS crisis in Africa. An understanding of historical and contemporary migration patterns, as well as associated phenomena, better explain the spread of the virus.

Grace Bunyi, a researcher at Kenyatta University in Nairobi, Kenya, argues that educating children in their own language is pedagogically more effective, and that mass education in indigenous African languages is the surest way to increase literacy rates and to further develop the human capital necessary for economic development. Véronique Wakerley, a professor of modern languages at the University of Zimbabwe, sees a role for local language instruction at the primary level, especially if the student is unlikely to go further in his or her studies. However, she asserts that instruction in a language such as English is more effective at higher levels because it allows access to the international community.

Richard A. Schroeder, an associate professor of geography at Rutgers University, presents a case study of a group of female gardeners in The Gambia who, because of their growing economic clout, began to challenge male power structures. Women, who were the traditional gardeners in the community studied, came to have greater income earning capacity than men as the urban market for garden produce grew. Furthermore, women could meet their needs and wants without recourse to their husbands because of this newly found economic power. Human Rights Watch, a nonprofit organization, describes how women in Kenya have property rights unequal to those of men, and how even these limited rights are frequently violated. It is further explained how women have little awareness of their rights, that those "who try to fight back are often beaten, raped, or ostracized," and how the Kenyan government has done little to address the situation.

Michael Bratton, professor of political science at Michigan State University, and Robert Mattes, associate professor of political studies and director of the Democracy in Africa Research Unit at the University of Cape Town, find as much popular support for democracy in Zambia, South Africa, and Ghana as in other regions of the developing world, despite the fact that the citizens of these countries tend to be less satisfied with the economic performance of their elected governments. Joel D. Barkan, professor of political science at the University of Iowa and senior consultant on governance at the World Bank, takes a less sanguine view of the situation in Africa. He suggests that one can be cautiously optimistic about the situation in roughly one-third of the states on the African continent, nations he classifies as consolidated democracies and as aspiring democracies. He asserts that one must be realistic about the possibilities for the remainder of African nations, countries he classifies into three groups: stalled democracies, those that are not free, and those that are mired in civil war.

Arthur A. Goldsmith, professor of management at the University of Massachusetts in Boston, examines the relationship between the amount of development assistance given to sub-Saharan African countries in the 1990s and the evolution of their political systems. He suggests that there is a positive, but small, correlation between donor assistance and democratization during this period. He views aid as insurance to prevent countries from sliding back into one-party or military rule. Julie Hearn, lecturer in development studies at the School of Oriental and African Studies, University of London, investigates democracy assistance in South Africa. She critically examines the role assigned to civil society by donors, questioning the "emancipatory potential" of the kind of democracy being promoted.

Robert I. Rotberg, director of the Program on Intrastate Conflict and Conflict Resolution at Harvard University's John F. Kennedy School of Government, holds African leaders responsible for the plight of their continent. While he admits that Africa's failure to develop in the postcolonial period has many causes, he suggests that "the visible hand of individual leaders can also be discerned." Rotberg concludes that a large part of the problem is that absolute power corrupts, and that there are limited checks and balances to curb this tendency in Africa. In this regard, he states that "Mugabe's mayhem" was aided and abetted by an underdeveloped civil society and the fact that the rest of the world has failed to judge Zimbabwe's president more harshly. Arthur A. Goldsmith, professor of management at the University of Massachusetts in Boston, suggests that African leaders are not innately corrupt, but are responding rationally to incentives created by their environment. He argues that high levels of risk encourage leaders to pursue short-term, economically destructive policies. In countries where leaders face less risk, there is less perceived political corruption.

Issue 20.   Are International Peacekeeping Missions Critical to Resolving Ethnic Conflicts in African Countries? 358

John Stremlau, professor and head of the Department of International Relations at the University of Witwatersrand, South Africa, argues that far too little is being done to check conflict in Africa and that the United States needs to do more in this regard. According to Stremlau, "[w]hile preventing conflict in Africa is primarily a task for Africans... the 1990s showed that outside help is needed." William Reno, associate professor of political science at Northwestern University, contends that no peacekeeping is better than bad peacekeeping. In his discussion of the failed Lomé Peace Accords, a settlement negotiated between warring parties in Sierra Leone, he notes that "[m]any Sierra Leoneans regarded positions taken by the UN and foreign diplomats who stressed reconciliation as offensive." As opposed to the more bureaucratic peacekeeping approaches taken by the United States and the UN, he lauds the hands-on tactics of the British.

Introduction

Interpreting African Issues: Commentators, Scholars, Policymakers

William G. Moseley

West Africa is becoming the symbol of worldwide demographic, environmental, and societal stress in which criminal anarchy emerges as the real "strategic" danger. Disease, overpopulation, unprovoked crime, scarcity of resources, refugee migrations, the increasing erosion of nation-states and international borders, and the empowerment of private armies, security firms, and international drug cartels are now most tellingly demonstrated through a West African prism.

Robert Kaplan, "The Coming Anarchy," *The Atlantic Monthly*
(February 1994)

This is our most popular African adventure safari, also ideal for adrenaline junkies! This African adventure starts in Cape Town and travels through the Cedarberg Mountains and onwards to the Fish River Canyon. In Namibia we visit the Sossuvlei Dunes: safari game viewing, dune boarding and quad biking are a few of the adventure activities that can be enjoyed at this stage of our trip. We spend the night in a Bushman Village where you can purchase local African crafts, before flying on to enjoy a classic African wildlife safari in the Okavango Delta & Chobe National Park. Then it's off to Victoria Falls, where the adventure continues with optional activities such as white-water rafting, micro lighting, bungi jumping, African elephant-back riding, canoeing, and a horse riding safari to name but a few! So join us for our trip between Cape Town and Victoria Falls: the ultimate African adventure safari.

WhichWay Adventure Company (2002)

Africa . . . has been running in reverse for the past two decades. Things are so bad . . . that Americans are even willing to allow themselves to be rescued by the French military. That edifying spectacle took place Monday, as French helicopters began airlifting stranded Americans from the besieged Liberian capital of Monrovia to a French ship offshore . . . Meanwhile, French troops are struggling to contain civil wars in Congo and

> *Ivory Coast, rioters are protesting the despotic rule of Robert Mugabe in Zimbabwe, and the president of Liberia has just been indicted for war crimes committed in neighboring Sierra Leone. And we still haven't mentioned the impact of HIV-AIDS, which has become the continent's silent Holocaust.*
>
> David Ignatius, "Turning Africa Around," *The Washington Post*
> (June 10, 2003)

In the Popular Press and Public Imagination: From Anarchy to Exotic Adventure

Images and descriptions in the popular press suggest that the African continent is a troubled land where corruption, ethnic warfare, poverty, hunger, environmental destruction, and pestilence prevail. Some have even suggested that Africa is a lost cause, asserting that the continent should be "written off" by international development organizations. Meanwhile, commercial tour operators also hawk the region as a place of high adventure and exoticism. Even quasi-scholarly publications such as the *National Geographic* magazine often promote a vision of a primitive or wild Africa. What these popular and commercial descriptions hold in common is the level of superficiality and one-sidedness. Yes, bad things do happen in Africa, and there is beautiful scenery to be seen, but this is only one part of a complex and highly varied picture. It is the apparent unwillingness (or laziness) of popular commentators to provide a more nuanced view of an enormous continent that is often frustrating to scholars of Africa (or Africanists).

Africa is, after all, a place of extraordinarily diverse, vibrant, and dynamic cultures. Since the early 1990s, no other continent has seen more dramatic improvements in human rights, political freedom, and economic development—from the overthrowing of apartheid in South Africa to the revitalization of economies in countries such as Ghana and Uganda. Although environmental threats are real, African societies have proven their capacity, when given a chance, to use resources with sustainability. Some conservation efforts in Africa have even become models for progressive community-based resource management in Western societies. The importance of human relations, family, and good neighborliness in many African societies also stands in stark contrast to the more closed and individualistic tendencies in a number of Western settings.

Contemporary African Issues and African Studies

While a variety of disciplinary departments offer courses that deal with contemporary African issues, it is notable that there are now a number of interdisciplinary programs and departments devoted to African and Afro-American studies. In the United States, the first African studies programs were established at Northwestern University (1946) and Boston University (1953). A number of other large programs were started in the early 1960s at universities such as Michigan State University, the University of Wisconsin, the University of California at Los Angeles, the University of Indiana, and Ohio State University (*African Studies and the Undergraduate Curriculum*, Patrica Alden, David Lloyd and Ahmed Samatar, eds., Lynne Rienner, 1994). By the mid-1990s there were approximately 330 programs of African and Afro-American studies in the United States (*Directory of African and Afro-American Studies in the United States*, African Studies Association, 1993). African studies is as or more developed at institutions in other areas of the world. While the names of these organizations are too numerous to list, there are over 1,800 academic institutions, research bodies, and international organizations involved in African studies research in all parts of the world (*International Directory of African Studies Research*, 3rd ed., Philip Baker, ed., Hans Zell Publishers, 1994). In the United States, over 1,300 scholars from a broad array of disciplines meet each year at the annual meeting of the African Studies Association to present their research findings and debate key African issues. African studies associations and societies also exist in a number of other countries (e.g., Australia, Canada, France, Germany, India, Japan, Netherlands, South Africa, Spain, Sweden, Switzerland, United Kingdom).

Prior to the establishment of African studies as a recognized interdisciplinary field of study, some scholars encountered resistance from quarters of the academic establishment that perceived Africa to be lacking in history or otherwise unworthy of academic investigation. African studies has grown to be a dynamic realm of inquiry that regularly contributes to the broader academic discourse. Africanist scholars have tested the validity of widely accepted notions in the African context. They also developed new theories that have influenced thinking in other regional contexts. Those disciplines contributing to African studies may roughly be divided between the humanities (mainly history and literature), the social sciences (anthropology, archeology, education, geography, political science, sociology), and the physical sciences (physical geography, ecology, botany). In the North American context, African studies meetings tend to be dominated by the humanities and social sciences, with historians and political scientists attending in the largest numbers (perhaps due to the sheer size of their disciplines). This volume largely deals with controversial African issues in the social sciences, although some of the questions have a significant historical or environmental dimension that pulls on literatures from the humanities or physical sciences. The number of historical issues that could be included in a volume of this nature are so numerous that they easily could constitute a separate book.

The burgeoning field of African studies has spawned a number of academic journals, examples of which include *Africa*, *Africa Today*, *African Affairs*, *African Studies Review*, *Cahiers d'Etudes Africaines*, *Canadian Journal of African Studies*, *Journal of African History*, *Journal of Modern African Studies*, *Journal of Southern African Studies*, and *Review of African Political Economy*. There are also a number of more policy-related or popular media magazines that focus on Africa, including *Africa Analysis*, *Africa Confidential*, *Africa Now*, *Afrique Express* (French), and *Jeune Afrique* (French).

The Academy and Area Studies

For roughly the past century, the North American academy has been organized largely along disciplinary lines. The disciplinary approach encourages focused investigations from one perspective across a range of geographies. Area studies complements disciplinary inquiry by promoting multi-perspective investigations of focused regions. In other words, the region or area represents a different framework around which to organize knowledge. Recent world events have revealed that there is a dearth of regional experts within the academy, government, and the private sector. The ideal scholar-practitioner may therefore be someone who not only has a firm grasp of the methods and theories of a particular discipline, but who also has a broader understanding of the issues and challenges (cutting across several disciplines) facing a particular region of the world.

Policymakers and Development Professionals

In addition to the media and the academy, the other major sphere where African issues are framed and examined is in the offices of government bureaucrats, global policymakers, and non-governmental organizations. These include individuals within the bilateral development agencies (e.g., U.S. Agency for International Development, British Department for International Development, Cooperation Française, Canadian International Development Agency), international financial institutions (e.g., World Bank, International Monetary Fund, World Trade Organization), UN agencies (e.g., Food and Agricultural Organization, World Health Organization), and nonprofit communities (e.g., CARE, Oxfam, Save the Children). While these organizations contribute to and influence discourse, their role is slightly different than the media and the academy in that they are where rhetoric is transformed into programmatic reality.

While I present the media, the academy, and development institutions as separate spheres of debate on Africa, there is, in fact, a considerable amount of crossover in thinking, not to mention personnel, between each realm. Policymakers and development professionals do not make decisions in a vacuum; instead they are influenced by public opinion (shaped by the media) and the latest academic research. Academics and policymakers attempt to influence public opinion through the media (via op-eds and interviews with the press). Governments may seek to shape academic findings by funding research projects or hiring scholars as consultants. Scholars may agree to consult on development issues because they believe this is a way for them to influence decision making. Of course, there is also a certain amount of personnel exchange between the different sectors (e.g., it is to think tanks, multilateral development agencies, and universities that a number of foreign service and aid advisors flee each time the party in power changes in Washington, Ottawa, or London). Despite these exchanges, there remains a difference in tone, timeliness, depth, and emphasis in analysis for each of these spheres as they cater to somewhat dissimilar audiences, standards of knowledge production vary, and the distribution of power (i.e., the power to have one's ideas heard) differs within each forum.

Major Themes

If this book has one overarching theme it is African development. I interpret development in the broadest sense of the term, going beyond conventional measures of economic progress, to embrace processes occurring at a variety of spatial and temporal scales that allow people to meet their full potential. When viewed in this manner, nearly all of the issues presented in this volume pertain in one way or another to development. In order to give the volume a more accessible format, it has been organized into four thematic categories of contemporary African issues that are most oft debated by Africanists in the social sciences, policymakers, and media commentators: 1) development; 2) agriculture, food, and the environment; 3) social issues; and 4) politics, governance, and conflict resolution. While I present these themes as distinct, the reader should understand that there are often a number of connections that exist between issues in different parts of the book.

Development

The nature of and approach to development in the African context is highly contentious. Commentators, scholars, and policymakers argue, for example, over the perceived failure of Africa to develop, the impact of the colonial experience on contemporary events, the influence of global economic structures on African development patterns, and the role of the state versus the private and nonprofit sectors in the development process.

The perceived lack of development in Africa is often articulated in terms of externalist and internalist explanations. Scholars evoking externalist (or structuralist) explanations suggest that contemporary development patterns cannot be properly understood without an understanding of historical patterns of resource extraction and political control, as well as the position of Africa within the global economic system. In particular, these explanations often look to problematic colonial legacies to elucidate contemporary economic distortions. Scholars emphasizing internalist explanations tend to look to local phenomenon (corruption, mismanagement, incompetence, nepotism, ethnic allegiances, regional ties, obligations to the extended family, and patron-client relationships) to explain difficulties on the development front. While a number of scholars draw on both sets of explanations, externalist/structuralist appraisals of internalist positions range from accusations of spatial and temporal myopia to "blaming the victim." Internalist leaning critics suggest that the structuralists are apologists who deny Africans' "agency" (i.e., the ability to influence contemporary events) and responsibility, thereby fostering a sense of victimization, helplessness, and further mismanagement.

Closely related to this externalist/internalist debate is one concerning the most appropriate approach to development in Africa. The structuralists evoke dependency theory and world systems theory to suggest that, even though the colonial era has ended, historical patterns of resource extraction persist. In many instances, African nations are supplying cheap commodities (minerals, oil, lumber, cotton, coffee, cocoa, tea, and sugar) to Europe and North America in exchange for relatively expensive manufactured goods. As such, the participation of African nations in the global economy under current conditions leads to a process of underdevelopment. The best way to avoid this trap, according to the structuralists, is to diversify the national economy by producing imported goods at home, an approach also known as import substitution.

The general failure of import substitution and the related Third World debt crisis led the international financial institutions (especially the World Bank and the International Monetary Fund) to begin pushing for the structural adjustment of African economies beginning in the early 1980s. With the basic aim of balancing the national budget and spurring economic growth, these programs obliged African government to privatize state-owned enterprises, devalue local currencies, cut government programs, and expand exports. The uneven success and social costs of these programs led to bitter debates in academic and policy circles. Many contend that the role of the state in the development process has been underestimated by neoclassical economists. Although contested, debt forgiveness increasingly is seen as one way to restart development and lift the burden of past indiscretions (on the part of African governments and donors).

Another key aspect of African development is the rise in the number and prominence of non-governmental organizations (NGOs) since the 1980s. Bilateral and multilateral donors increasingly bypass the African state by providing funds to NGOs for project and program implementation. NGOs (both local and international) are perceived to be more efficient and more in tune with the needs of the local population. Critics argue that NGOs can be equally autocratic, patronizing, and detached from the local population.

Agriculture, Food, and the Environment

Since the global media focused attention on large-scale droughts in the early 1970s and mid-1980s, famine and environmental destruction in Africa have loomed large in the public imagination. Key debates have centered on how best to increase food production, whether or not the continent's population is growing too quickly for its agricultural base, the role of mineral and energy resources in African development, the resolution of conflicts between local people and wildlife, and whether or not the continent is facing a deforestation crisis.

A fundamental debate persists as to whether or not famine and food insecurity in Africa is the result of underproduction or the maldistribution of food. Many development-assistance programs, including the international network of crop development institutes under the aegis of the Consultative Group on International Agricultural Research (CGIAR), are predicated on the assumption that Africa's food problems will be resolved by increases in food production. Economist and Nobel laureate, Amartya Sen, has argued that famine rarely results from an absolute shortage of food but instead evolves from the inability of poor households to access available supplies.

Related to the aforementioned debate is a long-standing discussion concerning the relationship between food production and population growth. The eighteenth-century parson, Thomas Malthus, asserted that human populations would inevitably grow more quickly than food production, ultimately leading to famine. Contemporary neo-Malthusians, such as Paul Ehrlich, have similarly argued that urgent measures are needed to control population in order avert disaster. Ester Boserup (1910–1999), who was a member of the UN Committee of Development Planning, countered that population density controls the level of food production, rather than food production quantities setting population thresholds. In recent years, some of the best case studies (both pro-Boserupian and pro-Malthusian) on the relationship between population growth and food production have come out of Africa.

The *tragedy of the commons* is another prominent paradigm that has been used to explain resource problems in Africa. According to this theory, commonly held natural resources will tend to be overexploited by individuals seeking to maximize personal gain. The solution advocated by economists is to privatize common resources, as it is believed that the private owner will more carefully husband environmental resources over the long term. The commons is actually a misnomer in the African context because many African communities have effectively managed commonly held natural resources through traditional mechanisms of control and enforcement. It is when these traditional mechanisms break down, and the commons become open-access resources (or resources where there is no effective management authority) that problems develop. The loss of wildlife in Africa has often been described as an open-access resource problem. Here it is assumed that communities bordering parks exploit wildlife because the state is seen as an ineffective or illegitimate manager. Integrated conservation and development programs seek to resolve this problem by giving local people a financial stake in the conservation of these animals. Critics argue that this approach further restricts the development of these communities. Deforestation in Africa has also been characterized as an open access resource problem. Environmentalists argue that drastic measures are needed to protect the continent's dwindling forest resources from overuse by poor households. Critics suggest that the assertion of widespread deforestation is based on bad science and a historical misreading of environmental change in many instances.

Finally, the notion that a rich natural resource–base may play a key role in the development of some African nations is hotly contested. Critics contend that bountiful resources may actually be a "curse" because they tend to inhibit economic diversification, fall under the control of unscrupulous government officials, and attract predatory multinational corporations. Proponents of mineral and energy resource development point to the positive contributions of mining in several developed-country economies.

Social Issues

Perhaps more than any other set of contested African issues, those pertaining to the social sphere tend to provoke deep-seated emotional responses. Different aspects of the AIDS crisis in Africa, a perception that the continent is overpopulated, female genital cutting, and the position of women in African societies all have attracted considerable media attention and scrutiny in recent years.

Although less noticed, there is also an ongoing debate about the use of local versus European languages in the classroom.

Many of these issues get at a deeper debate between those who assert that there are certain universal rights and wrongs, and cultural pluralists who believe that we need to evaluate practices within their own cultural context. Advocates of the universality of norms disparage defenders of certain African practices as cultural relativists. Cultural relativism is cast as problematic because it may be used as an excuse to say that anything goes. In contrast, cultural pluralists assert that there are separate and valid cultural and moral systems that may involve social mores that are not easily reconcilable with one another. Cultural pluralists are not necessarily cultural relativists as many would argue that everything does not go (e.g., murder is wrong). The challenge for cultural pluralists is to determine if a practice violates a universal norm when it is viewed in its proper cultural context (rather than in the cultural context of another). The result of this deep philosophical divide is that we often see Western feminists pitted against multiculturalists (two groups that frequently function as intellectual allies in the North American context) over some of the issues addressed in this section of the book.

Some of the debates in this section also pit political economic, or structuralist, explanations against those that emphasize internal factors. As such, AIDS or corruption may be seen as a result of broader economic processes (migration in former instance and economic incentive structures in the latter) or personal failings and problematic cultural practices. More positive, culturally imbued internalist explanations might assert that what is corruption in one culture is not necessarily corruption in another.

Politics, Governance, and Conflict Resolution

The terrain of politics, governance, and conflict resolution is simultaneously one of the most hopeful and distressing realms in contemporary African studies. While more contested elections have been held in the last 10 years than at any other time in the postcolonial period, the African continent also suffers from more instances of civil strife than other world regions. Key debates have focused on the success or nonsuccess of multiparty democracy in the African context, the role of foreign assistance in promoting democracy, reasons for corruption among African officials, and how best to resolve ethnic conflicts in African countries.

An underlying theme related to several of these questions concerns the most appropriate form of governance in the African context. Proponents of multiparty democracy assert that this form of government will promote economic growth and minimize ethnic tensions. They also believe that a healthy civil society will serve as a check on corruption and other government excesses. Increasingly, foreign assistance for a variety of projects is conditional upon certain types of democratic reform.

Other scholars and African leaders see the imposition of democracy in Africa as a form of neoimperialism. They suggest that the problem with multiparty democracy in many African countries is that it leads to the formation of too many political parties, each with a regional or ethnic outlook, and none representing the interests of the country as a whole. Furthermore, some Afrocentrists assert that the one-party state is more consistent with traditional consensus decision making that occurs at the village level. They maintain that the process of competitive elections is a foreign notion that is divisive in the African context. They have also argued that democracy may actually inhibit economic growth. According to this argument, the problem with democracy is that it does not allow leaders to make tough economic decisions, such as the austerity measures required under structural adjustment.

World Bank Group for Sub-Saharan Africa

The World Bank Group for Sub-Saharan Africa site includes annual reports, publications, speeches, and other sources of information about issues including rural development, education, and the incorporation of indigenous knowledge into development.

http://www.worldbank.org/afr/

United Nations Development Programme

The United Nations Development Programme provides information on UN development work in Africa. A wide range of topics is covered, including poverty and globalization.

http://www.undp.org/dpa/publications/regions.html#Africa

Economic Commission for Africa

The United Nations' Economic Commission for Africa site provides information on the regional integration of African economies, including information on regional economic organizations. The organization is attempting to reform and integrate African economies to improve the quality of life on the continent.

http://www.uneca.org/index.htm

United States Agency for International Development (USAID) in Africa

USAID in Africa details African development from the perspective of USAID, the bilateral development assistance arm of the U.S. government. Information is available by country, topic, or date and ranges from short articles and press releases to full publications.

http://www.usaid.gov/regions/afr/

PART 1

Development

*W*hat constitutes "development" in the African context very much
remains an open question. Since the majority of African nations gained
independence in the 1960s, how best to pursue this process has been an
oft-debated question. The role of the state versus the commercial sector
versus nonprofit organizations in development is highly contested. Con-
troversies have also raged over the extent to which Africa's progress on the
development front has been influenced by its precolonial and colonial histories, as well as its historic and current position in the global economic
system.

- Is Africa a Lost Cause?

- Has the Colonial Experience Negatively Distorted Contemporary
  African Development Patterns?

- Have Structural Adjustment Policies Been Effective at Promoting
  Development in Africa?

- Are Non-Governmental Organizations (NGOs) More Effective at
  Facilitating Development Than Government Agencies?

- Should Developed Countries Provide Debt Relief to the Poorest,
  Indebted African Nations?

ISSUE 1

Is Africa a Lost Cause?

YES: Gavin Kitching, from "Why I Gave Up African Studies,"
African Studies Review and Newsletter (June 2000)

NO: Jeff Popke, from "'The Politics of the Mirror': On Geography
and Afro-Pessimism," *African Geographical Review* (December 2001)

ISSUE SUMMARY

YES: Gavin Kitching, professor of political science at the University
of New South Wales, left the field of African studies because he
"found it depressing." According to Kitching, Africanist scholars
have failed to see Africa's own ruling elites as the principal culprits
for the continent's dire predicament. He suggests that we "have to
ask what it is about the history and culture of sub-Saharan Africa that
has led to . . . its disastrous present."

NO: Jeff Popke, a professor of geography at East Carolina University,
challenges the notion that "Africa's 'failure' is due to its backward
and uncivilized culture." He suggests that Afro-pessimists are prone
to ethnocentric assessments of Africa that judge the continent on
Western rather than African terms. While African realities may not
reflect our own assumptions about modernity, he argues that
Africans are pursuing their own vision of development with great
skill.

Perhaps more so than any other world region, the African continent suffers
from a public relations problem. The continent is portrayed regularly in the
Western media as an area plagued by wars, disease, famine, corruption, incom-
petence, and drought. As a result, Africa is often perceived by the rest of the
world as a "lost cause." Emblematic of this genre of portrayals is Robert Kaplan's
February 1994 article in *The Atlantic Monthly* entitled, "The Coming Anarchy:
How Scarcity, Crime, Overpopulation, Tribalism, and Disease Are Rapidly
Destroying the Social Fabric of Our Planet."

Many Africans and Africanist scholars are deeply disturbed by one-sided
journalistic representations of the region. While admitting that Africa suffers

2

from a number of pressing problems, they also point to several positive developments in recent years. They argue that Africa is not only a place of extraordinarily diverse, vibrant, and dynamic cultures, but that no other continent has seen more dramatic improvements in human rights, political freedom, and economic development in the last 15 years. Advances have ranged from the overthrowing of apartheid in South Africa, to the rise of multiparty democracy in Mali, to the revitalization of economies in countries such as Ghana and Uganda. In short, many Africanists assert that sweeping generalizations are made about the entire continent based on events in a few areas. These gross simplifications overshadow many of the positive developments that have occurred in recent years.

In his selection, Gavin Kitching riled a number of Africanist scholars by suggesting that they have failed to see Africa's own ruling elites as the principal culprits for the continent's dire predicament. He suggests that we "have to ask what it is about the history and culture of sub-Saharan Africa that has led to . . . its disastrous present."

As the title of his selection suggests, Kitching left the field of African studies in 1983 because he "found it depressing." Only after repeated requests from the editor of the *African Studies Review and Newsletter* did he agree to write his article in 2000. Initially a little-noticed piece in an obscure journal, the article began to attract attention when it was reprinted in the online journal *Mots Pluriels*. Kitching's ideas, and reactions to them, were most recently featured in a March 2003 edition of the *Chronicle of Higher Education* as well as a summer 2003 special issue of the *African Studies Quarterly*.

In his selection, Jeff Popke refers to those who traffic in negative stereotypes about Africa as "Afro-pessimists." According to Popke, many outsiders have assessed Africa in an ethnocentric manner. He notes that, "Africa is clothed in a set of meanings from outside the continent." In other words, Africa is often judged based on the cultural norms of Europe or North America rather than on its own terms. Europe, he suggests, has long used Africa as a "mirror." Through this mirror, things European are defined as positive and advanced in relation to a "backward" Africa. Popke states that even though African realities may not reflect our own assumptions about modernity, Africans are pursuing their own vision of development with great proficiency.

Gavin Kitching

Why I Gave Up African Studies

In a word, I gave up African studies because I found it depressing. But that is hardly an explanation, even if it is an emotionally precise description of what occurred. To explain my giving up African studies I have to say why I came to find the activity depressing. This requires a little history—both personal history, or autobiography, and history of Africa.

I began to be interested in Africa whilst an undergraduate student of economics and politics at Sheffield university (1965–8) and I began my doctoral work in African politics in Oxford in 1969. That means that I first entered this field of study when the hope and optimism generated by Africa's independence from colonialism was still in the air. My doctoral research was conducted in Tanzania, and that was not accidental of course. As an undergraduate I had been deeply moved and impressed by the writings of Julius Nyerere, and like many young intellectual radicals of that period I was eager to see Nyerere's experiment in a 'Third Way' African socialism at first hand. I undertook doctoral field-work in Arusha district, Tanzania between 1969 and 1971, undertook further field research on peasant agriculture and rural stratification in Kenya in 1972–3, and returned to East Africa at intervals between 1973 and 1983. I also conducted shorter research and consultancy visits to Ghana, Senegal and Zambia in the late 1970s and early 1980s. My last visit to Africa—to Kenya and Tanzania—occurred in 1983.

This history means that, like many other Africanist scholars of my generation, I lived and worked through the period when optimism and hope in and about Africa were replaced by pessimism and cynicism. A particular memory comes to mind. I am in a bar on the campus of the former Kwame Nkrumah university in Kumasi, Ghana in early 1980. A number of African colleagues are there too, including a lecturer from the university's Department of Law. He has had rather too much to drink. He is watching the news on the television above the bar. A news reader announces the declaration of the formal independence of Southern Rhodesia, now to be called Zimbabwe. There is the conventional film footage of these occasions—cheering crowds, brass band, new flag rising on floodlit flag pole. My colleague smiles drunkenly, murmurs "poor sods!" loud enough for all in the bar to hear, and staggers from his bar stool and out of the

From Gavin Kitching, "Why I Gave Up African Studies," *African Studies Review and Newsletter,* vol. 22, no. 1 (June 2000). Copyright © 2000 by Gavin Kitching. Reprinted by permission of the author. The *African Studies Review and Newsletter* is now retitled as *The Australasian Review of African Studies (ARAS).*

door. I am shocked. But, significantly perhaps (at least in retrospect) nobody in the bar, African or European, protests. And this is not an isolated occurrence. In Ghana in 1980, in Kenya and Tanzania in the late 1970s/early 1980s, it is not difficult to meet older African people who will tell one—and totally unbidden— that "we were better off when the British were here". I even hear it from middle aged men in the Murang'a district of Kenya's Central Province, men who had been Mau-Mau fighters.

Radical Perspective

Of course I had a ready-made radical perspective into which I could accommodate and by which I could explain away such uncomfortable experiences and observations—the dependency perspective. What after all was so surprising about all this? Had I not said, in my own book on Kenya, that the country was in the political grip of a dependent African 'petit-bourgeoisie' which, by its very dependent nature, was unlikely to be an effective agent of 'real' economic development in Kenya. And had not a raft of other scholars—European and African—said similar things about a raft of other newly-independent African states. So, in such circumstances, why was it at all surprising that popular disenchantment with such elites within Africa was so widespread—up to and including a retrospective nostalgia for colonialism?

Yes. But this same perspective had been predicated on the view that these 'dependent' or 'neo-colonial' governing elites were agents of 'imperialism' or of 'transnational capital' in Africa. And as the 1970s turned into the 1980s and the political fragility and economic involution of so many African states became palpable this notion itself seemed ever more questionable. As I put it in a number of public presentations in the early 1980s—"if the ruling elites of Africa are seen as managers or agents for western capitalism or imperialism, one can only say that the latter should get itself some new agents. For the ones it has seem remarkably inefficient." In other words, the decay of Africa's production structures and economic infrastructure, the continent-wide slump in investment (domestic and foreign), the endemic inflation and balance of payments problems, the sharp absolute falls in real standards of living of the mass of African people and (accompanying all this) the massive levels of governmental corruption and persistent breakdowns in civil peace and order in so many states—all this seemed hard to square with a basically functionalist perspective which had the governing elites of Africa somehow doing the bidding of international capitalism. And it was all the harder to maintain this position when so many of the official spokespeople for that capitalism (the IMF, the World Bank, corporate executives with African investments) far from endorsing the activities of their supposed 'agents' were endemically critical of the failure of African elites to provide domestic environments in which any form of capital investment could be secure and profitable.

This is not to deny, of course, that there were powerful factors beyond the activities of Africa's governing classes which also pushed Africa into political instability and economic decline from the mid-1970s onwards—everything from the oil shocks which brought an end to the post-war long boom in the world economy and led to at least the onset of Africa's debt problem, through the Cold War flooding of the continent with arms and various forms of military subversion, to the global arms trade and traders. But it is to say that such factors did not seem to account, either individually or even in conjunction, for the particularity of the African situation. For all these factors had impacted on other parts of the South or Third World too without effects as catastrophic as those to be observed in Africa. I came up with a hackneyed but useful analogy. The African ship of state was ploughing through heavy international seas, yes. But that only strengthened the need for an excellent captain and navigator at the helm and a well disciplined crew. But as it was the captain and all his officers seemed to be drunk or absent from the bridge and the crew engaged in various forms of mutiny. No wonder the ship had run aground.

So, and leaving aside hackneyed analogies, I was returned again and again to an overwhelming question, which can be phrased in various forms, but always remains essentially the same question. Why are some governing elites economically progressive and others not? Why are some ruling classes exploitative, selfish and corrupt but also genuine agents for national economic and social improvement, while others are just exploitative, selfish and corrupt? Why are some states 'developmental' and others not? Of course these issues have come to the forefront of African studies since I gave it up, even if they are dealt with in vocabularies remote from dependency theory (which was declining rapidly in popularity as I left the field). So now we hear a lot more about African 'cleptocracy' or 'modern patrimonialism', about 'rent seeking behaviour' among state elites and about 'state failure' generally in Africa and the need for market-led 'structural adjustment'. But though the question may be posed in new forms and though helpful new descriptions of African state functioning (mainly derived from the public choice paradigm) have emerged, I see no significant progress made in answering the question 'why?' Why have African governing elites been particularly prone to behaving in ways which are both economically destructive of the welfare of the people for whom they are supposedly responsible and which have led—at the extreme—to forms of state fission, (civil war etc) collapse or breakdown?

At this point I must make a confession. After over thirty years of studying this question (including the last ten years in Australia giving far greater attention to the 'developmental states' of SE Asia) my first and predominant answer to it is still that I do not really know (or not in any hard or definite sense of 'know'). I have some suspicions about where an answer might lie. The lack of economic or social 'depth' of the colonial experience over most of sub-Saharan Africa; the fragile and inexperienced stratum of educated African people which that 'shallow' experience left behind; the chronic lack of 'fit' between Africa's indigenous structures of ethnicity and the so-called 'nation-state' structures which colonialism bequeathed and which the first generation of nationalist leaders (probably very unwisely) opted to retain; the inability (probably derived from the above) of Africa's governing elites to identify themselves with, or as part of, an 'imagined community' of the nation states which they nominally superintend; their consequent failure to manifest any sense of loyalty or of a duty of care or responsibility to the people who make up these entities; their tendency to restrict such a sense of moral duty only to some particular ethnic or other sub-group of their citizens (thus increasing both economic and social polarities among citizens and the likelihood of the political fision of the state, especially in economically desperate times). I could add further suspicions to the list if asked. But none of them are certainties or anything like certainties, and again it is comparative study which muddies the water. Because, of course, you could have said all of the same things about the post-colonial elites of Indonesia or Malaysia and some of the same things about the elites of Thailand, South Korea or Taiwan. This reflection only leads me—and rather flatly—to the conclusion that it must be the concatination of these domestic elite characteristics with the particularly weak global economic situation of sub-Saharan Africa which was the fatal two-sided recipe for developmental failure. And that may be true, but, as I say, I still have no certainty that it is or about how precisely to weight the relative importance of the list of usual suspects above.

But why did all this lead me to give up African studies? Well it compounded the depression. I was depressed, that is to say, both by what was happening to African people and by my inability even to explain it adequately, let alone do anything about it. And also, I was depressed by the polarization, within the world of African studies as it was in the early 1980s, between those advocating what were called 'internalist' explanations of Africa's problems and those who continued to favour 'externalist' explanations. I was depressed because advocates of the latter view often charged advocates of the former with "blaming African people" for Africa's parlous state, a charge which seemed at once incoherent, even in its own terms (was colonialism in Africa an 'internal' or 'external' factor, for example?) and, above all, enormously hurtful. For of course the vast majority of African people are the victims, often the horrific victims, of Africa's plight, not its perpetrators in any sense, and I, at least, would never wish to deny that. But I would also wish to assert that, though the political elites of Africa change their social composition (often quite significantly and rapidly) their economically and politically destructive behaviour mostly does not change with their personnel. And that certainly suggests—at least to me—that there are broader social/cultural factors (in the mass milieu from which such elites are recruited) continually making for, and reinforcing, this behaviour. So, we do have to ask what it is about the history and culture of sub-Saharan Africa that has led (at any rate in part) to its disastrous present. But that can only be construed as "blaming African people" or, more broadly as "blaming the victim", if a guilt-ridden confusion is made between context and agency. That is, it may be in the broadest historical and cultural context of African society that we find the clue to destructive political elite behaviour. But that does not make that context an agent of that destruction. Nor (therefore) does it prevent political and moral responsibility for their actions being sheeted home to narrow and privileged sub-groups or classes of African people (and not 'African people' as a whole) classes which have been the agents—even if not the sole agents—of that destruction.

Role for Australasian Africanists

In short, and to conclude, I left African studies because what was happening to a continent and a people I had grown to love left me appalled and confused. But I also left it because I felt that the emotionally stressed and guilt-ridden debate which arose within the African studies community about the causes of Africa's decline was itself a powerful testimony to a fact even more depressing in its implications that anything that was happening in and to Africa. This fact is, to put it simply, that the most damaging legacy of colonialism and imperialism in the world has not been the global economic structures and relations it has left behind nor the patterns of modern 'neo-imperialist' economic and cultural relations of which it was the undoubted historical forerunner. Rather its most damaging legacy has been the psychological Siamese twins of endemic guilt on the European side and endemic psychological dependence on the African side, legacies which make truth telling hard and the adult taking of responsibility even harder. Imperialism fucked up the heads of so many people whom it touched—both colonialists and colonized (Frantz Fanon was absolutely and deeply right about that) and until that—ultimately depressing—legacy of its existence is finally killed, neither Africa nor African studies will be able to make real progress. It was that conclusion which led me—very sadly—to leave both behind.

The "Politics of the Mirror": On Geography and Afro-Pessimism

As we enter the new millennium, the immediate outlook for the countries of Africa appears decidedly mixed. The positive side has witnessed a return to democratic elections in Nigeria, Ghana, and Senegal; "consolidating" elections in Mozambique and Namibia; and sustained, if modest, economic growth in a number of countries, including Benin, Botswana, Cote D'Ivoire, Ghana, Mozambique, Senegal, and Uganda. Consensus is emerging around an alternative model of African development, first articulated in the so-called Lagos Plan of Action in 1980, and subsequently developed into an alternative to structural adjustment programs (United Nations 1990; Oloka-Onyango 1995; Mkandawire and Soludo 1999). The African Alternative Framework emphasizes popular participation in the development process, arguing that "adjustment must be for the benefit of the majority of the people and as such, adjustment programs must derive from within" (UN Economic Commission for Africa 1991, 12). Calls for debt relief, such as the Dakar Declaration of December 2000 assert the need for a truly African form of development (Dakar-2000, 2000). There is even recent talk of an "African Renaissance," articulated by new African leaders such as Yoweri Museveni (2000) and Thabo Mbeki (1999, 2000).

At the same time, there are reasons to be less sanguine. We see troubled elections in Cote d'Iviore and Tanzania; increasing civic unrest and state repression in Mugabe's Zimbabwe and Moi's Kenya; conflict and continued tension between Eritrea and Ethiopia and civil wars in Angola, the Congo region, Burundi and Sudan; and the continuing struggle of "collapsed states" such as Somalia, Rwanda, Liberia and Sierra Leone to regain some measure of stability. We can add to this crushing poverty in many parts of the continent, a staggering debt burden in most countries, recurring drought in the Horn and flooding in the south, and the devastating impacts of HIV/AIDS. Understanding the complex roots of these problems and posing potential solutions are challenges for all of us who care about the African continent and its inhabitants.

That task is made all the more challenging by a profusion of images, commentaries and assessments attempting to describe or explain "the African crisis."

From Jeff Popke, "'The Politics of the Mirror': On Geography and Afro-Pessimism," *African Geographical Review,* vol. 21 (December 2001). Copyright © 2001 by The African Specialty Group, The Association of American Geographers. Reprinted by permission. References omitted.

A host of travel narratives, policy analyses and popular commentaries articulate a view that Africa is a basket case, beyond help, a negative outlook known as "Afro-Pessimism" (Rieff 1998; for critiques of Afro-pessimism, see Gordon and Wolpe 1998; Borgomano 2000). One flagrant case of Afro-pessimism is *Out of America,* a soul-searching account by African-American journalist Keith Richburg, who spent three years covering conflicts in Somalia, Rwanda, Liberia, and Zaire. In the end, Richburg is happy to distance his American identity from his African roots:

> excuse me if I sound cynical, jaded. I'm beaten down, and I'll admit it. And it's Africa that has made me feel this way. I feel for her suffering, I empathize with her pain, and now, from afar, I still recoil in horror whenever I see yet another television picture of another tribal slaughter, another refugee crisis. But most of all I think: Thank God my ancestor got out, because, now, I am not one of them. In short, thank God that I am an American. (Richburg 1997, xiv)

Richburg is not alone: other commentators describe the African continent using terms like "anarchy" (Kaplan 1994), "chaos" (Ayittey 1998), and "hope-lessness" (*The Economist* 2000). What these assessments (and others like them) share is a view of the continent as a space outside the bounds of modernity and rational explanation, whereby Africa is considered unlikely to achieve any significant level of "development." One of the most troubling features of the recent Afro-Pessimism is the deployment of a startlingly simplistic notion of geography as a combination of country size and shape, resource endowment, transportation access, and climate. Africa's comparative underdevelopment has been recently described as "a case of bad latitude" because Africans are "prisoners of geography" (Hausmann 2001) and "enslaved by . . . [their] environment" (Etounga-Manguelle 2000, 69). Differences within Africa are explained by the fact that some countries have "favorable geography," while others are "challenged by geography" (Herbst 2000). Taken as a whole, this work augurs nothing less that the political acceptance of a renewed form of geographical or environmental determinism (Diamond 1997; Landes 1999; Herbst 2000; Sachs 2000; Hausmann 2001).

It is clear that geographers have an important role to play in complicating such crude formulations. Indeed, recent work in geography has contributed greatly to our understanding of a wide range of issues, including HIV/AIDS (Oppong 1998), structural adjustment (Konadu-Agyemang 2000; Owusu 2001), housing (Oldfield 2000), the informal sector (Aspaas 1998), citizenship (McEwan 2000), political ecology (Bassett and Zueli 2000; Awanyo 2001; Freidberg 2001), and refugees (Bascom 1998; Hyndman 2000). Taken together, such work has helped to produce a nuanced understanding of African issues, problems, and potential solutions. Still, the narratives of Afro-pessimism continue to circulate, and for this reason, I believe it is important to investigate the epistemological dimensions of this particular form of knowledge. In doing so, I want to draw upon what Patrick Chabal (1996) has called the "politics of the mirror." Chabal uses this term to describe the common tendency of Western commentators to view the African continent with assumptions that govern our beliefs about our own society. Such ethnocentric assessments frequently reflect deeply held assumptions about our own Western modernity. Chabal (1996, 46) describes the impacts of doing so:

> First, we have perennially been disappointed in that the reality of Africa has never matched our expectations. Second, and more ominously, we have failed to look at Africa as it is . . . rather than as we imagine it to be. Third, and as a result, we have confined Africa to the dustbin of history; that is, as a continent the history of which we cannot be expected to understand and on which we eventually "give up."

In what follows, I want to reflect on the implications of viewing Africa with an ethnographic mirror, and to suggest what this might mean. . . .

Colonial History . . .

Two fundamental kinds of processes—both of which are hallmarks of Western modernity—were used to operationalize European colonization. The first of these was the delineation of a subject position via processes of inclusion and exclusion and the second was the territorial structuring of colonial space. Taken together, the processes comprise an epistemology of modernity, which is wedded to a particular understanding of identity and space—identity as the Cartesian rational subject (Bauman 1991; Sibley 1995; Kirby 1996) and space as a form of "presence" (Shields 1992; Strohmayer 1997).

The civilizing project marked Africans as different, and in seeking to transform them, it also redefined the European as the privileged subject of modern history. This process of differentiating subject positions was abundantly clear to those on the receiving end of colonization. As Franz Fanon described it, "the colonial world is a world cut in two . . . inhabited by two different species . . . When you examine at close quarters the colonial context, it is evident that what parcels out the world is to begin with the fact of belonging to or not belonging to a given race, a given species" (1963, 38, 40).

This "parceling out" of the world was not only a discursive maneuver of the colonial project, but was actively produced through the structuring of colonial space. A whole range of material practices transformed the chaotic and premodern spaces of the colonies into structured and orderly territories: "reserves" were marked out for indigenous occupation; towns were surveyed and built; land was fenced and enclosed; transportation networks were laid out and built; administrative districts delineated. Such "spatial discourses" operated as a means to make the space of the colony legible to the European gaze, and hence subject to appropriation and control, as recent work has shown in places such as Egypt (Mitchell 1988), German Southwest Africa (Noyes 1992), South Africa (Robinson 1996; Popke 2001), Zanzibar (Myers 1998), and Tanganyika (Kopf 2000). In the process, geographical difference was domesticated, such that African space could be remade, or territorialized, in the image of Europe (Young 1995; Mbembe 2000). Those aspects of African subjects and societies that did not fit this rational, Cartesian model (e.g., polygamy, nomadism, witchcraft, communal land holdings) required erasure, if even in a violent form. *Through* such erasure and exclusion the European subject of culture and history could be affirmed as universal (Mitchell 2000). As Stuart Hall has put it, "the very notion of an autonomous, self-produced and self-identical cultural identity, like that of a self-sufficient economy or absolutely sovereign polity, had in fact to be discursively constructed *in and through 'the Other'; through a system of similarities and differences"* [*emphasis added*] (1996, 252). Going at least as far back as the mid-nineteenth century, the spaces of "darkest Africa" served as the unacknowledged "constitutive outside" against which the positive image of European culture and society could be maintained (Driver 2001). Africa, as the "other" against which the light of modern civilization could be projected, has been fundamental to the ways in which we have crafted and understood our modernity. It is in that sense that Charles Piot (1999, 21) reminds us, "Modernity's roots lie as much in Africa as Europe."

Africa . . . became a kind of "geographical unconscious" vis-à-vis the hegemony of a Eurocentric outlook. That Africa remains a backdrop for European modernity is reflected in the work of contemporary Afro-pessimists. But contemporary debates about African politics, economic development, and cultural identity challenge those who wield the African mirror as a way to legitimate a particular, European form of modernity.

Rethinking Politics

Perhaps no topic receives more attention from the Afro-pessimists than the failure of African politics. The history of one party rule, military coups, and despotic leaders like [Uganda president (1966–1971, 1980–1985) Apollo Milton] Obote and [Zaire leader Joseph Désiré] Mobutu are cited as proof that Africa is not ready for modern democracy. Keith Richburg's suggestion that "corruption is the cancer eating at the heart of the African state . . . " leads him to conclude that "this problem of corruption, from the president all the way down to the customs officials at the border posts, that seems to me about as good an explanation as any for Africa's plight" (Richburg 1997, 173, 175). George Ayittey (1998, 150–1) is even more blunt, describing what he christens the "vampire state":

> Dishonesty, thievery, and speculation pervade the public sector. Public servants embezzle state funds; high-ranking ministers are on the take. The chief bandit is the head of state himself . . . The African state has been reduced to a mafia-like bazaar, where anyone with an official designation can pillage at will. In effect, it is a "state" that has been hijacked by gangsters, crooks and scoundrels . . .

These characterizations, and others like them, reflect a Western model of the political sphere—election procedures, representative institutions, and bureaucratic apparatuses. In the African context, however, there is little practical distinction between "formal politics" and the realm of civil society. The political exists in a much more informal and personalized form.

Actors and "political" activities cannot be as easily separated from other realms of society. Instead, they infuse society through a vast array of channels, networks and associations. Such networks cut across dimensions of African identity and may include, for example, affiliations based upon kinship, clan, ethnicity, locality, religion, or age cohort.

The notion of *le polique par le bas,* or "politics from below" has important implications for how we understand the nature of politics in the African context (Chabal and Daloz 1999, 28–9). Political relationships are not rooted in Western conceptions of formal representation, in which individuals act to institutionalize rules and decisions for the benefit of some constituency or of society as a whole. Instead, politics in Africa are generally filtered through a complex set of patron-client relationships. Chabal and Daloz (1999, 15) underscore the importance of patrimonial affiliation: "the legitimacy of the African political elites, such as it is, derives from their ability to nourish the clientele on which their power rests. It is therefore imperative for them to exploit governmental resources for patrimonial purposes." Legitimacy in this context is based upon notions of reciprocity and redistribution, which regulate the ability to meet the expectations of one's clients. In some cases, it is also based upon ostentatious displays of wealth, which can serve to assure political clientele that existing political networks and affiliations will bear fruit. Given the porous nature of the boundary between the institutional sphere of "high politics" and the everyday forms of patronage characterizing "low politics," it is quite reasonable for both resources and power to flow between them. For those able to gain access to administrative resources, the state offers a vast array of opportunities to political entrepreneurs for the accumulation of economic and political capital, if properly channeled through socially sanctioned networks of patronage and redistribution.

This notion of politics also has important implications for how we interpret the widespread charge of corruption and mismanagement leveled against many African governments. Bayart (1993, 235) argues that:

> "corruption" as well as conflicts inaccurately described as "ethnic" . . . [are] no more than the simple manifestation of the "politics of the belly." In other words, the social struggles which make up the quest for hegemony and the production of the State bear the hallmarks of the rush for spoils in which all actors—rich and poor—participate in the world of networks.

Widespread social practices referred to in the West as fraud or embezzlement are publicly accepted in Africa, providing they do not reach the level of a Mobutu-style kleptocracy. In this way, the lack of formalized checks and balances can be viewed as part of a range of social practices, through which the redistribution of the political and economic resources of modern state activities are negotiated. Moreover, these are not "antiquated practices on their way to extinction but, much more realistically . . . codes of conduct which are at the heart of modern economic activities" (Chabal and Daloz 1999, 101).

The Afro-pessimists, of course, would not agree. For Ayittey (1998, 343) "the state vehicle that currently exists in many African countries cannot take

Africans on the 'development journey' into the twenty-first century." To assert this, however, is to assume that African politics are somehow a "corrupted" form of the Western norm, and thus, to rely yet again upon the politics of the mirror. As Chabal (1996, 46) notes, "the West demands a democracy in which it can recognize itself: party plurality, party competition, regular multiparty elections and parliamentary politics." In Africa, however, this form of political modernity was never successfully imposed, and the African state bears little resemblance to the formal, institutionalized apparatus of traditional political theory. The colonial state, and the post-colonial state that followed, were always hybrids of modern bureaucratic rationality and African forms of political expression. The space of the state thus remains in constant flux, constituted by multiple political repertoires, which combine the formal political culture of modernity with African custom and practice (Bayart 1993). Instead of measuring African politics against our own expectations, we should examine the ways in which the realm of "politics" has been vested with meaning, based upon a creative and innovative articulation of local African realities with the Western model of the state. In this way, an appreciation of African politics can serve to alter the ways in which we understand the political geography of contemporary states.

Querying Globalization and Development

A second manifestation of the African mirror can be seen in common understandings of economic development. Africa is generally depicted as a space outside of global interaction, an area "left behind" by the rapid economic and financial integration of the globe. William Coffey, for example, describes the seventh and "bottom" category in a "newer" international division of labor as follows: "approximately fifty 'least developed countries,' almost of all of which are in Africa, maintain very few economic connections with the rest of the world, except for exporting small quantities of natural resources. For all intents and purposes, they are external to the world economic system" (Coffey 1996, 60). In a similar vein, Ricardo Hausmann (2001, 53) writes "it is the absence of globalization—or an insufficient dose of it—that is truly to blame for . . . [the world's] inequalities. The solution for geography's poverty trap is for developing countries to become more globalized." As Grant and Agnew (1996) caution, however, this thesis that Africa is "falling out" of the world economy is an oversimplification. In reality, African states are engaging to an increasing degree in the world economy, and the trajectories of their political economy can be seen as an explicit response to the increasing globalization of economic activity (Chabal and Daloz 1999; MacGaffey and Bazenguissa-Ganga 2000).

The problem with many economic explanations, like those of politics, is that they rely upon a Western understanding of the space economy as comprised of formal networks of finance and trade, which can be accurately enumerated and compared across different economies. If, however, the division between "high" and "low" politics is untenable, so too is the distinction between the formal and informal sectors of the economy. Measured in terms of

GNP [gross national product], imports and exports, and debt levels, there is little doubt that overall, African formal economies are lagging behind most parts of the world. This does not indicate, however, that Africans are not engaged in the global economy. As the need to find new sources of accumulation to sustain Africa's diverse networks of client-patron relationships has become ever more pressing in the context of Africa's deepening economic crisis, African elites have become adept at creating new forms of rent out of new forms of modernization.

Flows of aid are one case in point, suggesting some of the ways that African governments skillfully exploit the resources of dependence (Chabal and Daloz 1999). African leaders, that is, have learned how to play the international aid game, to speak the language of NGOs [nongovernmental organizations], the World Bank, and the IMF [International Monetary Fund]: "the fact that some Africans are now prepared to support structural adjustment is not in itself an indication of its desirability. It is, rather, the proof that those Africans are adept at learning the language which will deliver the most financial aid from the West" (Chabal 1996, 47).

In this context, financial liberalization offers new opportunities to accumulate power and prestige within networks of client-patron relationships. Many of these activities lie outside of the juridical apparatus of the state. They include, for example, price fixing, banking fraud, money laundering, customs evasion, Nigerian "419 frauds," Ponzi schemes, the personal manipulation of privatization schemes, and so on. While Afro-pessimists decry such activities as "corrupt," they cannot be considered in isolation as simply illegal or criminal, but are better seen as "among a larger variety of techniques designed to exploit opportunities offered by the state and to gain access to the profits generated by operating between local and international sectors" (Bayart et al. 1999, 81). Included in these "international sectors" are the greatly expanded illicit, yet international flows of commodities such as diamonds, arms, drugs, and timber.

Dense African networks of illicit commodity trade, drawing on personal and ethnic ties, extend from Dakar to Paris, from Lagos to London and beyond. This "second economy," operating outside of formal state regulation, offers myriad opportunities, especially in regions where state authority has collapsed (MacGaffey 1991; MacGaffey and Bazenguissa-Ganga 2000). Here again, we can see the politics of the mirror at work. As Filip De Boek (1998, 779) suggests, such trade is generally interpreted as an illogical or criminal distortion of more "normal" capitalist processes, and thus "reflects the dictates of first-world imperatives and moralities." Instead, he argues, such activities are concerted attempts to take advantage of Africa's insertion into the global economy, to manage at a personal level the contradictions of neoliberal modernity. In this respect, it is no surprise that criminal networks have flourished in Africa at the same time that globalization has erased the importance of the nation-state in regulating economic affairs.

This is the case even in the most violent and anarchic "warlord states." Such places appear to be the paradigm cases of Kaplan's (1994) "coming anarchy," a morass of "tribal conflict" and primitive savage violence, representations of what Richards (1996) has referred to such the "New Barbarism" thesis.

In reality, however, conflicts like those in Sierra Leone or Liberia are fueled by their insertion in global flows of commodities, weapons and money, flows which often rely upon violence for their continued functioning (Richards 1996; Ellis 1999). While such violence can in no sense be condoned, they are the prevailing strategies for creating new opportunities for accumulation, opportunities that have come about as a direct result of the ongoing reorganization of global capital (Reno 1998). As Richards (1996, xvii) points out, "the Sierra Leone conflict is . . . moored, culturally, in the hybrid Atlantic world of international commerce."

All of this suggests the need to re-examine the ways in which we conceptualize the geographies of spaces and flows of the global economy. African economies *work,* often through what Chabal and Daloz (1999) call the "instrumentalization of disorder." In a world where globalization has eroded the role of the state, the "criminalization" of the African state and the rise of international networks engaged in elicit smuggling is, in this sense, a fundamentally modern phenomenon, emerging as a result of international economic integration, rather then signaling an evasion from it (Bayart et al. 1999). The entrepreneurial traders manage this integration by self-consciously straddling traditional boundaries of sovereignty and juridical control, and in so doing, creatively "seek to provide their own order and predictability in the midst of disorder" (MacGaffey and Bazenguissa-Ganga 2000, 169). . . .

Conclusion . . .

For [Halford] Mackinder—and no less for recent Afro-pessimists like Robert Kaplan, Samuel Huntington, and George Ayittey—Africa is "clothed" in a set of meanings derived from outside of the continent. Rather than try to allow Africa to speak to us, it has been constructed through the familiar teleological discourses of our specifically Western modernity. In the words of Chabal and Daloz (1999, 141), "our approach to the continent has been driven by a need to fit its supposed complexities—it's enigmatic psychology, as it were—into an explanatory scheme congruent with our view of Western development." This is not simply a matter of geographical location, for African scholars and political leaders are not immune to wielding this very same mirror. As Kom has recently argued (2000, 2), "many Africans . . . spend the greater part of their time looking at themselves in the mirror of the very people who imposed their condition on us in practically all areas of life. . . . " The same point was of course made by an earlier generation of African leaders and intellectuals—people like Franz Fanon, Aimé Césaire, Amilcar Cabral and Léopold Senghor. As Senghor (1965, 62, 97) put it, "we must be sure in our own minds that Africa is not just a series of geographical facts . . . Africa's misfortune has been that our secret enemies, in defending their values, have made us despise our own. And so we now go around shouting slogans from their ideologies which we are now naive enough to believe in."

POSTSCRIPT

Is Africa a Lost Cause?

In his selection, Kitching touches on a long-running debate within the field of African studies between "internalist" and "externalist" explanations of Africa's problems. The internalist arguments tend to focus on factors within Africa (e.g., poor leadership, corruption, immorality) to explain developmental difficulties, whereas externalist explanations emphasize exogenous factors, such as the lingering effects of colonialism or globalization.

Internalist explanations date to the colonial era but include a number of contemporary works as well, such as George B. N. Ayittey's *Africa in Chaos* (St. Martin's Press, 1998). The primary criticism of these works is that they overlook deep and long-standing connections between Africa and the global economy, connections that may have a profound impact on political leadership and resource flows within African countries.

Externalist explanations have often relied on dependency theory and world systems theory to situate constraints on African development. Andre Gunder Frank is credited with the development of dependency theory, a Marxist-inspired framework that conceptualizes elites in many developing countries as pawns of capitalists in the industrialized world. See his book entitled *Capitalism and Underdevelopment in Latin America* (Monthly Review Press, 1967). Immanuel Wallerstein's related world systems theory places dependency theory in a broader spatial context by conceptualizing developing, or peripheral, nations as hinterlands supplying raw materials to industrialized, or core, nations. See his book entitled *The Modern World-System* (Academic Press, 1974). What both of these theories do is place Africa in a broader global context in which African elites, and African economies more generally, are not serving the needs of their own peoples but rather the demands of industrialized nations. Examples of scholarly works from the externalist camp include Terence Ranger's *Peasant Consciousness and Guerilla War in Zimbabwe* (University of California Press, 1985) and Allen Isaacman's *Cotton Is the Mother of Poverty: Peasants, Work, and Rural Struggle in Colonial Mozambique, 1931–1961* (Heinemann, 1996). One problem with the externalist argument is that it may deny ordinary Africans "agency," i.e., a recognition that local people have the ability to influence events in spite of external influences. As Popke notes, "Africans are not simply passive victims of the capricious whims of despotic leaders" but resist in obvious and not-so-obvious ways.

ISSUE 2

Has the Colonial Experience Negatively Distorted Contemporary African Development Patterns?

YES: Marcus Colchester, from "Slave and Enclave: Towards a Political Ecology of Equatorial Africa," *The Ecologist* (September/October 1993)

NO: Robin M. Grier, from "Colonial Legacies and Economic Growth," *Public Choice* (March 1999)

ISSUE SUMMARY

YES: Marcus Colchester, director of the Forest Peoples Programme of the World Rainforest Movement, argues that rural communities in equatorial Africa are today on the point of collapse because they have been weakened by centuries of outside intervention. In Gabon, the Congo, and the Central African Republic, an enduring colonial legacy of the French are lands and forests controlled by state institutions that operate as patron-client networks to enrich indigenous elite and outside commercial interests.

NO: Robin M. Grier, assistant professor of economics at the University of Oklahoma, contends that African colonies that were held for longer periods of time tend to have performed better, on average, after independence.

The degree to which the colonial experience has impacted contemporary patterns of development in Africa is a major issue of discussion. Prior to 1880 roughly 90 percent of sub-Saharan Africa was still ruled by Africans. Of course slavery had had a profound impact on the continent, but the European presence in Africa was limited largely to coastal enclaves before this time. At the famous Berlin Conference of 1884–1885 (at which Africans were not represented), the European powers established ground rules to divide up the continent. Two decades later the only uncolonized states were Ethiopia and Liberia. The

18

principal colonial powers in Africa were Great Britain and France, followed by Portugal, Belgium, Germany, and Spain. Italy would also control some African countries in the late colonial period.

What were the objectives of European colonial endeavors in Africa? Of course the colonizers themselves suggested, or rationalized, that they were agents of progress who had a civilizing influence on the African people. The reality, widely accepted by scholars today, is that African lands and peoples were colonized to provide key raw materials for the European powers. While accepting resource extraction as the principal goal of colonialism, some scholars assert that the experience had some positive benefits, namely the development of infrastructure (roads, railroads, etc.) and educational and medical systems that benefit the African peoples today. Others suggest that the influence of colonialism has been overblown, and that African leaders merely use the colonial experience as a scapegoat for their own mismanagement of African affairs.

Those who suggest that colonialism has had an enduring negative influence on Africa point out that the infrastructure, as well as the educational systems left behind, nonsensical national borders, and the political legacy of colonial governance, all serve to distort, rather than to facilitate, contemporary economic and political development. For example, the roads and railways built during the colonial era often extend from the capital or port city to interior regions rich in resources, rather than connecting a country in a fashion that would promote national unity. Educational systems emphasized rote learning as they were developed to train low-level civil servants, rather than managers and directors. Enduring and nonsensical national borders have compromised the economic viability of many African nations (there are a large number of landlocked countries in Africa) and aggravated ethnic tensions (because they often do not respect ethnic boundaries). Finally, colonial governance was anything but democratic.

In the following selections, Marcus Colchester and Robin M. Grier present contrasting views on the enduring legacy of colonialism in Africa. Colchester examines contemporary patterns of resource extraction in equatorial Africa and draws a link between these and colonial practices. In Gabon, the Congo, and the Central African Republic, an enduring colonial legacy of the French are lands and forests controlled by state institutions that operate as patron-client networks to enrich indigenous elite and outside commercial interests. In many ways, Colchester is suggesting that colonial patterns of resource extraction never really ended; we simply have gone from overt colonial control to neocolonial regimes of extraction in the postindependence era. In contrast, Grier argues that on average, African colonies that were held for longer periods of time tend to perform better after independence. Furthermore, she asserts that the level of education at the time of independence helps explain the development gap between former British and French colonies in Africa.

Marcus Colchester

 **YES**

Slave and Enclave: Towards a Political Ecology of Equatorial Africa

The three countries of Equatorial Africa—Gabon, the Congo and the Central African Republic (CAR)—are among the most urbanized in Africa, largely as a result of the resettling of rural communities in colonial times. Of a total population of five million, only some 40 per cent live in rural areas, comprising about one hundred, linguistically closely-related peoples who are described in Western anthropology as the "Western Bantu", as well as some 120,000 "pygmies". Those remaining in the forests are reliant on self-provisioning economies, based on shifting cultivation, treecropping, hunting and fishing.

Some 47 million hectares of closed tropical forests in the three countries, combined with those of neighbouring Angola, Cameroon, Zaire and Equatorial Guinea, make up the second largest area of closed tropical forests in the world after Amazonia. These forests are some of the most diverse in Africa, and contain an abundance of wildlife, including forest elephants, lowland gorillas, various kinds of chimpanzees, forest antelopes and a wide variety of birds.

The social and political structures of Equatorial Africa have been markedly transformed by the European slave trade, the French colonial era and by the subsequent interventions of commercial interests and the new African states. Despite the current political liberalization taking place in Gabon, the Congo and the Central African Republic after three decades of single party politics, the forests and the peoples who rely on them are still being sidelined.

The Slave Trade

Long before Bantu society came into contact with colonial Europeans, its egalitarian traditions—a reflection of mobile, decentralized settlements which included an unease towards the accretion of power—had been progressively overlain during several centuries by more hierarchical forms of social organization, resulting from warfare and competition for land. This process had gradually reduced the accountability of leaders to their people with disastrous social and, later, ecological consequences. The European slave trade, however, is the most obvious example of this phenomenon, during which several millions of Africans died and millions more were transported overseas. The Portuguese began the trade in slaves on the coasts of Equatorial Africa around 1580, but it only became vigorous after about 1640. The European slavers themselves, however, were minimally engaged in raiding, never going far inland. Capture was carried out by Africans which not only intensified raiding and war between local African communities but also transformed previous systems of bondage and servile working conditions into those of absolute slavery.

This meant that slaving had a profound impact not only on those communities whose members were captured but also on those engaged in the trade. To gain control of the trading network, dispersed and differentiated social groups merged their numbers and identities to increase their power and domain. These groups were dominated by "trading firms" often comprising a chief and his sons. The increasing importance of inheritable wealth, capital accumulation and the corresponding need to resist redistributive customs led many matrilineal societies of the middle Congo to become patrilineal. Trading firms swelled their numbers by recruiting male and female slaves as labour and as wives to produce pliable heirs devoid of inheritance rights through the maternal line.

Colonial Repression and Resistance

Direct colonial rule of Equatorial Africa by the French from the 1880s onwards further exacerbated the tendency towards more hierarchical forms of social organization. As the French government was unable and unwilling to administer directly the vast area of more than 700,000 square kilometres, it allocated 80 per cent of the region to some 40 companies.

Within these vast concessions—the area controlled by the Compagnie Française du Haut Congo encompassed 3.6 million hectares—the companies had almost sovereign control including the right to their own police force and legislation, slavery, violence, killing and inhuman punishment were widely documented. Their express aim was to extract the natural resources—chiefly wild rubber, timber, ivory, and later coffee, cocoa and palm oil—as cheaply and as quickly as possible.

Labour shortages, rather than shortages of land, were the main constraint of these enclave economies. To extract labour for the concession areas, plantations, road-building, portage and, in the 1920s and 1930s, for building the Congo-Ocean railroad, "man hunts" became fundamental to the colonial economy. The military was used to support concessionaires who had insufficient labour. In the rubber regions of Ubangi (now the Central African Republic), for example, villagers who did not manage to flee from the troops sent in to ensure their "participation" were tied together and brought naked to the forests to tap the rubber vines. They lived in the open and ate whatever they could find. A French missionary who witnessed the scene wrote: "The population was reduced to the darkest misery . . . never had they lived through such times, not even in the worst days of the Arab [slaving] invasions."

The colonial authorities also introduced various taxes and levies to oblige local peoples to enter the cash economy; in practice, this meant to work for the concessionaires. Along the coasts, logging rapidly replaced the trade in non-timber products, becoming the foundation for the political economy of the region. The extraction was carried out with axes and handsaws, the huge logs being dragged by hand to the rivers and floated out to waiting trading ships. By the 1930s, an estimated third to one half of men from the interior villages of what is now Gabon had been brought down to the logging camps on the coast, where they worked for minimal wages and in appalling conditions for months at a time. Disease was rife, alcoholism rampant and thousands died. Meanwhile, the workload of the women and children who remained in the villages increased correspondingly and their health too was undermined by the numerous diseases—influenza, yellow fever, sleeping sickness and venereal infections—which returning labourers brought back.

The conscience of the colonial authorities was obviously pricked by the all too evident degradation and exploitation of the local people. As Governor General Reste noted in 1937:

> "The logging camps are great devourers of men . . . Everything has been subordinated to the exploitation of the forest. The forests have sterilized Gabon, smacking down the men and taking off the women. This is the image of Gabon: a land without roads, without social programmes, without economic organization, the exploitation of forests having sapped all the living force from the country . . . There is not a single indigenous teacher, doctor or vet, agricultural officer or public works agent."

Those making the profits—some two billion francs between 1927 and 1938—were a few French companies in whose hands the logging concessions were concentrated. By 1939, of the one million hectares of Gabon under concessions, 66 per cent was controlled by just seven companies with only some 84 others controlling the rest. These companies were to play a key role in the transition to Independence.

Resistance, however, was widespread throughout French Equatorial Africa; villages flared up against demands to yield labour and forest products to the concessionaires and taxation to the French administration. The result was a protracted and brutal war in which the colonial regime sought to bring the whole interior of the colony under its control to facilitate its commercial exploitation. The taking of hostages, including women and children, pillage, arbitrary imprisonment, executions and massacres, the torching of settlements and the sacking of whole communities were commonplace. By the 1930s, when the last areas of resistance were being quashed in what is now the west of the Central African Republic, the French were using planes to spot villagers hiding out in the forests before sending in the army to bring them out.

Regroupement des Villages

Resettlement of dispersed and shifting African communities into larger, permanent villages on roads and portage trails became a central plank of French administrative policy throughout its colony. One of the main aims of this *regroupement des villages* was to control the local people—to oblige them to render tax and labour and to prevent further rebellions.

Regroupement had devastating impacts on the local peoples. One missionary reported to André Gide that the local people:

> "prefer anything, even death to portage . . . Dispersion of the tribes has been going on for more than a year. Villages are breaking up, families are scattering, everyone abandons his tribes, village, family and plot to live in the bush like wild animals to escape being recruited. No more cultivation, no more food . . . "

In the early days of colonial rule, it was carried out with little if any consideration of customary land rights, causing conflicts over land between different social groups. Traditional institutions, residence patterns and ties with the land were overturned while overcrowding in the new villages led to declining standards of nutrition and exposed people to epidemics. Sleeping sickness increased throughout the region in the first half of the 20th century, causing a massive decline in population.

Even after Independence in 1960, Gabon and the Congo pursued the policy right into the 1970s. As a regional *préfêt* noted in Gabon in 1963:

> "We have had enough of these isolated hamlets, which, being so numerous and of eccentric location, lost in the vastness of the Gabonese forest, have never allowed the Gabonese government to have control of their populations or permanent contact with them and have thereby prevented an improvement of their standard of living."

Convinced that "no family head can cut himself off from the duty to modernize", the post-colonial government continued to tear villages away from their crops without any compensation, overturning residence patterns that reflected local ways of life and throwing into disarray, at least temporarily, traditional systems for allocating rights to land. In some respects, *regroupement* under the independent administration was more onerous than in the colonial era: houses in the new settlements had to be laid out in regular rows with defined sizes of houseplots and pathways.

The Indigenous Elite

The striking continuity between the colonial and independence policies was assured by an élite of French-educated Africans. Under French law, most Africans were subject to the *indigenat,* a set of laws which ascribed an inferior status to local people: "persons under the *indigenat* were subject to penalties and taxation without the legal protection afforded 'citizens'". However, a small number of Frenchified indigenous people were considered as *évolué* and thus accorded "citizen" status, becoming a local "élite attuned to the French presence and subservient to its interests".

Cooption of the indigenous leadership extended down to the community level through a hierarchy of *chefs du canton* and *chefs du village,* chosen to act as intermediaries between the villagers and the authorities. Dependent on the colonial administration for their positions and often resented and even secretly ridiculed by the villagers themselves, these leaders became ready tools in the colonialists' hands and assisted with the unjust exploitation in the concessions. Favoured ethnic groups emerged, considered to be more "evolved" and less "backward"; groups which had been intermediaries in pre-colonial trade now became an important support base for the administration. This structure and practice still persist; local leaders and chiefs, and favoured ethnic groups, continue to owe their primary allegiance to the urban élites and to the administration, not to the villagers or to the remoter, more traditional, rural communities.

Independence: *Plus Ça Change . . .*

After Independence from France in 1960, French Equatorial Africa was divided into the three countries of Gabon, the Congo and the Central African Republic, a division based on the previous colonial administrative divisions. The crucial concern of the departing colonial power was to ensure that the new "independent" governments supported French interests; President de Gaulle warned African states that "France would intervene if it considered its interests in jeopardy." In both Gabon and the Congo, France's principal aim was to guarantee that French logging companies were assured continued access to the forests. Maintaining access to strategic minerals, notably manganese and uranium, the latter being of critical importance to France's civil and military nuclear programme, was also central to French policy. In the Central African Republic, France's main preoccupation was to protect its cotton, coffee and diamond interests.

Gabon

French interests were decisive in selecting the future leadership in Gabon after Independence; French logging interests poured funds into the successful election campaign of Leon Mba, an *évolué* from the coastal region. After Mba's accession to power, the press was suppressed, political demonstrations banned, freedom of expression curtailed, other political parties gradually excluded from power and the Constitution changed along French lines to vest power in the Presidency, a post that Mba assumed himself. However, when Mba dissolved the National Assembly in January 1964 to institute one-party rule, an army coup sought to oust him from power and restore parliamentary democracy.

The extent to which Mba's dictatorial regime was synonymous with "French interests" then became blatantly apparent. Within twenty-four hours of the coup, French paratroops flew in to restore Mba to power. After a few days of fighting, the coup was over and the opposition imprisoned, despite widespread protests and riots. The French government was unperturbed by international condemnation of the intervention; the paratroops still remain in the Camp de Gaulle on the outskirts of Gabon's capital, Libreville, to this day, where they share a hilltop with the presidential palace, an unforgettable symbol of the coincidence of interests between the French and the ruling indigenous élite.

With the establishment of a one-party state, abuse of power and office became the norm. Wealth became more concentrated in the hands of the ruling élite, and the network of patronage became further removed from the concerns of ordinary citizens. The Presidency passed smoothly from Mba to Omar Bongo, "the choice of a powerful group of Frenchmen whose influence in Gabon continued after independence." Bongo and his cronies have since amassed substantial fortunes, having "transformed Gabon into their private preserve, handsomely enriching themselves in the process". They have transferred *billions* of French francs annually to Swiss and French banks. Opponents of the regime have been arbitrarily imprisoned and tortured, among other human rights violations.

Central African Republic

Post-Independence politics in the Central African Republic were not dissimilar. After the independent-minded first President Barthelemy Boganda died in an aeroplane accident, his successor, David Dacko, received strong French support. But Dacko's repressive policies and lack of effective economic reforms made him unpopular locally; he was widely perceived as a "French puppet caring only about cultivating French interests."

When Jean-Bedel Bokassa replaced Dacko on New Year's Day 1966, it seemed that the country, which had been virtually bankrupted by Dacko's regime, might be given a chance to recover. However, Bokassa's capricious and violent rule became synonymous with the worst excesses of African dictatorship—"the systematic perversion of the state into a predatory instrument of its ruler". Massive corruption was the norm, and Bokassa himself appeared to make no distinction between the revenues to the state treasury and his personal income.

France only withdrew its support for the regime in 1979, when it was revealed that Bokassa visited prisons personally to torture and kill those who had stood up to his whims. As in Gabon, French paratroops were sent in and Dacko restored to power, to be replaced on his death in 1981 by army strongman General Andre Kolingba, the current President.

Congo

Independence in the Congo pursued a different course. Initially, the post-Independence regime was modelled in the neo-colonial mould—servile to French political and economic interests. However, "it was swiftly corroded by venality and became an embarrassment not only to its internal supporters but also to its French sponsors". After a street uprising in 1963, the regime was overthrown and a Marxist-Leninist government assumed power. The French did not intervene.

French influence within the Congo remained strong, however, and, despite the Congolese government's rhetoric to the contrary, the role of foreign capital was scarcely diminished. Although the state created marketing monopolies for agriculture and forestry, and nationalized some other sectors—including the petroleum distribution network—the timber concessions, some of the oil palm plantations and many other import-export concerns remained in foreign hands. Foreign oil companies, too, were assured a satisfactory cut.

Logging Enclave Perpetuated

In all three countries of Equatorial Africa, logging has intensified since Independence. Mechanized logging extended the area of extraction during the 1970s and 1980s, up to the remote forests of the northern Congo and the south of the CAR; the rate of extraction increased some six-fold between 1950 and 1970. The industry has remained an enclave of foreign companies who enjoy the patronage of the governing élites. In the Congo, foreign companies, or joint operations dominated by foreign capital, produce the vast bulk of the timber—nearly 80 per cent of the sawlogs, 90 per cent of the sawn wood and 92 per cent of the plywood. In common with both the Gabon and the CAR, the Congolese government "largely lacks adequate technical and economic competence to control and rationally manage its forests".

During the 1970s, logging in the Congo was widely used as a fraudulent mechanism for capital flight, through false declarations of the quantity and type of timber being exported and through transfer pricing. The government attempted to curb this by creating a state monopoly to market timber, but the inefficient and ineffective agency ran at a loss due to collusion between loggers and officials.

A leaked report, prepared for the World Bank, reveals that the logging industry in the Congo is still swindling the government of millions of dollars. Unpaid taxes, stamp duty, transport and stumpage (duty payable on each tree cut) fees are estimated to exceed US$12 million on declared production alone, while huge quantities of timber are slipping across the border illegally into neighbouring Cameroon and Central African Republic. According to the report, "almost all the companies in the forestry sector are 'outside the law'" and "forestry administration is nonexistent". As a result, "the forest is left to the mercy of the loggers who do what they like without being accountable to anyone".

Companies are taking maximum advantage of this lack of supervision. For example, the French company, Forestière Nord Congo, has exclusive rights over 10 years to log some 187,000 hectares in the north of the country. Its contract obliges the company to process 60 per cent of the logs on site and to establish a major sawmill and woodprocessing works constructed out of new, imported materials. In exchange, Forestière Nord Congo has received generous benefits—substantially reduced import duties and a five-year tax holiday on wood production, including company taxes, property tax and stumpage fees. Yet, in complete violation of the agreement, the company bought a non-functioning second-hand mill, has not processed any timber, has exported sawlogs for six years through its tax loophole, and has not paid even the tax it should have rendered. In total, Forestière Nord Congo has cost the Congo some US$2.9 million in lost revenue.

Elsewhere, the Société Congolais Bois de Ouesso (whose board includes the Congolese President himself) received foreign "aid" in the late 1980s and early 1990s to install a highly-sophisticated saw mill and veneer producing works. With technical advice from the Finnish company, Jaako Poyry Oy, the World Bank backed the project with some US$12 million. But costs soon rocketed from an estimated US$39 million to US$63 million. Further loans were incurred from several African banks, but the mill was never completed; today only a small sawmill is working. The World Bank report notes "the situation is catastrophic and no further activity within the present arrangement is possible". The company's failure is attributed "quite simply to the overvaluation of the project which has allowed some vultures to enrich themselves immeasurably at the Congo's expense."

In the Central African Republic, where concessions have been granted in 48 per cent of the "exploitable" forests, the accountability of the logging industry is more lax. Illegal cross-border logging into the forests of north Congo was observed by FAO [Food and Agriculture Organization] technicians in the mid-1970s, and smuggling of timber down the Oubangi and Congo rivers to the Congolese capital, Brazzaville, on the Zairean border was normal. Today it is common knowledge in Bangui, the capital of the Central African Republic, that nearly all the forest concessions are being illegally logged, the regulations are flouted and much of the timber is clandestinely leaving the country via new road connections with Cameroon. CAR President Kolingba himself is alleged to be closely linked to businessman M. Kamash who owns SCADS, the company which processes timber illegally slipped across the border from north Congo. In common with all the concessionaires in the country, "SCAD carries out no management and employs no foresters—they are just timber merchants who mine the forests".

Since the Gabonese forestry service is funded from the central government budget rather than through stumpage fees and other tariffs on logging itself, there are few incentives for foresters to impose the many rules and regulations to control logging, while the loggers themselves make sure the foresters are provided with suitable incentives not to apply them.

Rural Stagnation

For the rural communities of Equatorial Africa, the long history of exploitative, extractive and enclavistic development has meant marginalization and poverty. Despite the statistically high per capita incomes of Gabon and the Congo relative to other Sub-Saharan countries, the rural people are poor and getting poorer.

The explicit aim of most government efforts concerning agriculture has been to replace itinerant, family-based, labour-intensive agriculture with fixed, capital-intensive, mechanized agriculture serviced by wage labour. Heavily-subsidized agribusiness schemes to promote large scale farming—cattle ranches, sugar plantations, battery farms of poultry, rice schemes, rubber and oil palm estates, and banana plantations—have undercut small farmers, destroying the last elements of a cohesive rural society.

A century of neglect and disruption of smallholder agriculture has had inevitable consequences. In the Central African Republic, only one per cent of the country is farmed. In Gabon, agriculture accounts for only eight per cent of the GDP [gross domestic product], occupies only 0.5 per cent of the land area and supplies only 10–15 per cent of the country's food needs, the remaining 85 per cent being imported; even traditional peasant crops such as taro, yams, mangoes, avocados and vegetables are imported from neighbouring states, particularly Cameroon.

In the Congo, agriculture yields only 5.9 per cent of GDP. Since Independence, there has been a massive migration to the cities. By 1990, 52 per cent of the total population, and 85 per cent of men aged between 25 and 29 years, lived in two cities alone, Brazzaville and Pointe Noire, although the Congo barely has any industrial base. As in colonial times, the lack of young men in the countryside means that 70 per cent of Congolese farms are managed by women. Education, preferentially given to young men, exacerbates this trend, resulting in the towns being considered the domain of men and the countryside that of women.

The marginalization of smallholder agriculture has also transformed local political institutions. Increasing mobility has weakened community ties and diminished customs which favour the redistribution of wealth and land. Echoing the political shifts which took place during the European slave trade, matrilineal groups have become more patrilineal and marriages with women classified as "slaves" (that is, without lineage and therefore without kin to make demands on agnatic inheritance) have been favoured. These internal trends have been reinforced by imposed national laws which favour cognatic succession, all tendencies which have further weakened the status and security of women.

NO

Robin M. Grier

Colonial Legacies and Economic Growth

Introduction

Development theorists have long hypothesized whether the identity of the colonial power mattered for subsequent growth and development. Many authors . . . once concluded that the colonial experience was insignificantly different under the major colonial powers. Recent research though has shown that colonialism did significantly affect development patterns. . . . In this [selection], I examine whether the duration of colonization has a significant effect on later development and growth and whether human or physical capital in place at the time of independence can help to explain why British colonies perform significantly better than French ones.

In the first empirical application of the [selection], I pool data from 63 former colonies and find that the length of colonization is positively and significantly correlated with economic growth over the 1961–1990 period. While the direction of causality is not conclusive, I find no evidence to support exploitation theory. Given that a country was colonized, since it would be impossible to test the counterfactual hypothesis, the longer it was held by the mother country, the better it did economically in the post-colonial era.

In the second empirical application, I reduce my sample to 24 countries in Africa. . . . I find that the length of colonization is still positively and significantly related to economic growth and that former British colonies still outperform their French counterparts. . . . I find that the newly independent British colonies were significantly more educated than the French ones. . . . I find that the inclusion of education at independence can explain the development gap between the former British and French colonies and the positive relationship between length of colonization and growth. . . .

Data and Variables

My empirical work addresses two questions. First, does the length of the colonization period matter for subsequent growth and development? Second, can education levels at the time of independence explain why British ex-colonies perform significantly better than their French counterparts?

From Robin M. Grier, "Colonial Legacies and Economic Growth," *Public Choice,* vol. 98, nos. 3–4 (March 1999). Copyright © 1999 by Kluwer Academic Publishers. Reprinted by permission of Kluwer and the author. Notes and references omitted.

In the first empirical application of the [selection], I perform a cross national study of 63 ex-colonial states and test to see whether the length of colonization is correlated with subsequent growth rates. . . . In the second empirical application, I reduce the sample to British and French Africa and test whether the duration of colonization matters for later development and whether human or physical capital levels at the time of independence can help to explain why British ex-colonies perform better on average than French ones. . . .

Econometric Results

Does the Identity of the Colonizing Power Matter for Economic Growth?

Five year averages are calculated on data from 63 countries for the years 1961–1990, resulting in 6 observations for each country and a sample size of 378 data points. . . .

Because of the vast differences in the length of colonization for different ex-colonial states, I add a variable called TIME to determine the effect of the duration of colonialism on subsequent growth, where TIME is the year of independence less the year of arrival. The most common argument against colonialism is that it exploited the native population, causing dependency and instability in the colonial states. The results . . . show that TIME is positively and significantly related to subsequent economic growth. I do not claim that longer colonization causes higher economic growth, but my findings do reject a crude form of the exploitation theory. While it is possible that countries in my sample might have had higher growth rates if they had not been subjected to colonialism, the results do show that colonies that were held for longer periods of time than other colonies have had more economic success in the post-colonial era. . . .

Does Colonialism Matter for African Development?

. . . I found that former French colonies perform significantly worse on average than British ones. It is possible that an African effect is exaggerating the development gap between the two. Because all of the French colonies were African (except Haiti, whose population is primarily African), and Africa has been characterized by poverty and underdevelopment in the last century, it may be that an African effect is biasing the earlier results. Limiting the sample to Africa also gets rid of the high-performing British outliers, like the United States and Canada, and helps to better distinguish the differences between French and British non-settlement colonies. . . .

French ex-colonies perform 1.38 percentage points worse on average than their British counterparts. That is, even when the sample is limited to Africa, the former French colonies still lag significantly behind the British ones. . . .

The results also show TIME to be positively and significantly related to economic growth, which implies that the length of colonization is positively related to growth and development. The next section, which examines the institutional legacies of colonialism, tries to determine why TIME is significantly related to growth and why the French ex-colonies perform worse than British ones, even when the sample is restricted to Africa.

The Colonial Legacy

To determine why ex-British colonies perform better than French ones, I look at the level of human and physical capital the colonizing power left at independence. . . . [E]ducation is an important component of growth and development. The British and French had contrasting philosophies of education, which translated into very different types of colonial education. The largest difference between colonial education policies is that the British made a conscious effort to avoid alienating the native culture, by teaching in the vernacular languages and training teachers from the indigenous tribes. While most of the teachers in French Africa were imported from France, the Advisory Committee on Native Education in the British Tropical African Dependencies recommended that, "Teachers for village schools should, when possible, be selected from pupils belonging to the tribe and district who are familiar with its language, tradition, and customs. . . . "

Students in British Africa were, for the most part, taught in their native language and in their tribal villages, which significantly eased the learning process. In contrast, in the French system, most students were boarded and only able to go home for the summertime vacation. Students were required to speak French, and all vernacular languages were forbidden. . . .

Conclusion

The literature on colonialism and underdevelopment is mostly theoretical and anecdotal, and has, for the most part, failed to take advantage of the more formal empirical work being done in new growth theory. This essay has tried to close that gap by presenting some empirical tests of oft-debated questions in the literature.

I find that the identity of the colonizing power has a significant and permanent effect on subsequent growth and development, which would deny the validity of a crude exploitation hypothesis. Colonies that were held for longer periods of time than other countries tend to perform better, on average, after independence. This finding holds up even when the sample is reduced to British and French Africa.

I also find that the level of education at the time of independence can help to explain much of the development gap between the former British and French colonies in Africa. Even correcting for the length of colonization, which has a positive influence on education levels and subsequent growth, I find support for a separate British effect on education. That is, the data imply that the British were more successful in educating their dependents than were the French.

POSTSCRIPT

Has the Colonial Experience Negatively Distorted Contemporary African Development Patterns?

Colchester describes a long process wherein relatively egalitarian and decentralized rural Bantu settlements in equatorial Africa were slowly transformed into more hierarchical forms of social organization. This process began with warfare and competition among Africans, was exacerbated by the slave trade and French colonial rule (for an excellent discussion of contemporary academic debates regarding the involvement of Africans in the trans-Atlantic slave trade, see Lassina Kaba's 2000 article in the *African Studies Review,* entitled "The Atlantic Slave Trade Was Not a 'Black-on-Black' Holocaust"), and continues today as the indigenous elite are more responsive to outside commercial interests than their own people. The limited accountability of leaders to their own people has led to exploitative logging with disastrous social and ecological consequences. As such, colonialism is not solely responsible for the lack of accountability among some African leaders in equatorial Africa today, but Colchester would argue that it definitely facilitated such behavior and created a set of relationships between French and African elites that persist today.

The process by which African elites were co-opted is worth highlighting. Colchester describes a Frenchified indigenous elite who were considered "advanced" enough to be granted French-citizen status. Through a process of acculturation, this elite came to share in and promote the interests of the French in their colonies. This approach, adopted by the French in their African colonies, is often referred to as the *policy of assimilation.* The policy of assimilation held out the carrot to all Africans (in French colonies) of potentially becoming French citizens. Although seemingly egalitarian on the one hand, it was a terribly ethnocentric policy on the other because it asserted that Africans only became civilized by adopting French culture. In many ways, the French policy of assimilation was an effective strategy for developing the group of "dependent elites" conceptualized by Andre Gunder Frank in his dependency theory.

The French policy of assimilation may be contrasted with the British policy of indirect rule. In nonsettler colonies such as Ghana and Nigeria, the policy of indirect rule meant that the British administered a country via its traditional leaders (up to a certain level in the colonial hierarchy). This difference in policy approaches is often used to explain why the British invested more heavily in education (because they relied on more Africans to run the colonial civil service). This is also a difference highlighted by Grier in her article for this issue. Grier further notes that the British encouraged instruction, at least at the lower grade levels, in local languages, whereas the French insisted that local people be educated in French (an approach consistent with their policy of assimilation).

Of course the French and the British were not the only colonial powers in Africa, but their approaches receive the most attention among academics because they had the largest number of colonies. For further information on the legacy of British and French colonialism, see a book chapter by the well-known Africanist Crawford Young, entitled "The Heritage of Colonialism," in John W. Harbeson and Donald Rothchild, eds., *Africa in World Politics: The African State System in Flux* (Westview Press, 2000). The Belgians and Portuguese have a reputation for having been particularly brutal colonial powers. See Adam Hochschild, *King Leopold's Ghost: A Story of Greed, Terror, and Heroism in Colonial Africa* (Houghton Mifflin, 1998) for a gripping account of conditions in the Belgian Congo, or Allen Isaacman, *Cotton Is the Mother of Poverty: Peasants, Work, and Rural Struggle in Colonial Mozambique, 1931–1961* (Heinemann, 1996), regarding the situation under the Portuguese in Mozambique.

ISSUE 3

Have Structural Adjustment Policies Been Effective at Promoting Development in Africa?

YES: Gerald Scott, from "Who Has Failed Africa? IMF Measures or the African Leadership?" *Journal of Asian and African Studies* (August 1998)

NO: Macleans A. Geo-Jaja and Garth Mangum, from "Structural Adjustment as an Inadvertent Enemy of Human Development in Africa," *Journal of Black Studies* (September 2001)

ISSUE SUMMARY

YES: Gerald Scott, an economist at Florida State University, argues that structural adjustment programs are the most promising option for promoting economic growth in Africa and asserts that mismanagement and corruption are responsible for prohibiting economic growth.

NO: Macleans A. Geo-Jaja, associate professor of economics and education at Brigham Young University, and Garth Mangum, professor emeritus of economics at the University of Utah, argue that structural adjustment programs and stabilization policies rarely have been effective. Rather, they contend that the available evidence indicates that these policies have "accentuated the deterioration in the human condition and further compounded the already poor economic conditions in many African countries."

$\mathbf{B}$oth the World Bank and the International Monetary Fund, or IMF, (along with the World Trade Organization) are sometimes referred to as Bretton Woods institutions because they (or their predecessors) were established during a conference of the major economic powers at Bretton Woods, New Hampshire, in 1944, as the end of World War II was in sight. The World Bank, or International Bank for Reconstruction and Development (IBRD), was established to rebuild Europe and Japan in the aftermath of the war, and the IMF was created to provide loans to help countries resolve short-term, balance of payment problems. The IMF and the World Bank have persisted as major multilateral economic development institutions, and they are particularly influential in Africa. These institutions are more influential in Africa because much of the debt incurred by national governments is public rather than private (i.e., loans from bilateral or multilateral development agencies rather than loans from commercial banks). In Africa, the IMF tends to focus on lending to resolve short-term problems (e.g., controlling inflation), whereas the World Bank has a slightly longer, "developmental" view.

Following the Third World "debt crisis" of the 1970s, the World Bank and the IMF initiated a form of policy-based lending known as structural adjustment. Prior to this time, much of the lending of the World Bank, in particular, had been project- or program-based. So, for example, in the 1960s the World Bank funded a number of infrastructure projects in Africa (such as dams), slowly transitioning to more programmatic funding related to basic needs in the 1970s (such as rural health care projects). The policy-based lending that began in the 1980s was somewhat different than traditional project- or program-based lending in that loans were held out as a carrot for countries that agreed to undertake a series of policy reforms. The basic aim of structural adjustment reform was to balance state budgets and, as either a cause or effect of the first, promote economic growth. According to the World Bank, the basic way to achieve such an end was to cut government expenditures and raise revenues, but not in a way that would encumber economic growth.

Policy reforms under the structural adjustment rubric in Africa include the privatization of inefficient state-run enterprises (which, if they are inefficient enough, may be a drain on the state treasury); a reduction of staffing and programming in state agencies in order to cut costs; the devaluation of national currencies in situations where they are deemed overvalued, and a redoubling of efforts in the export sector to foster the generation of foreign exchange. Overvalued currencies are seen as problematic because they may make a country's exports artificially expensive (and may reduce export potential) and its imports artificially inexpensive (and thereby encourage the overconsumption of imported products).

As evidenced by the readings in this issue, the effectiveness and appropriateness of structural adjustment in the African context is hugely controversial. In the following selections, Gerald Scott disputes the evidence used to suggest that structural adjustment programs have a deleterious effect on economic growth in Africa. Furthermore, he states that economic malaise in Africa is a result of corruption and mismanagement. In contrast, Macleans A. Geo-Jaja and Garth Mangum argue that structural adjustment programs and stabilization policies rarely have been effective. They maintain that the available evidence indicates that these policies have aggravated poverty and have failed to fulfill basic needs.

Gerald Scott

 **YES**

Who Has Failed Africa? IMF Measures or the African Leadership?

Introduction

Many writers have suggested that International Monetary Fund (IMF) Structural Adjustment Programs in Africa have not only damaged growth prospects for many countries, but have further worsened an already badly skewed income distribution. Some of these writers have claimed that IMF programs have ignored the domestic social and political objectives, economic priorities, and circumstances of members, in spite of commitments to do so. In a recent article, an African critic submitted that IMF measures have failed Africa. He claimed that "after adopting various structural adjustment programs, many [African] countries are actually worse off." Not unlike many, he seems to be suggesting that IMF programs have been somewhat responsible for the severe decline in economic conditions. Some critics of IMF programs have pointed out that the fact that economic conditions have deteriorated is not conclusive proof that conditions would be better without IMF programs. They do however stress that the developments associated with IMF programs have been extremely unsatisfactory. At the same time, this association does not necessarily imply that IMF programs cause economic decline in the region.

The main purpose of this [selection] is to argue that of all the feasible alternatives for solving Africa's current economic problems, IMF Programs are the most promising. The [selection] will not contend that the panacea for the seemingly unsurmountable problems rest with the IMF. However, it will argue that IMF programs are better poised to help Africa reach its economic goals, or improve economic performance. . . .

Why Has Sub-Saharan Africa (SSA) Performed So Poorly?

The problems of slow growth, high inflation, and chronic balance of payments problems continue to plague SSA well into the 1990s. In general these problems can be traced to international or domestic factors. During the last two decades a

From Gerald Scott, "Who Has Failed Africa? IMF Measures or the African Leadership?" *Journal of Asian and African Studies*, vol. 33, no. 3 (August 1998). Copyright © 1998 by Koninklijke Brill NV, Leiden. Reprinted by permission of Brill Academic Publishers. Notes omitted.

number of adverse events in the international economy have contributed to the economic decline in the region. These include oil crises, global recessions, deteriorating terms of trade, protectionism in the developed countries markets, rising real interest rates, and the lack of symmetry in adjustment to payments problems. In addition a number of adverse developments in the domestic economy have inhibited productive capacity and thwarted the attempts to initiate and sustain economic growth.

No doubt, many countries lack appreciable amounts of essential resources and adequate infrastructure for sustained growth. It is also true that growth and development in many nations have been set back by droughts, civil wars, and political disturbances. It may even be true that colonial economic structures still account for many inflexibilities that inhibit economic growth. However, many countries could significantly improve economic performance and reduce poverty significantly if they managed their economies more efficiently, controlled population growth, and abandoned those policies that are so obviously anti-developmental.

The major setback has been gross mismanagement, which has largely resulted from corruption, rather than from incompetence and absence of skilled administrators. In many nations, national resources for investment, growth and welfare have been consistently diverted into private hands and used largely for conspicuous consumption. Poor public sector management has resulted in large government budget deficits, which contribute to inflation, which in turn encourage undesirable import growth and serious balance of payments deficits. Quite simply African leaders, administrators, businesses, and political insiders have been engaged in corruption on a massive scale. The result has been almost complete destruction of the economic potential in many nations.

For the purpose of solving Africa's serious economic problems, there is need for the political will to attack the fundamental causes. If the present disquieting trends are not urgently tackled with the appropriate policies, then an even more somber future looms on the horizon for many Africans. Any package of measures should include policies designed to revitalize, expand, and transform the productive sectors into viable and self sustaining entities. In the absence of corruption public resources can be allocated efficiently to facilitate growth in the productive sectors. The microeconomic efficiency that results from efficient resource allocation, coupled with appropriate macroeconomic stabilization policies, would greatly enhance the prospects for economic growth and prosperity.

Assessing IMF Programs

Studies aimed at evaluating IMF programs in Africa conclude that the results are mixed, ranging from disappointing to marginally good. Inasmuch as it is difficult to assess the overall effect of IMF programs some studies have shown that they have been somewhat successful in terms of a number of key economic indicators. One main reason for the contention that IMF programs have been harmful is that many countries with programs have performed as badly as those without IMF programs. One must be cautious in examining the performance of key economic indicators following IMF programs because of the dynamics of the setting in which they are implemented. But let us suppose for the sake of argument that IMF programs actually result in deteriorating economic conditions immediately following the program. For example, suppose economic growth declines as a result of the program. Even though economic growth is perhaps the most important objective of national development policy, it is still reasonable to consider a program successful if it laid down the basis for future realization of economic growth, within some reasonable time period. In other words if it established the economic structure that promotes and facilitates long term growth, then it can still be regarded as successful. In addition it is possible that even though conditions did not improve, the program may have prevented economic conditions from deteriorating even more. It is not possible to subject IMF programs to controlled experiments. However, it seems more reasonable to argue that without IMF programs, in many countries, conditions would have been much worse, than one would argue that IMF programs cause conditions to worsen.

It has somewhat been fashionable, especially amongst those with very limited knowledge of the various economic rational behind IMF recommendations, to reject those recommendations without presenting a feasible alternative. Many object to the IMF and some regard it not only as a representation of western economic interest, but as too uncompromising and arrogant in its relationship with nations in crisis. The indications are that IMF is usually anxious to intervene even before conditions deteriorate into a crisis. But like any prudent banker it has to be concerned about repayment prospect, which is essential for its very own survival and continuous provision of its service to other deserving members.

Why IMF Programs May Be the Answer for Africa

IMF programs in the 1990s should have a major attraction for Africans genuinely concerned with the welfare of the people for a number of reasons. First, the programs are no doubt based on sound theory, always a useful guideline for policy-formulation.

The peculiar social, political and economic circumstances of African nations and the inability or refusal of the IMF to take them into account in the design and implementing of programs have been cited as reasons why IMF programs have "failed" in Africa, or are doomed to fail. On the contrary, these particular African circumstances are in fact another good reason why IMF programs may be the right answer to the problem. Because of the nature of African economic circumstances, particularly problems in economic administration, the conspicuous absence of commitment on the part of politicians and administrators to the development and welfare of the nations, the absence of institutional capacity and the weak civic consciousness, the best policy is to embrace IMF programs. IMF programs encourage the dismantling of controls and simplification of the bureaucratic process; emphasize the strengthening of institutional capacity; require public accountability and responsibility; emphasize efficiency and economic discipline; encourage private sector participation in the economy;

foster coordination in economic decisions and promote macroeconomic stability; and emphasize measures designed to expand aggregate supply.

The optimal policy intervention for dealing with an inefficiency or distortion is to seek the source of the problem. IMF programs are attractive because they are designed to attack the problems at their source. In African nations there are many problems that are outside the control of the officials and administrators. However corruption is not one such problem and it need not be so pervasive and economically destructive. Although it is very important not to under-emphasize the importance of many other problems of development, corruption is an obstacle that can largely be controlled, if the top leadership is committed to that objective. It is not the same problem as say drought or poor resource endowment, or an absence of a skilled workforce, that is largely outside the control of officials.

One major attraction of IMF programs is that they tend to remove all opportunities for corruption, i.e., they seek the source of the problem. For example, the suggestion that controls should be dismantled is in recognition that the reliance on physical controls for resource allocation is inferior to the market mechanism, especially in the absence of an efficient administrative machinery for the effective administration of controls. But perhaps the more relevant point is that a proliferation of controls usually lays the foundation for corruption, which has continued to destroy economic life in the region.

In reality it is not the IMF who has failed Africa, but the African leadership. The politicians and public sector officials have conspired with private businessmen and firms to adopt and implement policies that benefit themselves at the expense of national development and welfare. The inability or unwillingness of Africans to demand more accountability and responsibility from both politicians and public servants ensures that violations of the public trust are not treated as illegal, immoral, unethical or non-nationalistic actions. If the IMF has failed Africa, it has done so by failing to vigorously condemn or expose corruption or even assign it the prominent place it deserves in the design of programs.

Many who oppose IMF programs have argued that they impose severe economic harm on the deprived peoples of Africa. Whom are these deprived peoples and what is the evidence? The majority of them are rural inhabitants who have virtually been untouched by modernity. They are largely farmers, have limited participation in the modern economy, consume limited manufactured goods and have very limited access to basic social services provided by governments. In the urban areas there are Africans of diverse economic circumstances ranging from those in abject poverty and squalor, to those of enormous wealth. The urban population is usually more politically powerful and its views have been the barometer used to measure or assess the political climate. On balance IMF programs will tend to harm the urban poor given the structure of their consumption basket and their production pattern. On the other hand the rural population could benefit immensely from IMF programs for similar reasons, and the efficiency gain to the nation would more than compensate for the loss experienced by the urban population. No convincing evidence has been advanced to support the claim of impoverishment of the majority of rural African peoples.

Even though the urban poor could face the most severe hardship as a result of IMF programs, such adverse consequences could be mitigated even within the context of those same programs that supposedly impose such hardships. There is some empirical evidence that IMF reforms will improve the distribution of income and help the poor. There is also evidence that appropriate exchange rates and price incentives improve economic performance, and that private enterprises perform better than state enterprises.

What Africans need is a set of institutions that would enable them to effectively demand the very modest conditions the people deserve and subject all officials to full responsibility and accountability. Given the levels of ignorance, ethnic loyalties, poverty, disillusionment and despair, absence of strong nationalistic and patriotic attitudes, I shudder to imagine the difficulties associated with establishing such institutions. Notwithstanding, the task is possible if the leadership is committed to doing so. Based on the current structure of African institutions, and the record of policy makers, IMF programs are more likely to be effective than other possible alternatives.

Let us examine some of the recommendations and issues in IMF programs and discuss their effects on national welfare.

Devaluation

A devaluation increases the prices of traded (relative to nontraded) goods and will induce changes in production and consumption. First, as imports become more expensive less will be demanded, thereby curbing excessive import demand which is a major source of balance of payments deficits. At the same time production of import substitutes will be encouraged. Secondly, exports will become expensive so that less will be consumed locally and more will be produced. Exports will also be cheaper in foreign countries, so that more will be demanded. Foreign firms that split production into several stages will find the country attractive for their investments, and tourism will also receive a boost. The devaluation will therefore stimulate the export and import substitution sectors. The political concern usually is that the urban consumers whose purchasing power has already been eroded by inflation partly from excessive government spending, will have to pay more for basic manufactured goods, the bulk of which are imported. Not surprisingly, there is usually an anti-devaluation sentiment in the main urban areas. It is very important to emphasize that the devaluation by itself will not correct the problem of macroeconomic instability. It must be accompanied by sound fiscal management that complements rather than counteracts the effects of the devaluation. For example, if the government continues to maintain significant fiscal deficits after devaluation, then the devaluation would soon be reversed as the exchange rate becomes overvalued again. An overvalued exchange rate is subversive to long-term growth and balance of payments adjustment.

The African rural population consumes imported manufactured goods only in limited amounts, but could potentially benefit from devaluation because it will increase the price of agricultural exports. A program that prescribes a devaluation so that exchange rates are competitive, should ensure that the producers of exports are not unreasonably exploited by middlemen (including government) to the extent that they have no incentive to expand production.

A legitimate concern is that devaluation will raise the price of essential inputs and stifle the supply response as the cost of production rises. In the first place, as long as cost of production lags behind prices, producers will find it profitable to expand production. In any case the appropriate supply response could be encouraged by an appropriate production subsidy. This of course involves an additional strain on the budget, and the IMF insists on fiscal restraint as we will see shortly. Fiscal reform involves maximizing tax revenue and ensuring that it is used to maximize macroeconomic performance. This means that those who have been avoiding their tax burden, especially the self employed, must be made to meet their tax obligations, and that frivolous and wasteful expenditures must be avoided.

Government Budget Deficit

When IMF programs recommend reductions in government expenditures, the concern is not only with the adverse effects of budget deficits on inflation and the balance of payments, but also with bogus budgetary appropriations that benefit private individuals and deprive the nation of developmental resources. As a result of the pervasiveness of corruption, many governments typically appropriate funds for the salaries of nonexistent civil servants or for goods and services that are not received. Similarly it is common for governments not only to pay highly inflated prices for goods and services, some of which are totally inessential, but also for governments to receive far less than market value for goods bought by some individuals or firms. IMF prescriptions on the budget can be viewed as perhaps a subtle way of telling African leaders that from their past record they cannot be trusted to appropriate the nation's resources in the national interest. This appears paternalistic, but should be acceptable to all concerned with the welfare of the mass of African peoples.

Government budget deficits as a percentage of GDP [gross domestic product] increased sharply after independence in many countries, as the states intervened ostensibly to correct the perceived flaws of a market economy. The evidence indicates that throughout the region the states have failed to perform the role of a prudent entrepreneur, and government investments have resulted largely in considerable inefficiency. Public enterprises have been inefficiently operated, as they have largely been used as a way of providing patronage to political insiders.

Government budget deficits financed largely through money creation, have contributed to serious inflation and balance of payments problem. These deficits have not been consistent with other macroeconomic objectives of the government. The control of the deficit usually requires reducing expenditure, including the elimination of subsidies to consumption, and increasing taxes. In many African nations it is common for the government to subsidize the consumption of essential food items, gasoline, electricity, public transportation etc.

The major beneficiaries are the urban population and mostly political insiders who for example obtain goods at subsidized prices and resell at black market rates. The typical rural inhabitant, because of the structure of the consumption basket does not benefit much from government subsidies.

Market Prices

IMF programs attempt to promote a strong link between work effort and reward. This involves appropriate prices of goods and services, and factors of production. Prices not only provide information to producers but serve as an incentive that facilitate efficient resource allocation. The major problem in African countries has been inadequate production. Production has been constrained by a large number of factors including inappropriate prices. In many African countries the tax system has turned the terms of trade against agriculture and has resulted in very slow or negative growth rates in this sector. Overvalued exchange rates are an implicit tax on exporters since exporters receive the official rate.

The imposition of market prices for agricultural commodities typically results in higher food prices. Rural farmers benefit as producers, but as consumers they lose. However as long as they can respond sufficiently as producers, their gains will be more than enough to compensate for their losses and the nation as a whole will benefit. The challenge of reforming prices is to ensure adequate production response, which may require other complementary policies.

The continuous proliferation of price controls will only continue to stifle production, worsen shortages, and reduce incentive for investment.

Privatization

African governments have argued that they have an obligation to provide goods and services usually provided by private enterprises in developed countries, because too often the market fails to do so. Thus they are compelled to invest in capital formation that will increase output, improve efficiency in resource allocation, and make the distribution of income more equitable. Those are desirable objectives and any government that achieves them deserves widespread commendation. Unfortunately the record of the public enterprises which are usually set up to pursue these objectives, have been very disappointing. These public enterprises have been very inefficiently administered, and have been widely used by politicians as opportunities for patronage to their supporters.

In recommending privatization of certain public enterprises, IMF programs attempt to deal with two problems. The first is micro inefficiency in the productive sector, and the second is government budget deficits that result partly from the need to subsidize inefficiently run enterprises. Private enterprises that continuously make losses go out of business, but government enterprises with similar balance sheets receive political relief. By turning over certain enterprises to private institutions, the pressure on the budget eases, and there is a greater chance of increasing efficiency in production. . . .

Conclusion

African economic problems over the last two decades, can be traced mainly to a host of international and domestic factors. Many of the international factors and some of the domestic factors such as lack of suitable resource endowment, are outside the control of the governments and administrators. However, a significant part of the problems can be traced to corruption and other forms of inefficiencies. Instead of blaming the IMF for the dismal performance in Africa, we should focus on the African leaderships and their policies. The level of their commitment and the policics they have adopted and implemented increasingly seem to confirm only their deplorable lack of compassion for fellow Africans and a callous detachment from the people's welfare. The status quo must change to prevent further erosion of the economic base on the continent.

The best foreign assistance is one that has a lasting effect; it is one that would empower Africans to fully participate in the growth process, and provide them with the irrevocable ability to effectively demand the modest living conditions that they have been unjustly deprived of by their leaders for so long.

Macleans A. Geo-Jaja and
Garth Mangum

 **NO**

Structural Adjustment as an Inadvertent Enemy of Human Development in Africa

African economies have experienced numerous disruptions since the 1980s, while other parts of the world have significantly prospered. Sub-Saharan Africa (SSA) has become impoverished in absolute and real terms, and today faces the need for some form of an economic miracle. According to the harsh verdict of the International Monetary Fund (IMF) and the World Bank, Africa is unable to point to any significant growth rate or satisfactory index of general well-being in the past two decades. To a substantial degree, these two international organizations have themselves to blame. Following Africa's emergence as independent nations after the collapse of colonialism in 1960, too many African nations immersed themselves in debt, overspending on public enterprises, public employment, excessive military, and, all too often, cronyism and corruption. Their currencies collapsed, and debt repayment ceased. When these troubled nations sought assistance, the international lending agencies justifiably demanded reforms classified as structural adjustment. However, structural adjustment programs (SAPs) and stabilization policies have seldom delivered on their promises. Even worse, the available evidence suggests that they have accentuated the deterioration in the human condition and further compounded the already poor economic conditions (Adedeji et al., 1990; El badawi et al., 1992). The primary reason was that the structural adjusters took the easy way out. Instead of insisting that corruption be ferreted out or that military expenditures be cut, they slashed more visible expenditures on education or public health, as well as made more justifiable cuts in public enterprises and public employment.

As a result, previous gains made in human development and in employment, education, and health sectors during the immediate postindependence period have since eroded. The same could be said of economic performance, which was largely respectable, as demonstrated by growth in per capita income and saving, investment, and export earnings. Growth rates in nonadjusting countries have actually been better overall between 1980 and 1987 than in strong adjusting countries (United Nations Economic Commission for Africa, 1989). Even countries with a decade or more of adjustment behind them have very little to show for it in terms of human development (for more details, see United Nations, 1991; United Nations Development Program, 1994, 1996, 2000). According to the United Nations Development Program's (UNDP's) (1994) *Human Development Report,* there were 41 African countries in the category of countries with low human development and not a single one in the category of countries with high human development. Thus, the prevailing evidence is that adjustment policies seriously rolled back some of the socioeconomic progress that had earlier been recorded. The apparent social cost of SAPs begs reconsideration of the overall efficacy and appropriateness of economic reforms pursued to build institutional and human capacities and solve unemployment problems. To illustrate these facts, we focus on SSA with Nigeria as a case study.

In the context of this article, human development is synonymous with human resource development. It is considered in the context of empowering people and enabling them to participate economically to enhance their opportunities for productive employment and income generation, unleashing their creativity, building human capacities, and expending their entitlements for effective popular participation in the development process. Africa is deficient in all of these respects. The failure of SAPs to induce human development in SSA has more to do with their conditions than with their implementation. In light of these and other considerations, we study the impact of SAPs on human development and unemployment vis-à-vis budgetary allocation and the education system. We focus on education because it is widely regarded as the primary means to combat unemployment and the key to human development. We investigate the relationship between SAPs' impact on education, which directly or indirectly determines the quality of a nation's human resources and employability.

Structural Adjustment Conditions for SSA

Since the beginning of the 1980s, SAPs aimed at integrating developing countries into the world economic system have been introduced in many SSA countries. As a result, adjusting countries must export more and governments must cut back on social sector programs, reduce financial regulations, and keep their exchange rates favorable to the North. These policies encourage the aggressive opening of countries for unbalanced trade, accompanied by excessive deregulation and the removal of subsidies. These policies also set the conditions for downsizing public enterprises, reducing their staffing, intensifying their work requirements, and creating intrasystem competition between educational and health institutions. Only when governments sign these loan agreements, does the IMF agree to arrange a restructuring of a country's debt that includes a provision for a new loan package.

Post-SAP changes in SSA are ubiquitous; they permeate all aspects of society. But with much talk about strengthening productive capacity and restructuring fiscal and monetary equilibrium, and with the success stories of the World Bank, hopes were raised. Yet, there is no sign of an end to what has become a series of economic crises. The Asian crisis illustrates the point. The devastation that the crisis has wrought is a good indicator of the inappropriateness of the

IMF and World Bank's model of development. One need only to recall the criticisms leveled at the IMF and World Bank for their inability to either foresee the crisis or come up with appropriate policy response that could have mitigated its effects. Related is the disagreement between the two institutions, with Joseph Stiglitz, economic advisor to the World Bank, pointing to the flaws in the economic logic of SAPs. The main point here is that after 3 years, the Asian crisis is still taking its toll on human development. The most insidious effect of the crisis on human development is reflected in education (Stiglitz, 2000). What has occurred to date raises doubts about the validity of the adjustment model, which threatens the very foundation of human rights and national sovereignty. A reasonable test of that statement is the degree to which SAPs validate or invalidate the realities of human development in SSA. The following section examines the role of SAPs, per human resource development, and views their impact on capacity-building institutions.

Economic Adjustment and the Social Sector in Africa

The IMF and World Bank's recommended economic reform and stabilization programs have been highly controversial. They are not merely about stabilization or liberalization of programs for sustained development. Rather, they are prescriptions for the intensification of the existing conditions of maldevelopment (Lensik, 1996; Lopes 1999). The 1999 Asian monetary crisis—which plunged economies into recession or depression and subsequently led to adverse socioeconomic effects, such as intensification of unemployment, bankruptcy, high inflation, and high incidence of poverty—illustrates the profound effect of economic recovery programs. Such programs comprise another of a series of expensive, nonlocalized, inadequate, and inappropriate development strategies imposed as an experiment on SSA countries, leading to dependency and underdevelopment, and culminating in unemployment and impoverishment. Other outcomes of such programs are dislocation in the quality of education and the health system, which is of great consequence to capacity building and productivity.

Most relevant to this article, and perhaps to the bulk of Africa's masses, is the study by Cornia, Stewart, and Jolly (1987) for the United Nations International Children's Emergency Fund entitled *Adjustment with a Human Face.* This work gained wide publicity for uncovering the adverse consequences imposed on the masses by SAPs. Even in some of the countries with upward trends, growth rates are below what is needed to make a minimal difference in the accumulated problems of poverty and unemployment.

According to the secretary general of the United Nations, it is clear that insufficient progress has been achieved, and the response of the international community has been far from adequate for SSA's governments' commitment to undertake the SAPs. A number of other studies, including that by Lockheed and Verspoor (1990), both senior education specialists at the World Bank, attribute declines in education effectiveness in adjusting countries to the reduction in public expenditure on education. Blundell, Heady, and Medhora (1994) and Jayarajah and Branson (1995) have all consistently reported that education quality suffered as a result of education budget cuts required by adjustment conditions. As we can see, at the end of the day, the prescriptions have brought neither development nor economic relief. Rather, they have exacerbated unemployment, poverty, and decay in human resource development by requiring governments to cut back spending on social services.

Concerned with the intensification of the SSA crisis, the United Nations Economic Commission for Africa (1989) argued that there was little evidence of recovery and that growth rates in non-adjusting countries actually did better overall between 1980 and 1987 than in adjusting countries. In response, they issued *African Alternative Framework to Structural Adjustment Program for Economic Recovery and Transformation,* which placed significant emphasis on human-centered development and the maintenance and improvement of the social sector. Later that year, the president of World Bank, Barber Conable, admitted that structural adjustment without a human face had done much less than they had hoped. It had retarded efforts for human resource development and exacerbated unemployment and poverty, thus reducing the benefits connected to government expenditure (United Nations International Children's Emergency Fund, 1989). One wonders why the World Bank would request nations to cut back government expenditures to social services, the very same provisions that have facilitated their development in the past. The Marshal Plan, the American New Deal, and Europe's welfare states are good examples.

In light of the aforementioned factors, the aggregate evidence shows that neglect and underfunding of the social sector, particularly education and health, negate the development of a skilled labor pool, the capacity and capability build up in research and policy, and the provision of the management talents demanded in an adjusting economy.

These are a few of the lacking human capacities that inform the argument that the present socioeconomic conditions of SSA are not amenable to IMF and World Bank's model of development as currently constituted, because they compromise quality and the effectiveness of education. In some countries, quantity and quality are sacrificed for debt repayment, whereas in others, either quality is traded for quantity or vice versa. In summary, these adjustment and stabilization programs tend to endanger human resource development. Even more telling, social conditions and human development considerably worsened, with the deterioration in the areas of education, health, employment and income, with especially serious effects on children, school graduates and women (Cornia et al., 1987). . . .

The Economic and Social Cost of Adjustment

Before we proceed to the discussion on the impact of adjustment measures on employment, a look at some of the World Bank's compensatory programs intended to mitigate the social cost of SAPs on the population is important. These compensatory measures were advocated by the United Nations and the United Nations International Children's Emergency Fund to introduce a human dimension to IMF and World Bank adjustment programs. In response, the World Bank designed and implemented a number of social safety net programs to protect vulnerable groups affected by adjustment: the Social Action Program in Cameroon, the Grassroots Development Initiatives Project in Togo, the Social Development Project in Chad, the Program of Action to Mitigate the Social Consequences of Adjustment in Ghana, and the National Directorate for Employment Project in Nigeria, to mention a few. In the process, the World Bank has admitted that the social consequences of adjustment are real and that the necessity of designing safety nets in conjunction with SAPs is compelling. Remedial programs such as these imply that their parent policy, structural adjustment, failed to deal with the real structural problems and the overall issues that brought about the crisis in the first place. Thus, in spite of these social safety nets, available evidence points to unemployment increasing from 7.7% in 1978 to 22.0% in 1990, and unemployment is projected to increase to 30% in 2000, while underemployment is estimated to affect 100 million Africans (International Labor Organization, 1993).

The evaluation in the next section focuses on the social cost of adjustment programs on employment in the continent, using Nigeria as an example.

Social Cost and Unemployment in Africa

Clearly from experience and research, we are reminded that designing optimal macro policies is not a guarantee for employability nor a condition for improving the overall work force. We are reminded that the failure of programs to induce employment in Africa has more to do with SAPs' designs and conditions than with their implementation. Technological changes affect the world of work by restructuring the economy, demanding new sets of job skills, and de-skilling workers. The whole employment situation is further compounded by the failure of most bilateral and multilateral agencies to appreciate that basic education should respond to social and economic transformation more than it determines them. In other words, the gap between education and employability keeps widening.

Some key features of the sacrifices made by adjusting countries have been drastic budgetary setbacks to critical social sectors, which have intensified social and economic problems. Especially affected is the enclave of mass unemployment, which has been further complicated by retrenchment and underemployment of workers because jobs that materialized from this market-oriented policy could not replace those lost. The point is that those compensatory measures for the generation of employment built into the adjustment process cannot negate the externalities associated with the massive retrenchment and retirement of labor that almost invariably accompanies the implementation of SAPs. Hence, the need for outside intervention, a visible hand, usually that of government, to help the crippled invisible hand of market forces. Of course, these policy actions must be wise ones, but they are essential. The SAPs did not have built-in mechanisms to balance changes in occupational skill caused by structural shifts in the economy. In most adjusting countries, people, particularly women, moved into the informal sector to cushion adjustment impacts. Some impacts facilitated the expansion and dependence on the underground economy, while driving others to elicit activities such as corruption, prostitution, and drug dealing.

Tutu, Jubuni, and Oduro (1994) showed that with the introduction of economic reforms, workers in Ghana suffered unparalleled retrenchment, displacement, and hardship. Ghana's experiment with adjustment had a catastrophic impact on urban low-income groups, especially retrenched public sector employees, who were already living below the poverty line. Their purchasing power was drastically reduced, making life nearly unbearable for them and their families. These authors further illustrated that there were substantial shifts from wage work to self-employment. In the case of Zambia, unemployment stood at more than 60% for the whole period of the adjustment program, 1985 to 1990. In Côte d'Ivoire, adjustment emphasized expenditure cuts in social services, which were accompanied by poor growth and unfavorable impacts on the impoverished and led to decay in human development. Drastic cuts in education expenditures had tremendous effect on the employability of the masses. The urban poor were the hardest hit because of both employment and real wage reduction. Zimbabwe witnessed accelerated inflation and poverty even as unemployment increased. Approximately 11.7% of able-bodied Kenyans seeking employment in 1983 were not able to obtain any kind of employment. By 1985, the figure had risen to 16.5%, with unemployment highest among tertiary institution graduates. The high rate of graduate unemployment could be attributed to the poor quality of education, which resulted in a mismatch between educational outcomes and skill needs within the reengineered economy. These country case illustrations demonstrate that privatization of state enterprises and the weakening of governmental roles in the economy will only deepen the crisis of unemployment and human resource development unless clearly and effectively accompanied by private sector alternatives. . . .

Conclusion

The conclusion reached in this article is that the SAPs of the IMF and the World Bank have failed in SSA, when assessed by their own stated objectives. Scale destruction in the conditions of social and economic infrastructures, service dislocation of rural economies, retrogression in the quality of education and employment, decay in human development, and the failure of the majority of the citizens to have access to education and other basic needs have been the general patterns in adjusting countries. Core elements of SAPs' failures have been faulty resource allocation caused by cost-price distortions and the elimination of positive government activities in the economy. These factors have resulted in serious dislocations in the social sector, which have imposed public sector retrenchment, reduced educational and health services, and resulted in the intensification of unemployment and poverty. Thus, the call for an immediate increase in the proportion of total budgetary expenditure devoted to human development, including greater spending on education, vocationalization, and basic health. SSA countries cannot be expected to realize any substantial social or economic benefits with the IMF and World Bank model of development, which does not emphasize social investment in people, education, and health care; poverty reduction; fair employment; and broad-based development.

This does not mean, however, that unemployment, poor quality of education, and underdevelopment, which are outcomes of IMF and World Bank prescriptions, cannot be remedied in the long term. The arrival of the 21st century may form the impetus for countries to further reevaluate the potential achievements of adjustment and educational strategies. SSA education and training systems based on commitment to equity, exploration of the world of work, and sensitivity to the role of technology and globalization are most urgent. This new blueprint has the capability to remove adverse influences that might inhibit employability and development. The salvation of SSA from human development decay, unemployment, inadequate human resource development, and its underdevelopment is achievable by practical quality education accompanied by wise use of the government's role in the economy.

It remains that economic development depends on human resource development. Therefore, it is incumbent on the state to ensure that there is adequate funding for human resource development, not only to normalize and increase present income flows but also to ensure future high rates of growth.

POSTSCRIPT

Have Structural Adjustment Policies Been Effective at Promoting Development in Africa?

The type of development that structural adjustment policies are designed to facilitate is export-led, laissez-faire economic growth. The belief (under this conceptualization of development) is that if one gets the general policy environment "right" and minimizes interference of the state, then capitalist actors will make efficient and rational decisions that lead to the generation of wealth that will eventually trickle down to all members of society. Furthermore, as individual nation-states exist in a global economic system, it makes sense for each to specialize in the production of goods that each can create relatively cheaply and to trade for those goods that others can produce more efficiently.

Other than questioning the effectiveness of structural adjustment along the narrow lines of whether or not it promotes economic growth, critics of these policies and, more broadly, development conceptualized as export-led, laissez-faire economic growth, generally raise at least three issues. First, they assert that laissez-faire economic growth often does not help the majority of the population, but instead may serve to concentrate wealth in the hands of a few powerful individuals or entities. Konadu-Agyemang, in an article in an August 2000 issue of the *Professional Geographer* entitled "The Best of Times and the Worst of Times: Structural Adjustment Programs and Uneven Development in Africa: The Case of Ghana," highlights the very uneven pattern of development promoted by structural adjustment policies in Ghana, one of the World Bank's success stories. Second, critics assert that it is often problematic for African nations to engage in free trade with the rest of the world because they are relegated to the production of primary commodities for which prices are low and declining. See an article by Padraig Carmody in a March 1998 issue of the *Review of African Political Economy* entitled "Constructing Alternatives to Structural Adjustment in Africa." Finally, many critics would argue that an entirely different vision of development needs to be emphasized, a vision that prioritizes the fulfillment of basic human needs (e.g., adequate food, clean water, health care) and the development of human capital (via improved education).

It may appear that the two approaches in this debate are mutually exclusive. Others would contend that there is a third, hybrid approach, known as "structural adjustment with a human face," that is cognizant of the role of sound social services (and the development of human capital) in long-term economic development. Cynics view this change as superficial (and an approach that only was developed by the World Bank and IMF under pressure from its critics), and they assert that we are still dealing with the same wolf, albeit in sheep's clothing.

ISSUE 4

Are Non-Governmental Organizations (NGOs) More Effective at Facilitating Development Than Government Agencies?

YES: Maria Julia, from "One NGO's Contribution to Women's Economic Empowerment and Social Development in Zimbabwe," *Journal of Social Development in Africa* (1999)

NO: Giles Mohan, from "The Disappointments of Civil Society: The Politics of NGO Intervention in Northern Ghana," *Political Geography* (2002)

ISSUE SUMMARY

YES: Maria Julia, professor of social work at Ohio State University, argues that the work of non-governmental organizations (NGOs) is critical for facilitating the empowerment and development of poor women in Zimbabwe. She makes this point by examining the role of a micro-credit NGO that provides financial assistance, as well as educational and emotional support to female entrepreneurs.

NO: Giles Mohan, a lecturer in development studies at the Open University, presents a case study of non-governmental organization (NGO) intervention in northern Ghana. His examination reveals that tensions exist between the northern NGO and its partners, that local NGOs create their own mini-empires of client villages, and that some NGO officers use their organizations for personal promotion.

While non-governmental organizations (NGOs), formerly known as charities, have been a long-standing fixture of the post World War II development enterprise, these entities increasingly became important in the 1980s as bilateral donors and governmental development agencies (such as the U.S. Agency for International Development) funded them to implement local-level projects in Africa. Prior to this time, bilateral donors may have used their own staff to manage projects in Africa, or they granted or loaned funds directly to African governments for the implementation of projects. NGOs, as implementing agencies, caught the fancy of donors because they were believed to be more efficient and in touch with the interests and concerns of local people. Furthermore, donors saw NGOs as a conduit for delivering assistance to the local level that bypassed African governments, which increasingly were viewed as inefficient and corrupt. More recently, NGOs have been conceptualized as a critical component of civil society in the African context. Donors increasingly view a dynamic civil society as an important check on the excesses of government.

NGOs come in all shapes and sizes. Perhaps one of the most important distinctions is between international NGOs, which are often based in North America or Europe, and national NGOs, which operate in a particular African country and are staffed and directed by individuals from the nation in question. NGOs might further be broken down into those that are home-grown community organizations (frequently operating in a small area of the country) and their more professional counterparts that not only work in several regions of a country, but have paid staff (often well-educated urbanites from the capital city).

A key concern of some scholars is that NGOs may be undermining the role of the African state in some of the countries where they work (by diverting funding and power away from governmental structures). As mentioned, a key reason for the rise of NGOs is that donors sought an alternative project implementation avenue in lieu of their frustration with the inefficiency and corruption of the African state. Opponents of this approach would argue that Africa needs strong "developmental" states that will be around for the long term and effectively coordinate resources and development priorities. For an example of this perspective, see Abdi Samatar's book, *An African Miracle: State and Class Leadership and Colonial Legacy in Botswana Development* (Heinemann, 1999).

In the following selections, Maria Julia argues that the work of NGOs supports female entrepreneurs in Zimbabwe by facilitating empowerment and development. In contrast, Giles Mohan reveals the tensions that exist between northern NGOs and their partners, the tendency of local NGOs to foster clientelistic relationships with the villages they serve, and the inclination of some local NGO officers to use their organizations for personal promotion.

Maria Julia

One NGO's Contribution to Women's Economic Empowerment and Social Development in Zimbabwe

Introduction

. . . Women's poverty, due in part to their lack of economic self-sufficiency, has been recognised as one of the impediments in the empowerment of women in Africa. In 1998, as part of ten strategic actions, the United Nations Economic Commission for Africa proposed to reduce gender disparities on the African continent. The goal of the action is to expand women's opportunities for entrepreneurship by promoting access to and participation in economic processes and structures, including access to employment and control of economic resources.

People-centred socioeconomic development emphasises the need to strengthen and support the capacity and self-reliance of women. The opportunity for advancement of economic power can facilitate and result in empowerment and a sense of self-worth brought about by economic self-sufficiency and self-reliance, *"under conditions that give people both the opportunity and incentive to mobilise and manage resources in the service of themselves and their communities"* (Korten, 1987:147).

In Africa in particular, investments in women's micro-enterprises provide essential resources that women need in order to improve their productivity and capacity for self-reliance.

Although NGOs [nongovernmental organizations] have been frequently criticised (Zmolek, 1980; Korten, 1998; Lal Das, 1998; Kam, 1998), they have more often played an enabling and leadership role as catalysts and vehicles of empowerment and development in their responses to the economic needs of women. NGOs are considered a *"most promising solution"* for one of the most difficult problems women face, *"including the most intractable of all social problems—poverty"* (Bar-On, 1998:5–6). According to Korten (1987), in an *"era of declining financial resources and deepening poverty,"* NGOs have been looked upon as a means of getting women the economic benefits they need more directly and cheaply than governments have been able to accomplish. While many NGOs are involved in community development, only a relative few are *"active in giving small loans to entrepreneurs in the micro enterprise informal poverty economy for income and employment generation purposes"* (Waly, 1998:4).

I spent a week at the home of the Zimbabwe Women Finance Trust (ZWFT), one of the first and most successful women's NGOs of its kind in Zimbabwe. As has been traditionally the case with NGOs, ZWFT relies for its funding on a number of North American and European foundations, such as CARE. ZWFT opened in Harare in 1985 as a micro-credit programme, providing financial assistance and incentives in the form of loans to women entrepreneurs who are denied access to credit by the formal banking sector. Operating under the leadership of its founder and executive director, social worker Eisenet Mapondera, the Fund encourages the development of informal sector micro-enterprises for improving economic opportunity and activity among women.

Following a number of conversations with the ZWFT staff, I was allowed to observe interviews between ZWFT staff and loan applicants, conversations with ZWFT clients, and on a subsequent spontaneous role play between a staff member and the author, the following scenario captures a typical interaction between an African woman entrepreneur and the ZWFT staff. The italicised commentary indicates steps in the process used by the ZWFT to introduce a prospective loan applicant to the opportunities that ZWFT has to offer.

An Illustrative Scenario

It was 8:30 am when I arrived at the Zimbabwe Women Finance Trust (ZWFT). The offices are located in a suburb of Harare in a converted agency house. On this cool October morning, I am overtaken by the magnificence of the jacaranda in bloom and by the serenity of the surroundings in this sub-Saharan African city. As I enter the house, I hear murmuring in a back room. As I move toward what seems to be the living area (reception area), I realise that the staff has gathered to pray. I learned later that prayer is part of the staff's daily routine before they start their work activities.

The walls of the living/reception area are covered with familiar posters. *"If it is not appropriate for women, it is not appropriate;" "As a woman I have no country . . . my country is the whole world."* I especially like those posters targeting prevention of abuse of women. All of the areas of the house are simply furnished, but the handcrafted pieces made by some of the clients fill up the place with colour and, even more, with humanness. A cheerful woman with a big smile greets us. I must digress to emphasise these genuine, frequent, welcoming smiles that are so pervasive, and that make people feel so welcomed and give an almost immediate sense of belonging.

"Good morning. I am Kudzai, who are you?"

"Good morning, I am Maria."

"Where are you from, Maria?"

"I come from Marondera."

"What brings you here this morning?"

"I would like to receive a cheque."

"Do you mean a loan, Maria?"

"Yes that's it, I guess. A loan."

"That will take a while to process. There are papers to complete, and we'll need to talk about your ideas. Then, assuming you qualify to receive a loan . . . "

"Oh!"

"It's okay, Maria. Let's talk a bit. What do you want a cheque for?"

"For herbs. I want to grow and sell herbs. A woman in the waiting area was telling me that she sells "freezits" (plastic packets containing frozen individual servings of fruit juice), and she seems to be doing financially very well. She says you helped her get started by providing a cheque for the plastics and the machine to seal them. Maybe I should try working with "freezits" instead of with herbs . . . "

"There might already be too much competition for "freezits" in the Marondera region. I want to hear more about your original idea of growing herbs. Where are you planning to work?"

"At home. I want to be near my home."

"Do you have experience growing herbs? Do you know what you need to do to grow herbs?" *(Kudzai makes an initial assessment of skills).*

"Yes. I have about half an acre of good soil for planting herbs. I have grown small amounts for myself and my family."

"That sounds wonderful! What herbs do you think you would grow to sell?"

"I do not know yet, but maybe marjoram and mint—you know, maybe medicinal herbs."

"Why those particular herbs? Tell me more about them."

"I am not sure. I like growing herbs that cure ailments in people. I learned from my grandmother how to identify and use medicinal herbs, and how to grow them too."

"What will your herbs cure?"

"Everything! Tummy-ache, headache, burns and sores, colds, anything."

"Have you thought about the possibility of ending up growing herbs that no one will buy? Who are going to be your clients?" *(Here the staff member begins to create awareness of the need to assess competition in the market).*

"Women at the market."

"Have you checked who are the women that will want them and how much are they ready to pay for herbs?"

"Not yet."

"Have you checked which are the kind of herbs that the women at the market will want to buy? Have you met any of your possible buyers to see what herbs are they looking for?"

"Mmm . . . "

"It's alright. We are just trying to sort it out. Do you have an idea about how much money you will need to start the business?"

"No."

"Do you have any money saved to start the business?" *(Here Kudzai begins a credit assessment).*

"No. I have the land and my work. But listen, I did not think that this was going to be so complicated. I want to work. I want to start my own small business. I want to make money to help feed my family—a "push" is all I need. Do you know what I mean?"

"Yes, I know what you mean, but we need to talk more about this. Why don't you take a seat and we'll talk some more?"

"A woman from my community came here about a year ago and she told me that you sent her to sell peanuts before you gave her any cheques."

"Yes, you need some business experience before we give you a loan. Even selling peanuts for a while will give you some experience. You need to learn to speak the language of business. Maria, you have the most important part of what is needed—the interest—but let me explain this to you a little bit better. First, you are going to be invited to take part in training where, among other things, you can learn how to approach and get potential suppliers and clients, how to find out what clients need, where to get the best materials for the business, such as the seeds for the herbs for your garden, and how much it will cost. The training also includes information about how to manage money, how to prepare a budget, and how to save a part of the money you make."

"Mmm . . . "

"I notice that you are a bit hesitant. Not being timid or shy is very important in having a business. The training that we offer includes working on your shyness. *(Assertiveness training).* You will also have fun with other clients and our staff, sharing ideas and suggestions on how to deal with the business and carrying on with the responsibilities of the family and the house *(Balance between household and business demands).* Are there other women in your community who would like to get into business or who are interested in going into business with you?"

"Of course! Why?"

"You need to become a part of a group of about seven women with an interest in business. You can form this group with people whom you know and who are interested in a business, or we can put you in contact with some existing groups where the women are already in business. Only three members of a group receive loans at a time. *(Explaining the informal rotating credit system).* The remaining members of the group will not receive loans until after the first three loan recipients pay their first instalment." *(The group-based lending strategy results in peer/group pressure, and ordinarily group cohesiveness develops rather quickly).*

"But when do I get the cheque?"

"After you gather your group or join an already existing group, together we will assess the kind of training that you need before starting a business. You take the training, and in the process determine how much money it is going to take for you to get the business going, how much you foresee making after you start the business, and how you are planning to save the required 15 percent of the profit" *(Savings model).*

"Is that when I receive my cheque?"

"At that point, your group decides which members are ready for a loan. We also decide if you get your loan in a lump sum or if you receive it in instalments."

"Great! And then what?"

"You will have three years to pay your loan back, and you can also reapply in three years. After that, you would have become a business woman whom we will refer to a bank."

"Sounds a bit scary . . . "

"There's nothing to be scared about. We will be following up on how things are going with the business and will be providing you with consultation, assistance, and additional, ongoing training, if you need it and choose to take it. You need to understand that you, the staff who work here, the women whom you will join in the group—all of us become a part of this business of yours because we all have something invested" *(Partnership).*

"Where do I go to make payments? Here?"

"Not necessarily. There are some cooperative offices in different parts of the area where women gather either to form groups, to take training, or to conduct their business. You can pay in the office closer to you."

"What do you mean by saying that the women conduct their business in these cooperative offices?"

"There are women in the business of weaving, knitting, embroidering, all kinds of crafts. Many of them go to the cooperative offices and do their work there."

"I would not be able to do that. I have two children who are still small and require my staying home with them."

"That is not a problem. Women bring their children to the co-op office when they come to work. Maria, we have guests here today who we are taking to visit some of the women's businesses to whom we have given loans. We are going to the marketplace and to one of the co-op offices. Do you have time to join us?"

"Sounds like a good idea."

"Would you like to meet some of the women who are growing herbs? One of them might agree to help with your training and with starting your business."

"I'd love to!"

⚜

In the marketplace women are selling produce that they purchase wholesale from women who have received loans from ZWFT; others are selling what they have been funded to grow themselves. Caterpillars, corn, rice, mealie meal, and greens are some of the most popular products. The woman in the "freezits" business is there selling her product. We learn that she now employs six young men to help with her business. We visit the stalls of funded women who buy and sell crafts materials to women who are funded by ZWFT for craft-making and craft-selling businesses. Bangles, baskets, musical instruments, and stone figures are particularly popular items. Some of the businesses have became family enterprises, where the children and husbands are now employees of the women who started the enterprises.

In the route to the co-op office we stop at an herb-processing laboratory run by a university professor of pharmacy. He is buying from some of the women in the herb-growing business for his experiments with different kinds of indigenous medicinal remedies, as well as for his involvement with the growing international market for potpourri.

We also stop for lunch at a locally-owned and operated restaurant featuring typical food geared to tourism. This sort of establishment was virtually non-existent in Harare until ZWFT funded the owner of this restaurant. It is now one of the best of its kind in Harare. At the co-op office—one of five in Harare—a group of women is waiting to make their loan payments. Other women are talking about forming a group. Still other groups are weaving or receiving lessons from a master in the craft of their interest *(on-site training)*. Women are showing up with their babies "packed" (wrapped) on their backs. They are breast-feeding and telling jokes, and they are satisfied. They are essentially happy. I am impressed with the evident camaraderie, contentment, and fulfilment that they project. Despite the evident conditions of poverty, I sense hope and an implicit message of comfort and satisfaction that one does not easily find under the prevalent socioeconomic situation. I sense being in the presence of empowered women; I am experiencing a classic example of an NGO encouraging and strengthening existing communities.

While ZWFT leadership began to develop the idea of financing women's microenterprises in 1985, it did not start funding projects until 1991. More than 4 500 women have received loans of anywhere from a few Zimbabwe dollars up to between Z$4 000 and Z$10 000. Their repayment rate is 97%, and most of the women have been and continue in business for years.

The Role of NGOs

Nyang'oro (1993) argues that the economy of the sub-Saharan countries of Africa has performed in an extremely poor manner in the last decade: *"Bad government policy, natural disasters such as drought and famine, civil unrest . . . and the general structure and consequence of underdevelopment"* (citing Mittleman, 1988:283) characterise the African continent of today. According to the World Bank (1989), Africans are almost as poor today as they were 20 years ago. The economy has performed poorly, and foreign capital becomes less interested in investing in the region, causing peripheralisation of the country, the people, and the women in particular, from the global economy.

Zimbabwe is no exception to this discouraging picture. Although independence was obtained almost 20 years ago (1980) after 15 years of a struggle for freedom, the qualitative difference in living standards between the colonial period and post-colonist times has been minimal for the majority of the people. The economic development difficulties experienced by the country have resulted in the need for people to look for alternative means of meeting socioeconomic needs. The informal economy referred to as *magendo* has proliferated in response to the harsh socioeconomic situation.

In Zimbabwe, it is estimated that job creation in the informal sector may average 25 000 to 35 000 new employees a year. In general, however, *"women's lack of access to affordable credit, information, technical advice and services prevents them from expanding their enterprises"* (Manuh, 1998:7). In Africa in general, and in Zimbabwe in particular, women face the constraints of formal credit institutions demanding collateral in the form of landed property and male approval before loans can be made. They have to depend on money lenders whose loans are too expensive, and the repayment schedule is too short to allow for growth. Banking procedures require literacy, time, and documentation, not always possible or available, and the banking system believes that lending to the poor, particularly poor women, is risky (Waly, 1989).

In the context of the overall situation in Zimbabwe, NGOs have tried to help the plight of women in particular. Assessments of such NGOs range from praise to harsh criticism. The NGO practices in work with women that have been assessed as negative include ignoring the structure of households and of the social relations that influence women's roles in production, as well as the larger social processes that shape women's lives (Manuh, 1998). Other negative evaluations emphasise economic-related aspects, such as too early withdrawal of financial support, too many requirements before assistance is given, providing funding before adequate planning, giving inadequate explanations of implications of a loan, lack of following up and monitoring, inappropriate training, inadequate spelling out of mutual obligations, and handing over of financial management before the individual is ready (Hancock, 1996; Ziswa & Else, 1991; Bibars, 1998).

Despite the mixed, inconclusive, and frequently contradictory critiques resulting in lack of a consensus about the value of the role of NGOs in social development, negatively-ladened assessments of NGOs tend to "overpower" and "blur" the service/assistance provided by successful NGOs. There has been some tendency to ignore those NGOs that have made significant contributions toward *"making a better world"* (Nyang'oro, 1993:155). Waly (1998:6) makes it clear that:

> "successful programs depend on well trained . . . officers who would reach out for the poor, identify them, promote the program, train the beneficiary when needed and collect loan payments."

According to Waly (1998:7),

> "the best model for NGOs running a micro credit program would be lending cash and giving technical support in the form of advising on best suppliers, best markets, suggesting innovative activities to avoid market saturation and linking beneficiaries to each other so that they can complement each other's activities."

According to Bar-On (1998:3), the advantages of successful NGOs are well known:

> "They deliver services that governments find difficult to arrange and that the market is unable to deliver." He emphasises that, "it is mainly in the realm of moral values that NGOs are said to make their special mark in promoting social justice."

Looking Ahead

The illustrated example of ZWFT as an NGO speaks for itself in its upright ways of overcoming the earlier-mentioned criticisms. This is but one instance where an NGO is successfully making a difference in the lives of a segment of society in great need of assistance in Zimbabwe. The assessments of Ziswa & Else (1991:18) have concluded that NGOs are and have been an *"important element of Zimbabwe's future."* Several other writers and researchers (Clark, 1991; Bratton, 1990; Nyang'oro, 1993) go beyond the Zimbabwe situation and recognise the crucial role that NGOs have played in Africa, where they have been legitimised by their closeness and responsiveness to their constituencies, while governments have lacked the insight and flexibility to adapt to local conditions.

ZWFT illustrates that in a country generally characterised by, *"weak agricultural growth, a decline in industrial output, poor export performance, climbing debt, deteriorating social indicators, institutions and environment"* (World Bank, 1989:2), NGOs have the *"opportunity to exert badly needed leadership in addressing people-centred development"* (Korten, 1987:147). NGOs such as ZWFT have positively affected people by devising strategies to alleviate women's poverty, serving as a vehicle for empowerment, representing the interest of women at the grassroots level and developing a self-sustaining system of magendos that encourage and facilitate women's efforts to independently meet their needs.

According to Bratton (1990:91), developing organisations consist of *"efforts to maximise factors that affect realisation of goals."* ZWFT is playing a key role in women's realisation of goals, making a difference in the lives of people— differences that are and will be sustained beyond the period of NGO assistance (Korten, 1987). Projects can only be considered successful if they improve the overall quality of life of those involved, and if they contribute to the socio-economic development of society (Mapondera, 1998). Capable, committed individuals who combine a long-term vision with well-developed skills in leadership and management, all framed within altruistic desires, have relied on high moral purpose, good will, hard work, and common sense to achieve this success. Micro-enterprises funded through ZWFT have had long-term implications for improving day-to-day living conditions and social development in the Zimbabwean society.

By improving the situation of women, NGOs such as ZWFT are simultaneously strengthening the African society and enhancing the broader development prospects of the continent. We have seen that *"Africa is overflowing with women leaders. They lack only the training and the means to bloom"* (Manuh, 1998, citing Ndiaye Ba from the Women's Development Enterprise in Africa, p1). The key to what is needed for advancing the position for women in Africa is strengthening their capacities and skills and expanding their opportunities to fully develop their socioeconomic roles and power. As ZWFT is demonstrating, and as Waly (1998:1) emphasises very well in her work, there is evident need to *"back micro credit for macro changes."*

Giles Mohan

The Disappointments of Civil Society: The Politics of NGO Intervention in Northern Ghana

In recent years, and especially since the end of the Cold War in 1989, bilateral and multilateral donor agencies have pursued a 'New Policy Agenda' which gives renewed prominence to the roles of nongovernmental organisations (NGOs) and grassroots organisations (GROs) in poverty alleviation, social welfare and the development of "civil society" (Edwards and Hulme, 1996: 961).

Introduction

. . . [This selection] applies insights to an empirical study of state-NGO-society dynamics in northern Ghana. The study shows that strengthening civil society can create political tensions which ultimately undermine development. . . .

The Disappointments of Civil Society: Perspectives From Northern Ghana

The research upon which this section is based was enabled through contact with a UK-based NGO called Village Aid which works with partner NGOs[1] in northern Ghana, Sierra Leone, The Gambia and Cameroon. The research, undertaken within their northern Ghana programme, involved participating in programme evaluation and workshops, semi-structured interviews with their partners and reviewing project documentation. . . .

The External Determination of Local Agendas by Foreign NGOs

Within the realms of international civil society, a major line of tension exists between the northern and southern NGOs (N/SNGOs). Most donors and NNGOs work with local partner NGOs. Foreign interests may lack the local

From Giles Mohan, "The Disappointments of Civil Society: The Politics of NGO Intervention in Northern Ghana," *Political Geography,* vol. 21 (2002). Copyright © 2001 by Elsevier Science, Ltd. Reprinted by permission of Elsevier. References omitted.

knowledge or legitimacy to enter local communities so that partner NGOs are important gatekeepers in reaching the grassroots. Additionally, maintaining a fully staffed field office would be costly so that using local partners to deliver certain project elements is cheaper than using expatriates.

However, the notion of 'partnerships' is . . . a loaded process. Recent analysis (Edwards & Hulme, 1996; Bebbington & Riddell, 1997; Fowler, 1998) shows that the relations between partners is not even and that the funder tends to determine policy agendas to a far greater degree. As Nyamugasira notes, the NNGOs, despite commitments to participatory development, concentrate on ideas, networking, and education and leave the "time-bound, geographically fixed projects . . . to their Southern counterparts" (Nyamugasira, 1998: 298). While it would be tempting to place all the blame on an imperialising mission by NNGOs, the situation is far more complex. Fowler (1998) identifies various factors including paternalistic assumptions by the NNGOs, a bias towards their knowledge and procedures being superior, poor choice of field staff and a reluctance to release control of programmes that a true partnership requires. However, given the NNGOs' increasing reliance on official funding they too are pressed to show transparent success which breeds conservatism and a wariness to hand over the reins to local partners. On the SNGO side, many do lack capacity and transparency and react aggressively to any suggestions by the northern partners that this is the case. There are few transparent mechanisms for decision-making with limited methods for enshrining the principle of participation at all levels of the partnership. Partnerships are clearly suffused with political inequality which compromises the notion of an independent civil society emerging.

As donors seek out reliable and successful NGOs a market for development finance emerges where once small, agile, innovative and, at times, radical organisations quite rapidly become development 'success stories' and receive large inflows of foreign capital (Moore & Stewart, 1998). Such rapid growth is a problem for any organisation, but more importantly the funders tend to treat these organisations as infinitely flexible and capable of delivering any number of development competencies (environment, gender, water, health, etc., etc.). As a representative of NGO 'A' said "it's not easy to chew" (Interview with AS of NGO 'A', 4/2/00) implying that the organisation's 'mouth' became too full with demands from funders to deliver programmes at the grassroots. As the aid market contracts, the trend will be towards niched NGOs or larger, semi-commercial organisations which can deliver entire programmes for donors. In this sense civil society begins to massify and commodify with power increasingly concentrated in the hands of a few large organisations which may well be antithetical to a competitive and 'free' civil society.

Operationally, those NGOs receiving official aid have to be accountable to their funders which often brings conflict because the slow, flexible, culturally-specific processes do not translate easily into the fast turnaround demanded by log frame accounting techniques. On the other hand, the NGOs have a grassroots constituency, the supposed beneficiaries, who are increasingly alienated from the centres of decision-making. This double legitimacy bind leaves the

NGO somewhat stranded and far away from its developmental role (Bebbington & Riddell, 1997). This is clearly demonstrated in terms of 'capacity issues' where northern partners insist on transparent accounting procedures and systematic monitoring and reporting systems (Fowler, 1998; Moore & Stewart, 1998). For some organisations this is seen as an imposition since they were established precisely to break from this bureaucratic tradition. It also runs counter to much 'participatory development' which valorises local knowledge yet the organisations impose management systems drawn from the western corporate world. For Village Aid and its partners in northern Ghana, this has produced considerable tension over the past five years which has, at times, spilled over into outright confrontation. The following extracts both use the phrase 'dictatorship' when referring to Village Aid

> But there seems to be too much dictatorship from Village Aid. Information flow is not adequate, ideas from partners are not respected or not taken (NGO 'B', 1999: 2);

> dictatorship . . . sometimes Village Aid wants to be very strict . . . as if they have no confidence in us . . . Decisions taken at the Village Aid level before coming to the agency level . . . People have not been asked to participate at the planning level (Interview with NGO 'A', 4/2/00).

From Village Aid's position these issues related to inadequate capacity and commitment to participation shown by their partners. I return to this point below as we see how the partner NGOs negotiate and manipulate their own position. . . .

The Relationships Between NGOs and Village Organisations

By and large NNGOs tend to use partners for village level activities. The underlying assumption is that the northern NGO lacks the local knowledge and connections to represent the local communities so that intermediaries are needed. It is precisely this perceived 'closeness' to local communities and understanding of their cultures that gives the SNGOs their power. In practice, this assumption is not always borne out. In many cases the local NGOs behave in equally patronising, dictatorial and bureaucratic ways towards the villages they represent. The following extracts suggest that Village Aid's partners suffer from many of the supposed problems of inflexible state bureaucrats:

> 'B' and 'A' are still adopting old-style possessive tactics towards their client villages (Smith, 1999: 8);

> They (the villages) are our people (Interview with NGO 'A', 4/2/00).

The SNGOs are taking ownership of local culture and using it as a defence mechanism. The NNGOs realised they needed to have intermediaries, ideally working in a partnership relation, but the SNGOs use this powerful position to protect their constituency of villages. They claim to represent the local communities, but have rather patronising attitudes towards them, but know they are beyond reproach. In this way civil society organisations actually impede democratisation and good governance. . . .

Another manifestation of this problem is where partner NGOs intervene between the Northern NGO and the village organisations. Again, this represents a complex politics of knowledge generation and communication. In particular 'participatory learning and action' (PLA) has become a widely used research and conscientisation procedure whereby local communities generate their own knowledge which then helps in priority setting (Chambers, 1997; Mohan, 1999, 2000). A sustained critique is given elsewhere (Mohan, 1999), but what we see is SNGOs intervening in this process and transmitting alternative interpretations of reality which the NNGO takes as authentic needs.

> Participatory Rural Appraisals are undertaken by local NGOs as a duty in order to access funding. Consequently, village communities identify needs and problems and tailor their prioritisation of them to the services which they perceive the local NGO is offering. Subsequent to this, the local NGO amends the prioritisation and, more often than not, the nature of the project itself to what they believe its northern partner will fund (Waddington, 1997: 2);

> For the most part, projects are a response to the needs of a 'B' group within a village rather than the village in general (Village Aid, 1996b: 3).

The result is that the NNGO funds acceptable priorities which may not be the genuine priorities of the villagers. To combat such problems, Village Aid have begun a programme which works within existing cultural and linguistic systems so as to avoid imposing externally-driven practices.

The Autonomy of Both Foreign and Indigenous NGOs From the State

One obvious problem associated with strengthening the NGO sector at the expense of the state is that state institutions and actors feel threatened. The good governance and social costs of adjustment initiatives came in the late 1980s and early 1990s after almost a decade of adjustment. State officials had become used to loans and aid flowing through the state so that diverting much of this lucrative source of funds was bound to excite resentment. A countervailing problem is that under the adjustment process the state had to reduce its welfare bill while austerity created such hardship that any assistance in alleviating social problems was also a vital political resource. The compromise for many states was to welcome those NGOs which had a relatively circumscribed social welfare agenda and not those which might have more transformatory political agendas.

The mechanisms for dealing with NGOs reflects these shifting and paradoxical tensions. Most common is regulation, usually taking the form of official registration (Bratton, 1989; Gary, 1996). More 'arm's length' influence can be via co-ordination through an umbrella organisation or forum which seeds debates and is selective about which NGOs are members and, therefore, privileged in terms of resources and information. More heavy-handed influence has been via co-optation where the state takes over partial functioning of NGOs via QUANGOs [quasi-autonomous nongovernment organizations]. Additionally, as the frontier between state and society has further muddled, many politicians have established their own NGOs as patronage structures for capturing foreign aid and promoting themselves in their local constituency. Finally, dissolution and harassment is the final form of state influence, especially with those NGOs taking critical views of the state. In Ghana we have seen various strategies adopted by both the state and NGOs. In this sub-section I look at . . . key sites of interaction between the state and NGOs. . . .

NGOs as Patronage Structures and Party Political Vehicles

. . . [I]nsidious is the use of NGOs as vehicles for personal and party political gain by local officers. This is achieved through various mechanisms—petty corruption, largesse, interlocking political affiliations, and 'status'—and, as we have seen, the less obvious ways in which indigenous NGOs defend local culture in the face of 'outsider' intervention. In effect, some NGOs become fiefdoms for local élites to further their material and political status. As Hibou notes,

> The promotion of NGOs leads to an erosion of official administrative and institutional capacity, a reinforcement of the power of elites, particularly at the local level, or of certain factions, and sometimes stronger ethnic character in the destination of flows of finance from abroad. In many cases, these NGOs are established by politicians, at the national and local level, with a view to capturing external resources which henceforth pass through these channels on a massive scale (Hibou, 1999: 99).

Similarly, "the emergence of opportunistic organisations that call themselves NGOs but have no popular base at all. Many have been created as survival strategies for a professional middle class" (Bebbington & Riddell, 1997: 111). While 'A' and 'B' are by no means so opportunistic, both directors aspire to political power within Dagbon society. This is being pursued through various channels, one of which is the NGO. For example, the Director of 'A' is also a District Assembly member and the NGO is seen by villagers as indivisible from the former ruling party. Similarly, the Director of 'B' has been disciplined on various occasions for writing cheques without accounting for the destination of the money. His response was one of indignation in that Village Aid were behaving in a dictatorial and untrustworthy way. Recently, he was placed under investigation by the National Bureau of Investigation and the NGO's operations suspended pending the enquiry.

Decentralisation

Earlier work on local government showed that one of the key problems at the level of programme delivery is that NGOs have tended to set up parallel systems alongside a weak and under-funded local government system (Mohan, 1996b). On the other side, the decentralisation programme has been hampered by institutional dualism whereby local departments answer to central ministries and are not flexible with respect to local needs. The outcome is mutual mistrust and wasteful duplication of effort. This tendency is sufficiently widespread to exact comment from the outgoing Minister for Local Government who observed "the tendency of some of them (NGOs) to bypass laid-down structures and procedures at the district level and establish structures and programmes of their own without regards to their sustainability" (Kwamena Ahwoi reported in the Daily Graphic, 14/12/99). Within Village Aid's work similar problems occur:

> there existed a communication gap between the Project and the local extensionist in the sense that instead of communicating through District Officers the Project went straight to the local extensionist thereby marginalising the District Officers (Village Aid, 1999b: 3);

> District Agricultural Director's and District Forestry Officers appear to see themselves on the end of decentralisation policy and not a part of it (Village Aid, 1999a: 7).

As Aryeetey (Aryeetey, 1998: 308) comments "Neither the assembly members nor the technocrats in the district assemblies were seen to be in a position to make serious contributions towards strengthening consultation between the communities and the assemblies". The result is that contrary to the advocates of civil society, supporting NGOs does not lead to regularised interaction between society and state and in the process build the strength of both. In fact, it alienates the two even further and could undermine the longer term aim of building citizenship rights.

Conclusion: A Third Way for the Third World

. . . An assumption of the civil society route to development is that 'self-help' can reduce external dependency, because the local organisations more effectively 'own' the process. By multiplying the number of stakeholders it is assumed that more consensual and democratic development can be achieved which is ultimately more sustainable. In the Ghana case this clearly did not occur. While the number of development organisations mushroomed in the wake of structural adjustment and the drivers of this process championed partnership, the local 'partner' organisations and, more importantly, the rural poor were marginalised from decision-making. The paradox is that external NGOs, often heavily funded by their home governments, are charged with 'empowerment', but are so wary of upsetting their funders that they tightly circumscribe the activities on the ground and completely undermine independent development. Their partners are then trapped in an irreconcilable position of being the authentic representatives of their grassroots constituencies, but being accountable to organisations outside the locality. Squeezed in such a way they usually defer to the funder and present to them a relatively trouble-free view of local communities and their development needs, all of which further marginalises and alienates the rural poor.

The aid paradigm means that civil society organisations become more dependent on external funders as well as the market. The NGOs largely become service delivery mechanisms for pre-determined development agendas. In competing for these scarce aid resources, the NGOs position themselves strategically which creates tensions between organisations. There is nothing wrong with debate and contestation between organisations and political actors, but most of the conflict in northern Ghana was about scrapping over the spoils and not ideology. Indeed many NGOs have been set up precisely to divert aid for personal goals as opposed to responding to the needs of the poor. As the aid market has woken up to the opportunism of many so-called NGOs it has tightened up its funding criteria, which might alleviate some corrupt behaviour, but actually makes it more difficult for smaller, less professionalised organisations to succeed. It also works against the philosophy of much participatory development which seeks to valorise multiple local differences rather than impose a rigid model. This in turn undermines the democratic potential of civil society as only some interests are actively represented, whether they be ethnic, class or political party. Rather than reflecting social differences the uneven promotion of civil society covertly strengthens social divisions, promotes factionalism and deepens the marginalisation of some groups.

As we saw the real beneficiaries of strengthening civil society have been the local elites. Increasingly, we see a tier of professional NGO managers who use foreign aid and locally generated income as a means of achieving or consolidating their middle-class status. Similar processes have been observed in Britain with most social economy initiatives being "run by outsiders—professional social entrepreneurs who bring with them what amounts to an *ideology* of community empowerment which they can then set about enacting with local people" (Amin, Cameron & Hudson, 1999: 2041). In emphasising local knowledge, grassroots initiative, and community development this ideology of empowerment generates a discourse of discrete and bounded places amenable to a particular form of intervention that only they, albeit in partnership, can largely control. Again, the rural poor are only brought in as members of fictionalised 'communities' and are in practice denied any real voice.

In all instances the state, rather than being a detached political actor separated from society by a supposed ditch, is deeply implicated in these activities. The Ghanaian state has established its own 'non-governmental' organisations, confined registration to 'non-political' organisations, and engineered debate about the 'need' for economic liberalisation. At the local level, state agencies and NGOs are in some cases relatively separate but the civil society organisations do not channel opinion into government or scrutinise its operations, but treat it with profound mistrust and often duplicate its efforts. No synergy, in a formal sense, exists here between state and civil society. However, in more subtle ways the state and NGOs are mutually implicated. The central state in Ghana has used civil society organisations to drive local politics and actively promoted decentralisation as a means of consolidating rural support. On an even more subtle level we saw how some NGO officials purposefully misrepresent themselves to blur the boundaries between civil society and the state in an attempt to present themselves and the party in a positive light by utilising the financial resources of the NGOs. As Bebbington and Bebbington (Bebbington and Bebbington, 2001:9) comment the emphasis on society "diverts attention from the webs of relationships that link civil society organizations and the state and that may offer the prospect of changing forms of state action". I return to the possibility of changing the state below.

These developments within civil society are very much in keeping with Marx's theorisation in the 19th Century. That it is a normative concept whose realisation serves the interests of the (international) bourgeoisie. Hearn (Hearn, 1998, 2001) and Beckman (Beckman, 1993) suggest that the emphasis on civil society in Africa is central to modern imperialism in which new institutional actors are added to the array of players seeking to delegitimise the third world state and further erode what little sovereignty it has left. Hearn (Hearn, 1998: 98) comments that "In the immediate post-colonial period, the comprador class consisted of government official and private sector entrepreneurs and managers, in the 1990s it includes leaders of the voluntary sector". This neo-compradorism simultaneously fictionalises and factionalises civil society in a bold new experiment in socio-political engineering which aims to weaken the state, cheapen the cost of aid and promote market-based freedoms.

Specifically this imagining of civil society has a number of important political ramifications. First, it cheapens aid through match funding and the whole ethos of self-help. While self-help is more likely to embed ownership and inject greater relevance into projects it also serves to place the burden for poverty alleviation on the structurally poor which, in turn, leaves NGOs *de facto* legitimising SAPs [Structural Adjustment Programmes] by filling in the welfare delivery gap. As Hintjens (Hintjens, 1999: 386) comments "The state is no longer to be held accountable for ensuring that citizens' basic needs are met; instead private citizens, individually and collectively, are expected to provide for themselves, however poor or disadvantaged they may be". Also, as we have seen, those groups and institutions able to provide some match funding may well do so to ensure the favourable direction of aid which does not necessarily benefit the most poor and marginalised.

Second, and closely related to the first, is that 'partnership' and devolution might serve to spread the risk to 'locals'. Complex scenarios in which "social development is a matter of tidying up after the market" (Pieterse, 2001: 126) leaves development organisations in an awkward position, because there is little they can do to affect broader structures. In this sense, partnership becomes an insurance policy against lack of effectiveness. As with 'policy slippage' under SAPs, the donors can implicate the poor in the failure to achieve development which becomes a subtle form of blaming the victim. In the UK context Amin, Cameron, & Hudson (1999: 2049) observed "Very few social economy projects are underwritten by public authorities . . . (and) . . . Even less risk is borne by the private sector". Participatory development can be seen as a sensitive form of empowerment when it works or the result of grassroots incapacity when it fails.

Third, the question of risk, opens up the relationship of civil society to the market. As the Standard Chartered case in Ghana showed, the NGO effectively underwrites the risk on a multitude of small credit schemes. In this way the promotion of civil society is, as the Marxists have argued, very much about creating the conditions in which private capital and entrepreneurialism can flourish. Additionally, the NGO sector in Ghana has been at the forefront of the negotiated consensus around further liberalisation while the veneer of stable processual democracy acts as a major stimulant to inward investment from multinationals.

Finally, the emphasis on localism has a number of effects. On the one hand it factionalises and fragments political opposition. As Mamdani (1996: 300) argues "all decentralized systems of rule fragment the ruled and stabilize the rulers" so that the emphasis on atomistic civil society repeats many of the problems of governance laid down under Indirect Rule. Central regimes are often happy to promote development programmes which seek to build upon local energies because this absolves them of responsibility for welfare provision, earns political capital by being sensitive and dialogic, and disaggregates society into a series of unconnected, both spatially and politically, 'issues'. Potential alliances and solidarity against the structural forces generating poverty are undermined as civil society actors literally scrabble for the pickings of the aid regime. Where networks of local NGOs develop it is often to enhance efficiency of delivery, such as avoiding spatial overlap or sharing 'best practice', than it is to actively lobby the state or international organisations. So, localism diverts attention from the structural causes of poverty and feeds into the belief that market-based globalisation can and should be harnessed to work for the poor. . . .

Donor and NGO support for civil society, and 'localism' in general, keeps at bay debates about more fundamental structural changes to, say, unequal property rights or despotic, but economically useful, host governments. One thing that has emerged from discussions is that 'local' action must simultaneously address the non-local. As Nyamugasira (Nyamugasira, 1998: 297) observes NGOs "have come to the sad realization that although they have achieved many micro-level successes, the systems and structures that determine power and resource allocations—locally, nationally, and globally—remain largely intact". Recent efforts have begun to deal with these limitations by looking at strategies for 'scaling up' local interventions (Blackburn & Holland, 1998; Whaites, 1998). Only by linking participatory approaches to wider, and more difficult, processes of democratisation, anti-imperialism and feminism will long-term changes occur. Crucially, greater and more critical engagement with the state is required although this is incredibly difficult where states, donors and other aid organisations delimit the political space open to civil society. One route for this is more accomodatory via the recent emphasis on citizenship and rights which seeks to generate greater 'synergy' between state and society through the promotion of social capital and civic engagement. A second route is more radical and involves civil society actors opposing the dominant development discourse and challenging local, national and global structures. While the Zapatistas [Mexican rebels] have become the *leitmotif* [recurring theme] of this form of political action, it remains to be seen whether a new cohort of political leaders and agents emerge which actively reject the present neo-liberal consensus.

Note

1. Given the political nature of the inter-NGO relations discussed in this section, the names of these NGOs have been concealed. Village Aid works with two main SNGOs which I have re-named 'A' and 'B'.

POSTSCRIPT

Are Non-Governmental Organizations (NGOs) More Effective at Facilitating Development Than Government Agencies?

There are a number of sub-issues that flow out of the discussion regarding the role and efficacy of NGOs in fostering the development process in Africa. They include those related to the participation of local communities in development projects, the influence of donors on NGOs, and the limitations of NGO success when broader structural problems persist.

Participation has been a buzzword in NGO circles since at least the early 1990s. The emphasis on participation came about in reaction to a history of top-down development projects that left local people with little to no sense of ownership of the projects undertaken by outsiders in their communities. In order to foster community participation, NGOs have employed a range of techniques (e.g., participatory action research (PAR), participatory rural appraisal (PRA), and participatory learning and action (PLA)), to facilitate community problem identification and group problem solving. See numerous publications by Robert Chambers, such as *Whose Reality Counts? Putting the First Last* (Intermediate Technology Publications, 1997). The problem, and a latent contradiction, is that communities may identify problems that NGOs are not prepared to solve. The result is that communities often identify problems and select programmatic solutions that they think an NGO can deliver on.

As NGOs increasingly receive funding from bilateral donors, there is a concern that they may become extensions of donors rather than autonomous development agents. This could mean that NGOs actually change the nature of their programming to suit the desires of donors rather than catering to the needs and wants of the communities they are serving. In contrast, and somewhat counterintuitively, it could be argued that overattention to donor priorities might actually increase participation in some instances if the donors themselves are insisting on participatory approaches. See an article by Alan Fowler in a 1998 issue of *Development and Change* entitled "Authentic NGDO Partnerships in the New Policy Agenda for International Aid: Dead End or Light Ahead?"

One of the potential limitations of NGO projects, and the project approach to development in general, is that local social change may be circumscribed by broader structural problems. Even more troubling is the possibility that "micro-level successes" might actually work against broader-scale social change if they make people content with a bad situation, that is, have a palliative effect.

ISSUE 5

Should Developed Countries Provide Debt Relief to the Poorest, Indebted African Nations?

YES: Dorothy Logie and Michael Rowson, from "Poverty and Health: Debt Relief Could Help Achieve Human Rights Objectives," *Health and Human Rights* (1998)

NO: Robert Snyder, from "Proclaiming Jubilee—for Whom?" *Christian Century* (June 30–July 7, 1999)

ISSUE SUMMARY

YES: Dorothy Logie, a general practitioner and active member of Medact, and Michael Rowson, assistant director of Medact, argue that debt is a human-rights issue because debt and related structural adjustment policies reduce the state's ability to address discrimination, vulnerability, and inequality. Debt relief, if channeled in the right direction, could help reduce poverty and promote health.

NO: Robert Snyder, an associate professor of biology at Greenville College, counters that debt cancellation will only work if the factors that created debt in the first place are addressed. He uses a case study of Rwanda to demonstrate why political and social change must occur for debt forgiveness to work.

As Dorothy Logie and Michael Rowson note in their selection, 34 of the 41 most indebted nations in the world are found in Africa. Furthermore, these same authors note that 23 percent of new aid receipts in Africa are spent on debt repayment. Responsibility for the debt crisis in Africa is complex and fairly difficult to attribute to any single group of individuals or a single entity. Resolution of this problem is hugely controversial, and it has been a matter of intense debate over the past several years.

Debt has accumulated for African governments for a number of reasons. First, many African governments pursued a policy known as import substitution in the 1960s and 1970s. Rather than importing manufactured and processed goods from abroad, the idea was that African governments should foster enterprises within their own national borders in order to produce goods that would "substitute" for imports. This approach, conceptualized by the Latin American economist Raul Prebisch, was designed as an antidote to the perpetual quandary of developing countries trading inexpensive commodities for expensive manufactured goods. The problem was that these state-run enterprises were often relatively uncompetitive at best (i.e., they produced goods at a cost higher than the imported competition) or a serious drain on state resources at worst (especially if they were used as sources of patronage by the ruling elite). As a consequence, the privatization of inefficient state-run enterprises was an important aim of structural adjustment policies.

Second, while many African economies grew rapidly in the 1960s and early 1970s, this growth began to temper significantly in the late 1970s and 1980s. Part of this slowdown in growth has been attributed to a global rise in petroleum prices at the time—a problem for the majority of African nations that were (and are) petroleum importers. Even after African economies began to slow down, commercial banks and public lenders often lent to African governments based on previous levels of growth—contributing to unhealthy levels of indebtedness.

Third, the loan officers at commercial banks, and even at multilateral institutions like the World Bank, were (and are) often under a certain amount of pressure to make loans. This meant that creditors encouraged African governments, in some instances, to obtain loans they really should not have been taking. Finally, some African government officials inappropriately used borrowed monies, making it difficult to pay these loans back in a timely fashion.

The current high level of indebtedness in many African countries has led NGOs, think tanks, and religious organizations to call for debt forgiveness. Others have warned about the dangerous messages that debt forgiveness will send to future borrowers. In the following selections, Dorothy Logie and Michael Rowson contend that debt is a human-rights issue and that debt relief, if channeled correctly, can help the poorest, indebted African nations. In contrast, Robert Snyder states that the factors that created debt in the first place need to be addressed in order for debt cancellation to work.

Dorothy Logie and Michael Rowson  **YES**

Poverty and Health: Debt Relief Could Help Achieve Human Rights Objectives

UDHR [Universal Declaration of Human Rights] Article 25

1. Everyone has the right to a standard of living adequate for the health and well-being of himself and of his family, including food, clothing, housing and medical care and necessary social services, and the right to security in the event of unemployment, sickness, disability, widowhood, old age or other lack of livelihood in circumstances beyond his control.

2. Motherhood and childhood are entitled to special care and assistance. All children, whether born in or out of wedlock, shall enjoy the same social protection.

. . . [O]n the fiftieth anniversary of the UDHR [Universal Declaration of Human Rights], human rights activists still confront the bitter reality of widespread hunger, disease and discrimination in the context of a global economy characterized by widening inequality and poverty. Although the preamble to the UDHR reaffirms the worth of each individual person and the equal rights of men and women to progress towards a better standard of living, and although the idea of freedom from poverty has been in human rights law and discourse since the adoption of the UDHR, poverty-related issues have not been given priority in the human rights agenda. This anniversary represents both an opportunity to push poverty to the front of that agenda, and to promote action that might enable its reduction.

Poverty and Ill-Health

Poverty, in both its absolute and relative forms, is the single most important driver of ill-health in the world today. This is hardly surprising. In developing countries, the pathways by which low levels of economic and social well-being affect health are easily found and are the daily reality of hundreds of millions. These pathways include: lack of access to safe water (experienced by 1.2 billion people); lack of adequate sanitation; poor housing; low income (at least 1.3 bil-

From Dorothy Logie and Michael Rowson, "Poverty and Health: Debt Relief Could Help Achieve Human Rights Objectives," *Health and Human Rights,* vol. 3, no. 2 (1998). Copyright © 1998 by The President and Fellows of Harvard College. Reprinted by permission of The FranÁois-Xavier Bagnoud Center for Health and Human Rights and The President and Fellows of Harvard College. Some references omitted.

lion people live on under $US1 per day), and lack of access to health services (faced by 800 million people). Discrimination against women, the elderly, ethnic minorities, refugees, the disabled and other marginalized groups both causes and magnifies these problems. Many other factors, such as conflict and adverse climate changes, for example, can also contribute to poverty and ill-health. Another contributor can be the skewed spending priorities of governments which may have high levels of military expenditure or may devote most of their health budgets to secondary health care for urban citizens, at the expense of primary and preventative health services.

Human Rights Response

Faced by such diverse causes of poverty and ill-health, human rights activists obviously need to work at international, national and community levels, and both within and outside the health sector, to make their anti-poverty agenda effective and credible. . . . Here we will focus our attention at the level of the international economy.

The anniversary of the UDHR is taking place in the midst of a widespread economic crisis, which threatens to lead to a global recession. . . . For many countries, a new recession may be only the latest stage in a downward economic record—since 1980, 100 countries have experienced economic decline or stagnation, and 1.6 billion people have seen their incomes reduced. This decline has been most obvious in sub-Saharan Africa. . . .

Debt and Economic Adjustment

Today a large burden of foreign debt is carried by over 40 of the world's poorest countries. This potentially has multiple effects on their economies and acts as a significant constraint to improvements in international health. First, it undermines prospects for economic growth (and thus poverty reduction) by discouraging public and private investment; private investors, in particular, fear the higher taxes, higher inflation and currency speculation that an unsustainable debt burden can bring. Second, debt repayments siphon away precious foreign exchange needed to buy the imports essential for economic growth and the maintenance of health systems. Third, debt repayments divert money from government budgets which could be used for health, education and poverty reduction initiatives. And fourth, they use up sources of foreign exchange (other than export revenues) such as grants and loans from government and multilateral donors. For poor countries, the accumulation of a large foreign debt can thus be the prelude to a catalogue of economic disasters.

The Scale of Sub-Saharan African Debt

> Many states in Africa lack the financial capital needed to address basic expectations and fundamental needs. This is one of the central crises in Africa today, and one that is due in large measure to the problem of African public sector debt.[1]

In the early 1980s, the developing world as a whole faced foreign debts of around US$800 billion—the results of a decade of irresponsible actions on the part of both creditors and borrowers. Today, total debt stands at over two trillion US dollars, despite large repayments in the intervening years. Sub-Saharan Africa, containing 34 of the 41 most heavily indebted poor countries, is worst affected. Taken together, its US$230 billion debt is small compared to debt in other parts of the world, but compared to its earnings from exports or its total GDP [gross domestic product], the debt represents an overwhelming burden. More money is spent on interest payments than on health and education, and many countries are unable to afford even the most basic annual health package, estimated by the World Bank to cost around US$12 per capita. In Uganda, with one of the highest infant and maternal morality rates (and an AIDS epidemic), the government spends annually only US$2.50 per capita on health. Several other heavily-indebted countries are spending equally small amounts on health, leading to an increasing reliance on cost-recovery mechanisms (such as user charges) to cover the widening gap between health needs and budgetary capacity.

Many countries in the region are literally bankrupt, and are in the process of building up huge arrears of debts that can never be repaid. What repayment they do manage is financed by revenues from exports and other money from the domestic budget, new loans (thus incurring further debts), and grants from aid donors. It is sobering to note that in 1996, twenty-three percent of all aid given to sub-Saharan Africa was spent on debt repayments to financial institutions and governments in the North.

Structural Adjustment Policies

In return for either delaying debt repayments or lending more money, international donors such as the World Bank and IMF [International Monetary Fund] have demanded that countries undertake policy reforms, known as structural adjustment programs, aimed at reducing domestic demand by raising interest rates, devaluing currencies and reducing public expenditure. They also recommend that the state's role in public life be reduced by, for example, eliminating subsidies for food and fertilizers, relaxing foreign investment regulations, privatizing state-run industries and services, and cutting government bureaucracy (including in the health sector). . . .

Economic Crisis, Adjustment and Vulnerable Groups

Article 25 of the UDHR specifically points to mothers and children as social groups needing special attention. Today we might extend this concern to other vulnerable groups such as women, the disabled and people living with HIV/AIDS. These groups are at greater risk of suffering during economic crises for a range of reasons, most notably related to discrimination. Debt and inappropriate economic adjustment can reduce the state's capacity to intervene to correct discrimination, vulnerability and inequality, thus putting these groups at even greater disadvantage in difficult times. . . .

Due in part to the lack of funds which can result from the constraints debt and adjustment impose on social sector spending, international and national plans for improving the health of women and children remain to be fully carried out. For example, 40 African countries have prepared National Plans of Action in response to the global Plan of Action emanating from the 1990 World Summit for Children which incorporated priority health goals for children and for women. These plans include raising immunization rates, improving oral hydration, eliminating iodine and Vitamin A deficiencies and encouraging breast-feeding. But implementation has been slow or nonexistent due to lack of both money and political will.

. . . Africa is the only part of the world in which the number of children out of school is increasing. In Niger, for example, fewer than one-quarter of children attend primary school, and less than 20 percent of these are girls. Where it exists, school often means a mud hut with a leaking roof, classes of 40 or more students and a chronic lack of teaching materials. In Zambia, expenditure on primary schools is now at less than half its mid-1980s level. This is a denial of the right to education as stated in Article 26 of the UDHR and also of Article 15 of the Convention on the Rights of the Child (which has been ratified by all of the governments of Africa except Somalia).

Reproductive Health

Article 25 throws the spotlight on motherhood as a condition during which women require "special care." However, in what has been described as "the health scandal of our time," 585,000 women still die each year from pregnancy-related causes and many times that number are incapacitated as a result of child-bearing. In parts of sub-Saharan Africa maternal mortality is still rising and, a decade after the introduction of the 1987 Safe Motherhood Initiative, its implementation remains frustrated not only by lack of funds and political will but also by continuing discrimination against women.

. . . Low levels of spending on health can result in inadequate medical and physical infrastructure, maternal malnutrition, user charges, poor-quality staff training and motivation, unreliable blood supplies and lack of drugs. . . .

HIV/AIDS

AIDS has become a marker for injustice, discrimination and lack of realization of human rights. The virus thrives on poverty, social disruption, ignorance and accelerated urbanization: its spread is encouraged by lack of resources, lack of clean water, commercial sex, and rapidly declining health services. Each of these factors is potentially exacerbated when adjustment policies lead to budget cuts, unemployment, and migration. . . .

Using Debt Relief to Promote Gains in Health

After a decade of vociferous pressure from NGOs [nongovernmental organizations] and others, both creditor governments and the international financial institutions have moved to a point where they have agreed to put a mechanism into place (the Highly Indebted Poor Countries Initiative) which will cancel some low-income country debts. It is believed that by relieving a proportion of foreign debt, economic growth will be encouraged and, if the relief is deep enough, this should free resources that governments could then use to tackle some of the enormous health problems outlined above. However, the process is slow, the relief is offered to too few countries and, even where provided, the relief is often not extensive enough to make a significant impact on human development problems. Oxfam has put forward a scheme which would require countries, in return for faster and deeper debt cancellation, to use the gains from debt relief to implement national development priorities agreed to by governments, civil society, and donors. They have also shown that for some highly-indebted poor countries, debt cancellation could provide a significant portion of the external finance needed to implement the National Plans of Action agreed to after the World Summit for Children.

In the long-term, debt relief could reduce the need for countries to undertake economic adjustment—with its technical-fix approach to health as a commodity and the resulting emphasis on introducing user charges, competition, and other market mechanisms into the delivery of health care. Countries could then be free to reorient their vision of health towards universal primary health care (as contained in the Alma Ata Declaration) and to emphasize a preventative public health agenda.

The Problem of Conditionality

. . . [H]uman rights activists and health professionals might link up to ensure that debt relief does help to reduce poverty and promote health, and that it does so in a fair way. For example they might:

- work to ensure that the debt relief process is open and accountable and not dominated by the interests and agendas of creditors, but also includes the concerns of governments and civil society representatives;
- help to highlight and build upon schemes proposed by developing countries themselves—schemes that use debt relief to achieve human rights objectives. In fact, several poor countries (such as Uganda) already have human-development-oriented schemes waiting in the wings to be financed by the proceeds from debt relief;
- make sure that debt relief is not conditional on health or economic system reforms that are potentially harmful to health outcomes; and
- press for governments that have ratified human rights instruments (such as the Convention on the Rights of the Child) and the World Bank and IMF (which are components of the UN system) to modify their policies in accordance with these international obligations.

Conclusions

Debt relief is not a solution on its own, but if channeled in the right direction, and with the active involvement of civil society, it can achieve progress in reaching human development targets. There is a responsibility, and a legal obligation, for governments and international financial institutions to address the determinants of health (including income, education, safe water and food, and all human rights) in order to achieve lasting health improvement.

Human rights activists and health professionals should work toward addressing these issues by pressing for swift and generous debt relief. . . .

The fiftieth anniversary of the UDHR is a time to call again for poverty reduction and health equality to be at the center of development strategies. In the meantime, the UDHR remains a living document, a standard by which we can measure the world's imperfections and work towards righting wrongs—past, present and future. The more people—health professionals, nongovernmental organizations and activists—understand what human rights instruments have to say about poverty and health, the more these instruments can be used to lever change.

Note

1. K. Annan, from the "Report to the Security Council on the Causes of Conflict and the Promotion of Durable Peace and Sustainable Development in Africa," April 1998, quoted in Oxfam's submission to the UN Committee on the Rights of the Child, *Violating the Rights of the Child: Debt and Poverty in Africa* (Oxford: Oxfam, May 1998).

Robert Snyder  **NO**

Proclaiming Jubilee—for Whom?

Jubilee 2000 is gaining momentum. Centers for the movement have arisen in more than 40 countries, and numerous churches and nongovernmental organizations have signed on to the campaign. The goals of this movement, which seems to have originated with the All Africa Conference of Churches and is now centered in the United Kingdom, are best summed up in the apostolic letter issued by Pope John Paul II in 1994. It states: "In the spirit of the Book of Leviticus (25:8-12), Christians will have to raise their voice on behalf of all the poor of the world, proposing the Jubilee as an appropriate time to give thought, among other things, to reducing substantially, if not canceling outright, the international debt which seriously threatens the future of many nations." . . .

The idea is appealing. After all, there is no such thing as an international bankruptcy court which allows hopelessly indebted countries to declare themselves insolvent. Countries that have no hope of ever paying off their debt languish in a state of perpetual penury. The people of these countries barely eke out a living, while the banks owned by the wealthy prosper.

The world's financial institutions have recognized that something needs to be done to change this situation. The International Monetary Fund (IMF) recently started the Heavily Indebted Poor Country (HIPC) initiative, which singles out countries undergoing extreme financial stress. On the list are many African nations, including Rwanda, Burundi, Kenya and the Democratic Republic of Congo. Each country must pass a second screening to be eligible to receive some debt relief.

The Jubilee 2000 people claim that the relief proposed by the IMF is not enough. It does indeed seem to fall far short of what is needed. However, the concept proposed by Jubilee 2000 is riddled with pitfalls; to apply it universally would be naïve.

The economies of the heavily indebted countries would clearly benefit from debt relief. In countries with benevolent governments, the citizenry on the whole would gain. However, the socioeconomic structure of some of the heavily indebted nations is such that, in the long term, debt relief might only aggravate the condition of the poor.

As a former agricultural missionary in east and central Africa, I've learned that quick fixes can sometimes become excuses for not dealing with the more

From Robert Snyder, "Proclaiming Jubilee—for Whom?" *Christian Century,* vol. 116, no. 19 (June 30–July 7, 1999). Copyright © 1999 by The Christian Century Foundation. Reprinted by permission.

painful fundamentals of international and national problems. A poorly executed act of sympathy can exacerbate the problem that it is meant to solve. Consider Rwanda.

Until 1994 Rwanda was under the rule of President Juvénal Habyarimana.
Generally, Westerners liked him. From the perspective of international agencies, he was at worst a benevolent dictator, at best a progressive peacemaker
promoting development. Compared to many African countries, Rwanda experienced a time of stability and growth during Habyarimana's rule. We now realize, however, that he was a cunning power broker and, to a certain degree, a
racist. He made sure that the benefits of international aid projects accrued
mainly either to his extended family or to the northwestern region of Rwanda
from which he came.

The people of Rwanda's southern half were well aware of this inequity. All
Rwandans had to carry identity cards that showed their ethnicity. If you were
Tutsi, you faced discrimination whether you were from the north or the south.
Though 10 to 15 percent of the population was Tutsi, no Tutsi was allowed to
hold a leadership position in government or the military. A small group of Tutsi
ran profitable business enterprises, but they were well aware that the price for
the freedom to carry on business was not to interfere with or criticize Habyarimana's dictatorial hold. Rwanda's leaders drained the economy into their own
bank accounts, while making sure that no opponent could get enough political
strength to challenge the status quo. Habyarimana manicured his image for
Western donors, and aid dollars poured in. The government and the army put
on a friendly face to those of us working in the country.

The Rwandans were not fooled by this political masquerade. They understood the rules of the game, according to the former Rwandan minister of
defense, James Gasana, who escaped from Rwanda in 1993. An insightful moderate, he would probably have been killed for his political stance by the powers
that eventually led to the 1994 genocide. In a paper presented at the Ecumenical
Institute in Bossey, Switzerland, in 1996, Gasana stated that the Rwandan army
served only one purpose: to protect the power elite. This is not unique to
Rwanda. Says Steven Were Omamo of Kenya's leader: "[Daniel arap] Moi's government . . . is widely viewed as an engine of domination instead of the agent of
the popular will, more interested in maintaining old forms of influence and patronage for a minority than in expanding opportunity for the majority. This, I
believe, is the root of our current troubles." Wangari Maathai, the legendary
leader of the Green Belt Movement in Kenya, states: "Leadership in Africa has
been . . . concerned with the opportunity to control the state and all its resources. Such leadership sees the power, prestige and comfortable lifestyles that
the national resources can support. It is the sort of leadership that has built
armies and security networks to protect itself against its own citizens."

⌖

In countries such as these, the army and secret service are part of the political
machine. They silence their opposition and prevent any broad-based power
sharing. When I lived in Rwanda, one of my employees told me that his elderly

mother had tried to vote against the continuation of the Habyarimana regime and been prevented from doing so. When she then stated that an old woman with mud on her feet from the fields ought to be allowed to vote against the official who drives his Mercedes Benz to the polling booth, she was arrested.

Though it is hard to prove, it is widely accepted that some African leaders promote ethnic violence during election times or when their power is challenged. The powerful are willing to injure and kill people so that they can continue to feed unhindered on the country's resources. Mobutu Sese Seko, the former president of Zaire (now the Democratic Republic of Congo), so ferociously plundered his country's resources that at his death his estimated worth stood at between $5 billion and $10 billion. His country's national debt was $14 billion.

Even some of the church leaders in such countries become involved in power games and ethnic divisiveness instead of serving as champions of justice. They, too, may have a vested interest in maintaining the status quo. We only need to consider our own history of race relations to understand how this can happen. Sometimes the flow of international charitable aid into the church attracts self-interested people into the institution; not all church leaders are oriented to serving the people. Many courageous men and women of the church have fought for justice, but many others have manipulated the system for their own gain.

Do we need to do something to help deeply indebted countries? Absolutely. Is the industrialized world partly responsible for their plight? Absolutely. Do we want to encourage corrupt leaders by giving them money that will enable them to pretend to be benevolent lovers of the people? Absolutely not. If we are going to forgive debt, let us not fool ourselves into thinking that we can outsmart the cunning men and women who are experienced at manipulating the international community for their own benefit. These leaders who are so good at sleight of hand will empty our pockets while they throw a few crumbs to the poor, and then laugh as their own bank accounts grow.

If a country is governed by a small, corrupt power elite and the national debt is really the debt of that elite, then let them face their people without foreign aid. The international community placed strong economic sanctions on the former white South African government. Even though those sanctions also impacted the poor, no one called for their discontinuation. Everyone agreed that ending the evil of apartheid required stern measures. Why can't we see that apartheid-like policies also exist in other countries? The world has shut its eyes to the racist policies of Rwanda and Burundi. Instead of imposing sanctions, we want to forgive their debts. When Kenya's leaders stir the country's racial tensions into riots, we look the other way and then talk about forgiving the government's debts.

ᴇ⟨◉⟩ᴏ

Some will accuse me of paternalism and of ignoring our own guilt. But anyone who has lived among the people of countries with corrupt regimes has seen what happens when money comes in from the outside. The Jubilee 2000 campaigners claim to be aware of dictatorial and international power cliques. They state, "Jubilee 2000 calls for co-responsibility of debtors and creditors for the debt crisis. Remission of debt should be worked out through a fair and transparent process ensuring full participation of debtors in negotiations on debt relief." But can there be such a thing as "transparent processes" in countries where spies and guns counter any threat to the status quo? Why does it take a coup d'etat to change most African governments?

We will only increase our guilt if we inhibit necessary, fundamental changes from occurring in these countries. We recognized this in dealing with the former Rhodesia and South Africa. But not with Rwanda. We seem to be blind to black-on-black racism and corruption. Only fundamental change would have prevented the genocide in Rwanda. Only fundamental change will stop the incessant coups d'etat in nations where one group after another seeks to grow fat on the country's resources.

A groundswell of opposition to corrupt leaders is rising in several African nations. The West must not provide the leaders of such nations with the means to mollify their populations temporarily while they solidify their positions of power. Where the church is in bed with the government, it should also be considered suspect. At the same time, the church in the West must educate itself about our history of foreign political manipulation focused on protecting our own self-interests. This understanding should be a prerequisite to joining campaigns like that of Jubilee 2000.

Forgiving debts is a worthwhile enterprise, consistent with biblical teachings. But the admonition to fight for the oppressed must equally be kept in mind. Forgiving a national debt and freeing the oppressed are not necessarily the same thing. In fact, they may be opposites. Let us proceed cautiously. We should not help any poor country that has a large, internally focused military or secret service. We must deal with more than the superficial issue of debt relief. The West must acknowledge its role in creating and supporting corrupt dictatorships. The economic powers need to help poor countries ruled by benevolent governments to get a sure footing in the international economic system.

Ultimately, we must realize that we in the West can not "fix" the problems of the poor countries. The people themselves must rise up and say no to their corrupt power elites. They must say no to the petty corruption that occurs at every police station and customs office. They must say no to benefiting from the ill-gotten funds of family members with access to power. They must say no to preying on ethnic groups who are outside of the power clique. They must say no to corrupt spiritual leaders. Until this is done, debt relief will provide only a temporary respite, a time when leaders can rest more peacefully in their expensive villas. It will only camouflage the slow, under-the-surface boil in countries ruled by corrupt dictators and their minions.

The church must not look to economic cures while ignoring systemic disease. We must not swing the odds against our brothers and sisters who are fighting for change. They understand the need for changed hearts. To paraphrase Bakole Wa Ilunga's book *The Paths of Liberation: A Third World Spirituality:* The path of liberation is long and winding, but it always must go through the heart of humanity.

POSTSCRIPT

Should Developed Countries Provide Debt Relief to the Poorest, Indebted African Nations?

Snyder argues that one of the main problems with debt relief is that it does not make African governments accountable to their own people. In his case study of Rwanda, he suggests that, more often than not, excessive debt exists in African countries because of corruption and mismanagement. Other arguments against debt relief tend to focus on the turmoil that debt relief would create in the international financial system. There is a fear among public and private lenders that if one country is allowed to default on its loans without serious repercussions, then there will be a raft of other countries that will quickly follow suit. Proponents of debt relief would argue that both of these perspectives ignore the role that lenders have played in exacerbating the African debt crisis.

While Logie and Rowson certainly do not agree with Snyder, it is interesting to note that the former only see debt relief as useful given certain circumstances. Similarly, Snyder seems willing to accept debt relief if certain conditions are met. In other words, both sides find debt relief useful given the right circumstances (although the conditions advocated by both sides are fairly different). According to Logie and Rowson, the debt relief process must be transparent, and not "dominated by the interests and agendas of creditors," or "conditional on health or economic system reforms that are potentially harmful to health outcomes." Snyder calls on developed countries not to provide debt relief to nations with corrupt governments, a problem he hints at as being pervasive throughout much of Africa. He fears that debt relief for such nations will only allow their corrupt leaders to remain in power longer.

One major debt relief program under way since 1996 is the World Bank and the International Monetary Fund's (IMF) Heavily Indebted Poor Countries (HIPC) initiative. This program is significant because it is the first time in the 50-year history of these institutions that some of the debt is allowed to be written off. The program also allows for debt to be reduced from bilateral donors on a block basis (rather than forcing debtor nations to negotiate individually with each creditor). The Paris Club, an informal group of bilateral creditors, helped to develop this program along with the World Bank and the IMF. Of the 42 countries eligible for HIPC debt relief, 7 African countries have qualified to date, including Benin, Burkina Faso, Mauritania, Mali, Mozambique, Tanzania, and Uganda. Since its inception, the HIPC initiative has come under heavy attack from critics because of the conditions it imposes on debtor nations wishing to receive relief. According to the World Bank (http://www.worldbank.org/

hipc/about/hipcbr/hipcbr.htm), in order to receive relief, countries must "face an unsustainable debt burden, beyond available debt relief mechanisms", and "establish a track record of reform and sound policies through IMF and World Bank supported programs." It is the second condition in particular that has angered many debt-relief advocates.

One of the largest debt-relief groups is Jubilee Research (formerly Jubilee 2000). The Jubilee 2000 movement originated in the early 1990s with the All Africa Conference of Churches and slowly gained momentum since that time, especially in the United Kingdom. The Jubilee 2000 Petition was launched during the 1998 G8 Summit in Birmingham, England, where 70,000 people linked in a chain of arms to persuade leaders to provide debt relief to the world's poorest countries. See http://www.jubilee2000uk.org for more information on the Jubilee 2000 movement. The five-year anniversary of the Birmingham human chain was recently marked in June 2003, at the G8 Summit in Evian, France. Critics of the HIPC initiative continue to call for the relaxation of privatization and liberalization conditions that are linked to debt relief.

On the Internet . . .

AFROL News—Agriculture

AFROL News of the Heifer Project International provides information on agriculture from across Africa, including general agricultural trends for the continent and country-specific news.

http://www.afrol.com/Categories/Economy_Develop/
Agriculture/msindex.htm

African Conservation Foundation

The African Conservation Foundation Web site allows one to search its databases of African environmental topics by country, organization, type of information, and environmental category.

http://www.africanconservation.com

African Data Dissemination Service

The United States government's African Data Dissemination Service offers a warning system for possible famine or flood conditions. It also allows one to download data, both map based and tabular, regarding crop use, hydrology, rainfall, and elevation (as well as several other datasets).

http://edcw2ks21.cr.usgs.gov/adds/

United Nations Environment Programme: Regional Office for Africa

The United Nations Environment Programme: ROA provides information on UN environmental initiatives in Africa, both in restoring the natural environment and harnessing the environment for human use in an environmentally friendly way.

http://www.unep.org/ROA/

Food and Agriculture Organization of the United Nations

The Food and Agriculture Organization of the United Nations Web site contains expansive news articles and other information about agriculture, food, and environmental issues. This site also contains information with both African and international emphases.

http://www.fao.org

PART 2

Agriculture, Food, and the Environment.

*W*ith *more rural inhabitants as a share of total population than any other world region, questions regarding agriculture, food production, and environmental management have loomed large in debates concerning Africa. As a region rich in wildlife and natural resources, local, national, and global actors struggle over access and control of these environmental assets. Real and imagined crises have also fueled discussions over how to prevent famine and augment food production without undermining the integrity of the ecosystem.*

- Will Biotech Solve Africa's Food Problems?

- Is Food Production in Africa Incapable of Keeping Up With Population Growth?

- Are Abundant Mineral and Energy Resources a Catalyst for African Development?

- Are Integrated Conservation and Development Programs a Potential Solution to Conflicts Between Parks and Local People?

- Is Sub-Saharan Africa Experiencing a Deforestation Crisis?

ISSUE 6

Will Biotech Solve Africa's Food Problems?

YES: Florence Wambugu, from "Why Africa Needs Agricultural Biotech," *Nature* (July 1, 1999)

NO: Brian Halweil, from "Biotech, African Corn, and the Vampire Weed," *World Watch* (September/October 2001)

ISSUE SUMMARY

YES: Florence Wambugu, CEO of Harvest Biotech Foundation International, argues for the development and use of agricultural biotechnology in Africa to help address food shortages, environmental degradation, and poverty. She asserts that only wealthy nations have the luxury of refusing this technology.

NO: In a case study examining attempts to control the parasitic Striga weed, Brian Halweil, a research associate at the Worldwatch Institute, questions whether producing maize that is bio-engineered for herbicide resistance is really the best approach in the African context. He suggests that improved soil fertility management practices and mixed cropping are more appropriate and accessible strategies.

Humans have long sought to increase agricultural output through the use of improved seed. Initially this was done by repeatedly (over generations) saving seeds from plants with the most desirable characteristics. African farmers have been found to maintain and utilize an amazing number of crop varieties. Even when fields are not intercropped (i.e., several different crops planted in the same field), African farmers will often plant different varieties of the same crop in accordance with soil and moisture conditions that may vary throughout the field. Preserving the rich genetic diversity among crops at the local level in Africa has been the concern of some environmentalists.

Formal plant breeding emerged at the end of the eighteenth century. Crops were systematically cross-fertilized in hopes of obtaining plants with a desirable mix of characteristics. A significant advance was the development of a technique known as *hybridization* in the early twentieth century. Normally, a plant must be cross-pollinated with another. A plant that mates with itself is known as an inbred, and it usually produces offspring that perform less well than the parent. Hybridization involves cross-breeding two inbreds from desirable parentage, a technique that produces highly productive plants. See Robert Tripp, *Seed Provision & Agricultural Development* (ODI, 2001). In the 1960s a concerted effort, known as the green revolution, was undertaken to boost food crop production in the developing world. The green revolution, involving highly productive hybridized crops in conjunction with pesticides and inorganic fertilizers, largely benefited Asia and South America because it devoted most of its attention to food crops prevalent in these regions, mainly rice, wheat, and, to a lesser extent, maize. Africa was bypassed by the green revolution for the most part, with a few significant exceptions such as the production of maize in Zimbabwe. While the green revolution did boost food production, it has been criticized for not really resolving the hunger issues that it was designed to address, tending to favor wealthier farmers and spawning a host of new environmental problems related to chemically intensive agriculture.

The debate over the use and proliferation of agricultural biotech in Africa evokes many of the old pro and con arguments related to the green revolution, not to mention a host of new issues. The key advance of biotech, over hybridization, is that plant breeders are now able to insert genes from other (unrelated) species to effect desirable characteristics in a food crop. In the following selections, Florence Wambugu voices her concern that Africa will miss out on the agricultural biotech revolution as it missed out on the green revolution. She argues that the development and use of agricultural biotechnology in Africa will help to address food shortages, environmental degradation, and poverty. Brian Halweil is concerned that most Africans cannot afford bioengineered crops and that reliance on such solutions transfers problem solving from the farmer to the labs of multinational firms. Using control of the parasitic weed *Striga* as an example, he argues that more appropriate and accessible solutions exist in many instances.

One of the fundamental questions in this debate is whether food insecurity in Africa is a question of underproduction or maldistribution. Until quite recently, there was a consensus among food security specialists that hunger was more often the result of conflict, mismanagement, or poor distribution than underproduction. This pre-existing consensus was catalyzed, in large part, by the pioneering work of the Nobel laureate economist, Amartya Sen, who showed how national markets could be replete with grain, yet poor households might still not have the means to access this food. See *Poverty and Famines* (Clarendon, 1981). Wambugu might contest the maldistribution thesis by noting that the African continent imports 25 percent of its grain (suggesting that there is an absolute shortage of food at the continental scale).

Florence Wambugu

Why Africa Needs Agricultural Biotech

African Perspective

The critics of biotechnology claim that Africa has no chance to benefit from biotechnology, and that Africa will only be a dumping ground or will be exploited by multinationals. On the contrary, small-scale farmers in Africa have benefited by using hybrid seeds from local and multinational companies, and transgenic seeds in effect are simply an added-value improvement to these hybrids. Local farmers are benefiting from tissue-culture technologies for banana, sugar cane, pyrethrum, cassava, and other crops. There is every reason to believe they will also benefit from the crop-protection transgenic technologies in the pipeline for banana, such as sigatoka, the disease-resistant transgenic variety now ready for field trials. Virus- and pest-resistant transgenic sugar-cane technologies are being developed in countries such as Mauritius, South Africa and Egypt.

The African continent, more than any other, urgently needs agricultural biotechnology, including transgenic crops, to improve food production. African countries need to think and operate as stakeholders, rather than accepting the 'victim mentality' created in Europe. Africa has the local germplasm, some of it well-characterized and clean, being held in gene banks in trust by centres run by the Consultative Group of International Agricultural Research. It also has the indigenous knowledge, local field ecosystems for product development, capacities and infrastructure required by foreign multinational companies.

The needs of Africa and Europe are different. Europe has surplus food and has never experienced hunger, mass starvation and death on the regular scale we sadly witness in Africa. The priority of Africa is to feed her people with safe foods and to sustain agricultural production and the environment.

Africa missed the green revolution, which helped Asia and Latin America achieve self-sufficiency in food production. Africa cannot afford to be excluded or to miss another major global 'technological revolution'. It must join the biotechnology endeavour. Transgenic food production increased from 4 million to 70 million acres worldwide from 1996 to 1998 with measurable economic gains and with sustainable agricultural production. It would be a much higher risk for Africa to ignore agricultural biotechnology. Africa's crop production per unit area of land is the lowest in the world. For example the production of sweet potato, a staple crop, is 6 tonnes per hectare compared to the global average of 14 tonnes per hectare. China produces on average 18 tonnes per hectare, three times the African average. There is the potential to double African production if viral diseases are controlled using transgenic technology.

The African continent imports at least 25 per cent of its grain. The use of biotechnology to increase local grain production is far preferable to this dependence on other countries, particularly as the population growth rate exceeds food production. The inability to produce adequate food forces Africa to rely on food aid from industrialized nations when mass starvation occurs. Although biotechnology is not the only answer to this problem, Africa should certainly benefit in many ways from its use, for example in improved seed quality and resistance to pests and diseases.

The average maize yield in Africa is about 1.7 tonnes per hectare compared to a global average of 4 tonnes per hectare. Some biotechnology applications can be used to reduce this gap, for example in the case of the maize streak virus (MSV), which causes losses of 100 per cent of the crop in many parts of the continent. A biotechnology-transfer project is under way to develop MSV-resistant varieties. The project is brokered by the International Service for the Acquisition of Agri-Biotech Applications (ISAAA), and involves the collaboration of the Kenya Agricultural Research Institute (KARI), the University of Cape Town, the International Centre for Insect Physiology and Ecology in Kenya, and the John Innes Centre in the United Kingdom. Funding is coming from the US Rockefeller Foundation, and Novartis in Europe has donated some technology to KARI.

Researchers at KARI are studying the mechanism of MSV resistance and trying to map the genes responsible. Advanced biotechnology skills, including the use of advanced agroinoculation techniques and molecular markers, is at the core of this effort. A priority in Kenya is also to produce high-yielding, drought-tolerant crop varieties to boost food production in the 71 per cent of the country that is arid or semi-arid.

Africa needs biotechnology to solve its environmental problems, and there is unlimited public demand for agricultural biotechnology products and services. In Kenya, the demand for tree seedlings reaches 14 million per year, whereas the country can only supply 3 million, a clear indication of the need for tissue-culture and cloning techniques to curb deforestation and boost reforestation using indigenous species threatened with extinction. These technologies are being successfully used in South Africa, and ISAAA has facilitated a project for application in Kenya. There are issues of intellectual property rights and patents that require hard work to develop or acquire, and advanced agricultural biotechnology skills will be needed. There may also be a need to work out collaboration agreements with the private sector or with companies that already have patents.

Biotechnology in Africa is needs-based. After working at KARI for nearly a decade to help improve sweet-potato production using traditional breeding and agronomy methods, I made no progress. An opportunity to work in the private biotechnology sector abroad resulted in the development of a transgenic variety that is resistant to sweet-potato feathery mottle virus, which can reduce yields by 20–80 per cent. Control of this disease will improve household food security for millions. This project involved collaboration between KARI, a project called Agricultural Biotechnology for Sustainable Productivity, funded by the US Agency for International Development, and Monsanto. The work by Kenyan scientists focuses on local varieties, and there will be a smooth and sustainable transfer of the technology, which will be shared with neighbouring countries. Kenyan scientists have been trained in gene technology techniques. ISAAA has been asked to help with the transfer and licensing agreement. Similar projects are under way for bananas, sugar cane and tropical fruits.

Remaining Problems

Needless to say, Africa has many problems—a shortage of skilled people (especially in biotechnology), poor funding of research, lack of appropriate policies and civil strife. Nevertheless, countries such as South Africa, Egypt, Zimbabwe and Kenya are taking practical steps to ensure that they can use biotechnology for sustainable development.

African countries need to avoid exploitation and to participate as stakeholders in the transgenic biotechnology business. They need the right policies and agencies, such as operational biosafety regulatory agencies, breeders' rights and an effective local public and private sector, to interface with multinational companies that already have the technologies. Consumers need to be informed of the pros and cons of various agricultural biotechnology packages, the dangers of using unsuitable foreign germplasm, and how to avoid the loss of local germplasm and to maintain local diversity. Other checks and balances are required to avoid patenting local germplasm and innovations by multinationals; to ensure policies on intellectual property rights and to avoid unfair competition; to prevent the monopoly buying of local seed companies; and to prevent the exploitation of local consumers and companies by foreign multinationals. Field trials need to be done locally, in Africa, to establish environmental safety under tropical conditions.

The main goal is to find a balanced formula for how local institutions can participate in transgenic product development and share the benefits, risks and profits of the technology, as they own the local germplasm needed by the multinationals for sustainable commercialization. New varieties must not simply replace local ones. The removal of genes that were in the public domain into the private sector raises concern in Africa.

All these issues mean that Africa must strengthen its capacity to deal with various aspects of biotechnology, including issues of biosafety, creating and sustaining gene banks, and encouraging the emergence of a local biotechnology private sector. The great potential of biotechnology to increase agriculture in Africa lies in its 'packaged technology in the seed', which ensures technology benefits without changing local cultural practices. In the past, many foreign donors funded high-input projects, which have failed to be sustainable because they have failed to address social and economic issues such as changes in cultural practice. The criticism of agribiotech products in Europe is based on socioeconomic issues and not food safety issues, and no evidence so far justifies the opinion of some in Europe that Africa should be excluded from transgenic crops. Africans can speak for themselves.

Brian Halweil

 **NO**

Biotech, African Corn, and the Vampire Weed

A parasitic weed is sucking the life out of East African corn. One way to deal with it would be to engineer corn for herbicide resistance, so that herbicide could be sprayed on the corn to kill the parasite—even though the corn seed and the herbicide would probably be too expensive for poor farmers, the herbicide would pollute, and the weed would likely become resistant. Another way would be to improve soil health. Tough call.

I was hot on the trail of the infamous Striga weed. Though I'd never confronted a live specimen, the plant had been hounding me *in absentia,* and often in public, for several years. Again and again, in various panel discussions on biotechnology, I have listened to my debating opponents hold up Striga as proof that genetic engineering could one day eradicate hunger and poverty in the Third World. The particulars varied, of course, but I had heard the same basic Striga argument from biotech executives, from industry-funded scientists, and from the industry's advocates in academia and government.

Striga hermonthica is a member of the Scrophulariaceae family, a widespread group of about 4,000 plant species that includes a couple of "heirloom" garden favorites, foxglove and snapdragon. But it also includes several important Old World plant parasites—plants that live off other plants. *S. hermonthica* is one of these. It's common in East Africa, where it's called "witchweed" or in Swahili, "buda." It looks innocent enough, standing about 15 centimeters (around 6 inches) tall, and bearing little lance-shaped, pale-green leaves that make a pleasing contrast with its pink-purple flowers. But below ground, the plant is a monster. Its root-like organs, called haustoria, seek out the roots of nearby crops, then rob them of water, nutrients, and life. And those pretty flowers can set as many as 20,000 seeds per plant. The seeds are easily dispersed and can lie dormant in a range of soil conditions for decades.

In a badly infested field, Striga can destroy most of the harvest, perpetuating not only poverty and hunger, but also gender inequity, since it's usually women who must undertake the largely futile task of disentangling Striga from

From Brian Halweil, "Biotech, African Corn, and the Vampire Weed," *World Watch,* vol. 14, no. 5 (September/October 2001). Copyright © 2002 by The Worldwatch Institute. http://www.worldwatch.org.. Reprinted by permission.

the crop. Throughout East Africa, Striga causes several billion dollars in losses each year.

The solution, according to the biotech advocates, is to engineer varieties of African staple crops to resist herbicides, so that farmers can spray their infested fields—and kill the Striga without killing the crop. This is an extension of agricultural biotech's dominant commercial application: the engineering of soybeans, corn, cotton, and canola by Monsanto to withstand the company's best-selling herbicide, glyphosate ("Roundup").

An anti-Striga niche would make it easier to claim the moral high ground for herbicide-tolerant crops. Such products might eventually seem as humanitarian as the industry's "golden rice"—the beta-carotene enhanced rice variety that is being developed to combat vitamin A deficiency. Beta-carotene is the precursor of vitamin A, an especially important nutrient for children. Worldwide, nearly 134 million children suffer some degree of vitamin A deficiency, a condition that can suppress immune system function, cause blindness, and in extreme cases, even kill. Little wonder that golden rice has become the emotionally compelling hook for a $50 million public relations campaign launched by the Biotechnology Trade Organization. (It's true that not everyone is sold on this idea. Some nutritionists argue that it would make more sense to help poor people grow green vegetables, which produce more beta-carotene than golden rice—along with various other nutrients completely lacking in rice, golden or otherwise.)

◦◦◦

But in any case, I decided that the time had come for me to get a first-hand sense of this expanding moral high ground. From a political point of view, the Striga issue looked especially interesting because Striga, unlike vitamin A deficiency, is almost exclusively an African problem, and Africa is ground-zero in the global food debate. Although hunger is sorely persistent throughout much of the developing world, Africa is the only region where it is actually getting worse. In Latin America and Asia, the past two decades have seen a modest decline in malnourishment among children, in terms of both the percentage of children affected and their absolute numbers. In Africa, however, the share of children who are hungry has risen from 26 to 29 percent over the past 20 years, and the absolute number of hungry children has doubled. It now stands at 38 million. That helps explain why, sooner or later, almost any major agricultural development will have to justify itself in an African context.

That, in turn, explains why I was looking at cornfields outside Maseno, Kenya last February. Maseno is a small but rapidly growing town near the Ugandan border, about 30 kilometers northeast of Lake Victoria. And from what I could see, the local harvest was going to be an uncertain affair. In some fields there seemed to be more Striga than corn, which is often called maize in Africa. Spindly corn stalks with pitiful, dried-up ears stood above a carpet of purple flowers. But in other fields, the corn looked good and there was no Striga at all. Several farmers and ag extension agents helped explain what seemed like pure chance. One of them, a farmer named Paul Okongo, put it categorically: "Striga is only a problem in overused and depleted soils."

Striga thrives where farmers have grown nothing but corn for decades, especially where fallow periods have been shortened or eliminated. (During a fallow period, land is allowed to "go wild" or soil-building fallow crops are sown. The practice helps maintain long-term productivity by reducing weed and pest infestations and by allowing soil nutrient levels to recover.) But as Kenya's population has grown, the size of the average family field has declined, and farmers have become increasingly reluctant to take land out of production. Of course, fallowing isn't the only way to renew soil; another standard approach is fertilizing. And where farmers are able to do this—whether in the form of manure or artificial fertilizer—Striga is rarely a problem. But most East African farmers can't afford commercial fertilizer. And manure is often scarce because livestock are not generally penned, where their droppings could be collected, but grazed out in the open. No fertilizer and little fallowing: the result is depleted soil, and then Striga moves in. Striga infestation, in other words, is a kind of second-order effect: it's what happens *after* soil health declines. It was obvious that a herbicide-resistant fix wasn't going to get at this problem.

It was also obvious that Paul knew what he was talking about, and I began to wonder what local expertise might have to offer in lieu of a prepackaged, imported solution. And indeed, it turns out that some African farmers have found a way to suppress Striga with a widely available, home-grown technique: planting leguminous tree crops—that is, tree species that are members of the legume family. Plants in this family often have certain microbes on their roots that can "fix" nitrogen: the microbes withdraw elemental nitrogen from tiny air pockets in the soil and bond it chemically to hydrogen, producing compounds that plants can metabolize. (Pure, elemental nitrogen is useless to plants; fixed nitrogen, on the other hand, is the principal component of fertilizer.)

These nitrogen-fixing trees are grown as a fallow crop, during the one more-or-less obligatory fallow period of the year: from February to April. This is the dry spell between the two rainy seasons—the long rains of late spring, which yield the main corn harvest in July, and the erratic, short rains of fall, which produce the smaller, January harvest. Of course, three months isn't long enough for the trees to get very big; they're usually 2 or 3 meters high when they're pulled to sow the fall corn. But that one season of tree growth can cut Striga infestations by over 90 percent. This is not just the effect of the added nitrogen. Some of the preferred tree crops are native to the region and have co-evolved with Striga; their evolutionary defenses to the parasite apparently include chemicals that they exude into the soil and that disrupt the Striga lifecycle.

Control of Striga is not the only benefit of nitrogen-fixing crops, according to Bashir Jama, a scientist with the Nairobi-based International Center for Research in Agroforestry (ICRAF), the main promoter of these "improved fallows." Where farmers can be persuaded to employ longer fallows, leguminous trees can accumulate 100 to 200 kilograms of nitrogen in 6 months to 2 years. (The results depend on the species used, soil quality, moisture levels, and whether the fallow is taken all at once or in intermittent periods.) These fertilization rates would satisfy most U.S. and European farmers. In severely depleted soils, they generally increase corn yields two to four times—not bad for adding just one crop to the rotation.

Bashir reports other benefits as well, such as improved soil structure, better water retention, and higher levels of other nutrients besides nitrogen. The fallow crop also shades out weeds—which means fewer weed seeds to cause problems when the field goes back into production. And because it adds diversity to the agroecosystem, the fallow tends to suppress the most serious insect pests of corn as well. (Monocultures foster very high insect infestations; in more complex systems, the pests encounter more predators and less food so their populations tend to be much lower.) Fallow greens are also nutritious livestock feed; the wood is useful for fuel. A six-month tree crop fallow on as little as half a hectare can meet a family's cooking needs for an entire year.

◦✦◉✦◦

Another Striga control has emerged as the byproduct of an ingenious local response to the most important insect pest of corn, the stemborer. The borer is thought to chew up between 15 and 40 percent of Africa's corn harvest each year. In other corn-growing regions of the world, the borer is the target of extensive insecticide spraying. African farmers generally can't afford insecticide, but they may not need it anyway, to judge from the work of another Nairobi-based organization, the International Center for Insect Physiology and Ecology (ICIPE). ICIPE's work is founded on the idea that pest problems are caused by ecological imbalances; progress is therefore a matter of correcting the imbalances, rather than simply trying to poison the pests. "The long-term solution is generally more insect diversity, not less," says Hans Herren, ICIPE's director. This attitude helps explain why the gates at ICIPE headquarters bear a sign reading "Duduville." "Dudu" is Swahili for bug. Kenyans don't generally seem to like bugs any more than Americans do; I saw billboards throughout Kenya that read, "Raid kills dudus dead."

ICIPE scientists have developed what they call a "push-pull" strategy for dealing with the stemborer. The corn is sown with certain plants, such as molasses grass (*Melinis minutifolia*) and silver leaf desmodium (*Desmodium uncinatum*) that repel, or "push" the borer. The field perimeter is sown with other plants, such as napier grass (*Pennisetum purpureum*) and Sudan grass (*Sorghum vulgare sudanense*), that attract, or "pull" it. Most of the larvae end up trapped in the gummy substances produced by these grasses. This system can cut stemborer losses from 40 percent to below 5 percent. It also offers substantial relief from Striga, because desmodium secretes a chemical that interferes with the parasite's ability to tap into the roots of other plants. (Desmodium is a legume genus from the New World, so this mechanism is not an adaptation to Striga specifically, but perhaps the genus coevolved with a similar parasite.)

Ironically, a major U.S. foundation, which has been funding the "push-pull" research, has asked ICIPE to isolate this chemical—and the related gene. "Part of their interest is to use that gene to come up with molecular biology solutions to the Striga problem," says Zeyaur Khan, who directs work on the push-pull strategy at ICIPE. In other words, they might be interested in funding the development of a corn variety engineered to produce this substance. How would that relate to ICIPE's interest in promoting ecological balance? Presumably, it would tend to undercut it, because it would push the system back towards monoculture—and it would likely do nothing for the borer problem. The full benefits of ICIPE's system are only available if you buy in at the ecological level—not the molecular level. And of course, the ecological level is the one that's available to farmers.

⚬❦⚬

After several days of farm visits in western Kenya, I was beginning to get a better sense of how the biotech approach to Striga compared with the improved fallow technique. The biotech fix would be costly for the farmer, would increase chemical use, would add no other benefits to the system, and in any case, does not yet even exist. On the other hand, improved fallowing is extremely low-cost and confers all the benefits mentioned above. It's also readily accessible. In at least a rudimentary form, the technique is already being used by tens of thousands of farmers in eastern and southern Africa. Pedro Sanchez, who recently stepped down as director of ICRAF, sees improved fallowing at the early stages of an exponential growth curve similar to what happened with Green Revolution rice varieties in Asia. Sanchez envisions 50 million farmers using improved fallowing within the next five to ten years.

One of the most interesting features of the improved fallow system is that it allows for forms of R&D that farmers can do on their own. In 1997, for example, a moth infestation began to threaten a popular East African tree crop known as sesbania (*Sesbania sesban*)—until some farmers discovered that occasional rows of tephrosia trees (either *Tephrosia vogelii* or *T. candida*) would keep the pest in check. This field-level innovation is very far removed from the biotech paradigm, where innovation occurs, not on the farm, but in million-dollar laboratories, and where the principal actors are not farmers, but Ph.D. biologists and patent attorneys.

This distinction is critical, and not just for the harvest. The ability to innovate could be crucial to the future of farmers themselves. But if innovation is to contribute to the welfare of farming, it will have to extend beyond issues of yield. After all, many U.S. and European farmers have been teetering on the brink of economic extinction for years, and a substantial number have gone over it—even though they produce some of the highest yields in the world. In most developing countries, agriculture is still the predominant way of life, so the economic health of farming is a basic social issue. This is why the agricultural status quo is a dangerous absurdity. The multinational corporations that sell farmers seed and pesticide are ringing up tens of billions of dollars in sales each year; the multinationals that distribute, process, and retail the harvests are ringing up hundreds of billions. But farmers themselves are now members of the poorest—and ironically, the hungriest—occupation on Earth.

This is the problem that has come to dominate the agenda of another Kenyan NGO, the Association for Better Land Husbandry. ABLH was set up in 1994 to promote a variety of conservation farming techniques, including biointensive farming, a form of organic production popularized by one of organic agriculture's best-known proponents, John Jeavons of Willets, California. When I visited farmers working with ABLH in the Vihiga district of western Kenya, I instantly recognized the neatly laid out "beds" that characterize the biointensive method. In 1996, I had attended a three-day workshop taught by Jeavons, and I knew from direct experience how much effort went into those beds. They are prepared through a deep-digging technique, called "double-digging," which aerates the soil, and permits better nutrient circulation. Jeavons places tremendous emphasis on soil health. "Feed the soil, not the plants" is one of his mantras.

Biointensive farming can boost yield dramatically, and it packs a one-two punch against Striga by enriching soils, then producing a dense, diverse plant canopy that shades out most weeds. But by themselves, such improvements aren't going to bring prosperity to farmers. "Doubling maize yields and tripling kale yields doesn't make much of a difference if you can't get your product to market, or if a flood of cheap imports squashes your local market," says Jim Cheatle, founder and director of ABLH. To get at these issues, Cheatle expanded the group's agenda to include a kind of farmer empowerment. ABLH now coordinates seven farm cooperatives so that local growers can capture the marketing and distribution advantages that come with scale. (Nine more co-ops should be in operation by the end of the year.) "Instead of each of several thousand farmers buying their own delivery truck and setting up their own marketing offices," says Jane Tum, an ABLH extensionist, "the cooperative can pool its resources for a much larger delivery truck and a marketing staff." This might seem like kind of an obvious thing to do, but *any* concern with marketing is still unusual in places like Kenya. A recent survey of over 200 sustainable farming projects in the developing world found that only 12 to 15 percent had tried to improve marketing or processing.

Co-op produce is now selling in both local and national markets under the "Farmer's Own" brand name. Among the products bearing this label are Mr. Brittle, a macadamia nut energy bar, and Mchuzi Mix, a soup and sauce thickener made from locally grown beans and corn. "We're competing with the big boys," says Francisca Odundo, the marketing manager for Farmer's Own. Francisca is trying to cultivate allegiance for the new brand, which now sits on shelves alongside the Cadbury and Nestlé labels.

Farmer's Own products also bear the label, "Conservation Supreme," a quasi-organic designation that permits limited use of approved biopesticides, like the insect-killing "Bt" bacterium. The designation is a kind of marketing tool to help small-scale farmers make the transition to certified organic production. (Before it can pay off, the transition usually involves several years of uncertainty, in which both the farm and the farmer must adjust to the new regime.)

I got a sense for the tangible results of these efforts when Jane Tum took me to a farm to pick up produce for a Farmer's Own vegetable stand. The farm was run by Flora Mwoshi, a young mother of three. On the day I visited, she was harvesting bright green bell peppers. With a young child resting on her hip, Flora balanced a five-gallon plastic bucket of peppers on her head and bore it to our truck, where it was weighed. Several more buckets followed. Her six-year-old daughter, barefoot but wearing her best white dress because the family had heard that white people were coming to visit, watched in awe as her mother became the center of attention.

Jane paid the woman 600 Kenyan shillings, about $10. That doesn't sound like much, but that money went directly into her pocket—no middleman to pay, no bills for agrochemicals or expensive seeds. And many more veggies remained to be harvested. This simple transaction was the most inspiring moment of my trip. We, the "white people," hadn't arrived with some foreign technology of highly dubious potential. We were there as witnesses and—in a broad sense—colleagues. What we witnessed was a local response to a local problem. And we could see that the response worked, because the produce was beautiful, and the farmer got paid.

POSTSCRIPT

Will Biotech Solve Africa's Food Problems?

There are a number of health and environmental concerns related to the use of agricultural biotechnology in Africa. Perhaps one of the most controversial concerns is the apprehension about food safety. While there is considerable unease among some consumers regarding GM foods, not to mention a ban on the use and importation of GMO (genetically modified organisms) crops by the European Union (EU), studies definitively linking these foods to health risks are still lacking. Nonetheless, whether or not the health concerns are justifiable, EU restrictions on these crops have genuine implications for African farmers who may wish to export their crops to Europe.

Some of the most widely used GMO crops in North America are the Roundup Ready varieties of corn/maize, soybeans, and canola and Bt crops produced by the Monsanto Corporation. The Roundup Ready crops are resistant to the popular herbicide, Roundup. As such, weeds may be eradicated through spraying with no negative effect on crop growth. Bt crops contain a gene from the soil bacterium *Bacillus thuringiensis* (Bt) that acts as a toxin to insects that prey on the crops.

First, Roundup Ready and Bt crops may cross-pollinate with weeds or neighboring vegetation and thereby release herbicide resistance (in the case of the Roundup Ready varieties) or insect resistance (in the case of Bt crops) into the broader plant population. Second, because the Bt in plants does not discriminate between the insects it kills, useful or benign insects (such as butterflies) may be destroyed. Furthermore, some insects may more quickly develop a resistance to Bt (a problem known as pesticide resistance) through constant exposure to the toxin and through evolutionary adaptation.

A third issue that one may want to consider when evaluating the usefulness of GM crops in Africa is that of "terminator technology." It has been pointed out that terminator technology is disruptive to traditional African farming systems where seeds are saved from one year's crops to be planted the following season. Some may find it problematic, even if yields are higher, that farmers would be dependent on the market for their seed supplies each year. Others suggest that this is simply part of the agricultural modernization process in Africa.

It is finally worth noting that this debate took on added prominence in 2002 when some southern African countries rejected food aid containing bio-engineered crops. This occurred at a time when millions of people were at risk of starvation in these countries. See Ruth Gidley, "African Crisis Fuels Debate Over GM Food," *AlertNet* (July 19, 2002). For more general information on this debate, see Hannah Hoag, "Biotech Firms Join Charities to Help Africa's Farms," *Nature* (March 20, 2003) and Natasha McDowell, "Africa Hungry for Conventional Food as Biotech Row Drags On," *Nature* (August 8, 2002).

ISSUE 7

Is Food Production in Africa Incapable of Keeping Up With Population Growth?

YES: W. Thomas Conelly and Miriam S. Chaiken, from "Intensive Farming, Agro-Diversity, and Food Security Under Conditions of Extreme Population Pressure in Western Kenya," *Human Ecology* (2000)

NO: Michael Mortimore and Mary Tiffen, from "Population and Environment in Time Perspective: The Machakos Story," in Tony Binns, ed., *People and Environment in Africa* (John Wiley and Sons, 1995)

ISSUE SUMMARY

YES: W. Thomas Conelly and Miriam S. Chaiken, both professors of anthropology at Indiana University of Pennsylvania, examine an area in Western Kenya that has very high population densities. Despite the wide variety of sophisticated practices that maintain a high level of agro-diversity, they conclude that intense population pressure has led to smaller land holdings, poorer diet quality, and declining food security.

NO: Michael Mortimore, a geographer, and Mary Tiffen, a historian and socio-economist, both with Drylands Research, investigate population and food production trajectories in Machakos, Kenya. They determine that increasing population density has had a positive influence on environmental management and crop production. Furthermore, they found that food production kept up with population growth from 1930 to 1987.

There is a long-standing debate about the ability of agriculture to keep up with population growth in Africa. Those who are concerned that population growth will outstrip agricultural production are often referred to as neo-Mathusians. This perspective is designated as such in deference to the eighteenth-century British clergyman, Thomas Malthus, who posited such a scenario as inevitable in his famous 1798 treatise entitled *Essay on the Principle of Population.* The neo-Malthusians have generally dominated contemporary debates concerning global population growth and food supplies, and they are led by such figures as Paul Ehrlich (see *The Population Bomb* (Simon and Schuster, 1968)) and Lester Brown (see numerous World Watch Institute publications). The major contrarian perspective in African studies is the Boserupian viewpoint, named after Ester Boserup (see *The Conditions of Agricultural Growth* (G. Allen and Unwin, 1965)). She established that increasing population densities may induce farmers to intensify their efforts and thereby produce food at a rate that keeps pace with population increase.

A third perspective, often referred to as the technocratic or cornucopian viewpoint, is somewhat akin to the Boserupian population thesis. This view, most commonly associated with the late Julian Simon (see *The Great Breakthrough and Its Cause* (University of Michigan Press, 2000)), asserts that human ingenuity will resolve resource constraints created by population growth if the free market is allowed to operate and appropriate price signals that are transmitted to producers. The major difference between the Boserupian and technocratic viewpoints is that the former tends to emphasize indigenous methods of adaptation, whereas the latter stresses the importance of market-led change.

To some extent, both of the perspectives in this issue are sympathetic to the Boserupian view, but they arrive at very different conclusions in the final analysis. W. Thomas Conelly and Miriam S. Chaiken state that intense population pressure has led to smaller land holdings, poorer diet quality, and declining food security in an area they examine in Western Kenya. Michael Mortimore and Mary Tiffen assert that increasing population density has not negatively affected environmental management and crop production. They maintain that food production kept up with population growth from 1930 to 1987.

W. Thomas Conelly and
Miriam S. Chaiken

 **YES**

Intensive Farming, Agro-Diversity, and Food Security Under Conditions of Extreme Population Pressure in Western Kenya

Introduction

In 1882–84 the Scottish explorer Joseph Thomson, with a caravan of African companions, journeyed by foot across the width of modern day Kenya. One of his main goals was to find a new route from the Indian Ocean to the great Lake Victoria, located between modern day Kenya, Uganda, and Tanzania. In late 1883, after weeks of travel in the dry and sparsely populated Rift Valley, Thomson climbed the western escarpment onto the well-watered highlands of what is today Western Province. As he passed through the current northern Kakamega District, Thomson was amazed to find the area heavily populated by the Luhya people and the land intensively cultivated (1962, pp. 156–61):

> We passed over a fertile, rolling country, watered by a perfect network of rivulets. What most impressed me was the surprising number of villages and the contented well-to-do air of the inhabitants. . . Almost every foot of the ground was under cultivation. . . We passed along a perfect lane of people, all carrying baskets of food. . . There was honey, milk, eggs, fowls, and beans. . . The whole country was remarkable for its poverty in trees. . . There is in consequence a great dearth of wood, and firewood had to be bought. . . The extraordinary density of the population was to us a matter of great wonder. They streamed forth in thousands to see us, amid yells and shouts of the most deafening character.

A century later, the dense population and remarkably complex agricultural system of the Luhya continue to inspire great wonder. According to the 1979 census, population densities in southern Kakamega District ranged from 600 to more than 1000 people per square kilometer and landholdings were often less than one hectare in size (CBS, 1981). These circumstances pose numerous problems for successful farm production. As fallow periods decline in response to

From W. Thomas Conelly and Miriam S. Chaiken, "Intensive Farming, Agro-Diversity, and Food Security Under Conditions of Extreme Population Pressure in Western Kenya," *Human Ecology,* vol. 28, no. 1 (2000). Copyright © 2000 by Plenum Publishing Corporation. Reprinted by permission. Notes and references omitted.

land scarcity, the maintenance of soil fertility and the control of weed and insect pests become increasingly difficult. Crop yields steadily decline and labor efficiency typically plummets. Overworked soils with little or no fallow cover are easily eroded by heavy rains, especially in hilly terrain. As more and more land comes under permanent cultivation, grazing land for livestock gradually diminishes and animal products in the diet such as milk, blood, and meat become scarce. In response, many Luhya farmers have developed ingenious and complex systems of intensive agriculture that attempt to maintain soil fertility and maximize production on small areas of land. Nonetheless, because of minuscule farm size and a scarcity of alternate economic opportunities, many families in the region face the threat of chronic food scarcity and poverty and the "contented well-to-do air of the inhabitants" is now difficult to find.

The process of intensification experienced by the Luhya is interconnected with many other transitions that take place in rural societies as populations increase. In a recent overview of intensive agricultural systems in Africa, Hyden *et al.* (1993) note that rural population pressure is likely to lead to some or all of the following changes: (a) intensified inputs into agriculture (more labor, greater cash investments, increased erosion control measures, and the adoption of new cultigens); (b) intensification of outputs (greater cropping frequency and productivity per hectare); (c) expansion of land under cultivation; (d) change in the proportion of crops marketed; (e) greater economic diversification (including non-farm occupations) and migration for wage employment; (f) changes in physical well-being in terms of diet quality and environmental degradation; and (g) changes in social relations, such as growing socioeconomic inequality and a re-alignment of gender roles.

Virtually all of these changes have taken place in western Kenya as population has increased and agriculture has become more intense over the years. This [selection], which analyzes the farming system of the Luhya living in Hamisi Division in southern Kakamega District, focuses on two of these issues: the role of agro-diversity in the process of agricultural intensification and the impact of population growth and intensification on food security and the quality of diet. . . .

Historical Changes in the Production System

In describing the agricultural system and evaluating nutritional status in the community, it is important to understand historical changes that have occurred in the Hamisi agro-pastoral production system over the past century. One of the most significant changes has been the introduction of new varieties of higher-yielding maize. Beginning in the early colonial era, government research stations started to produce improved maize varieties for distribution to African farmers. By the 1930s, the new varieties had largely displaced local maize (originally grown only on a small scale) as well as sorghum and finger millet that were the traditional staple grains in the Luhya agricultural system (Wagner, 1970). Several high-yielding, hybrid maize varieties were also introduced in more recent decades. Today, although farmers still cultivate various local or "traditional" types of maize that have differing desirable properties, it is estimated that over half of all maize planted in the region consists of improved hybrid varieties from research stations (Jaetzold & Schmidt, 1982, pp. 387–88). Maize is now the dominant crop in Hamisi and is used to produce the staple food ugali, a polenta-like dish made from maize meal. Sorghum is still cultivated but it is clearly secondary to maize, although sorghum production increases during the second "short rains" cropping season.

A second important historical change in Hamisi has been a steady loss of grazing lands and a decline in the size of livestock herds resulting from population pressure and changes in land tenure. Since the colonial era (which ended in 1963), there has been a transition from individual use rights of land controlled by lineages to private ownership of land that can be bought and sold. Individual tenure and land scarcity have led to a decline in common lands that were once open for all families to use for grazing their animals. As a result of these factors, typical herd size in Hamisi today is quite small, averaging 2.5 cattle and 0.6 smallstock per household. . . .

A third major innovation has been the increasing frequency of male migratory wage labor. This process was stimulated initially by the colonial demand for taxes (Kitching, 1980), but later migrants were also motivated by a growing desire for a cash income necessary to purchase consumer goods and to pay for medical care and the education of children (Sangree, 1966). Wage labor, both local and migratory, continues to be an important part of the Hamisi economy today. In the baseline sample of households surveyed by the SR-CRSP [Kenya Small Ruminant Collaborative Research Support Program], 54% were headed by women whose husbands were working away from Hamisi. We were never able to obtain reliable information on the amount of money migrant workers send home to their families in the form of remittances. However, many migrant workers are unable to obtain regular and high-paying employment, and our impression is that remittances are not a reliable source of income for most families.

A final important change has been the introduction of export cash crops into the local economy. Coffee and tea cultivation were first permitted by the government in the late 1950s and early 1960s. The coffee and tea income that growers receive is dependent on unpredictable world market conditions and the efficiency of the parastatal organizations that market the crops, factors completely outside the control of farmers. Tea has proved to be a generally reliable if modest source of income over the years, but establishment costs have been estimated to be as high as Ksh 20,000/ha (U.S. $1400). As a result, tea is generally cultivated only by more prosperous households and growers usually plant only very small areas (Argwings-Kodhek, 1995). The coffee market was unpredictable through much of the 1970s and 1980s (Castro, 1995) and many farmers in the region have begun to abandon coffee production. Another cash crop, the French bean, was introduced in the early 1980s and is now cultivated on a small scale by many households. The beans are produced through contracts with a private agricultural corporation in the region that then markets the crop in Europe (Argwings-Kodhek, 1995). . . .

Population Pressure, Intensive Agriculture, and Food Security

In admiring the ingenuity and hard work of Hamisi farmers in coping with land scarcity it is tempting to see their agricultural system as highly "adaptive" and assume that it is successful. Mortimore and Tiffen (1994) assert that the similar (though somewhat less intense) agricultural system in Machakos District in central Kenya is an example of just such a successful response to population pressure. They argue that "increasing population density has had positive effects in Machakos. The increasing scarcity (and value) of land promoted investment, both in conservation and in yield-enhancing improvements. Integrating stock and livestock production improved the efficiency of nutrient cycling and, thereby, the sustainability of the farming system" (1994, p. 30). They conclude that population pressure and increasingly intensive agriculture in Machakos have led not only to the adoption of improved conservation measures, but have also permitted food production to keep pace with population growth and made possible higher per capita income and an improved standard of living in the Machakos District as a whole (Tiffen *et al.*, 1994, p. 261).

We agree that there can be many positive aspects of agricultural intensification. The complex and highly diversified farming system in Hamisi, combined with the income generated by migrant employment and local wage labor, have allowed the continued survival of the population under conditions of extreme resource scarcity. Despite intense pressure on the land, Hamisi farmers have avoided the threat of severe environmental deterioration that is often seen as a consequence of rapid population growth. The construction of simple terraces and the use of intercropping and cover crops have prevented serious soil erosion (Clay & Lewis, 1996). The incorporation of trees into the farming system has preserved an adequate supply of fuelwood and building material for the community's needs. The use of livestock manure has helped maintain crop yields that would otherwise have collapsed under the system of permanent cultivation. Though their return is often erratic, the adoption of export cash crops has helped to supplement the income families need for market purchases, school fees, medical care, and the numerous other expenses associated with life in modern Kenya.

In contrast to Mortimore and Tiffen's Machakos study, however, we are less optimistic that population pressure in Hamisi has led to an improved standard of living and we question the continued viability of any intensive agricultural system that has experienced such relentless population growth. Quality of life is difficult to measure and, in western Kenya at least, we found that estimates of household income and crop yields are often inaccurate and can be misleading if used as indicators of general community welfare. An alternate measure of the standard of living in a community is the level of household nutrition and the degree of food security, issues that many social scientists overlook (Frankenberger, 1987; Martorell, 1981). Dietary evidence from Hamisi indicates that most of the families have a low standard of living. People in the community are poorly nourished, consuming remarkably little protein and suffering seasonal scarcity of basic carbohydrates, which in turn necessitates a heavy dependence on purchased foods, and the children of Hamisi exhibit clear signs of chronic undernutrition (Onyango *et al.,* 1994; CBS 1983, see also Howard & Millard, 1997, for a comparable case). We conclude that, despite the resourcefulness and diversity of their agricultural system, the overall picture in Hamisi is of a people who have marginal diets and lack basic food security. . . .

Food Consumption Patterns

The typical diet for all households is composed primarily of carbohydrate-rich starches. Ugali is the most desired food and is the foundation for most meals. Ugali is made from coarse grain meal, most commonly maize, and is cooked with water to form a thick paste. Unlike true polenta there is no addition of fats, broth, or other condiments, so the nutritional value of ugali is rather low. Ugali is usually eaten in large quantities and gives a feeling of satisfaction, but fails to provide adequate nutrients. Because of the very small farm size in Hamisi and the allocation of 25% of farm land to cash crops, many households consume the grain from their fields within a few months of harvest. In our survey, maize meal used for making ugali was purchased from the market on 55% of the days over the year. . . . The harvest of the main long rains maize crop takes place in July and the short rains harvest is in December–January.

Other starchy foods such as cooking bananas, roasted or boiled maize cobs, and sweet potatoes are also eaten to supplement the consumption of ugali. However, some households produce these crops only in small quantities. In addition, some women (especially in families without significant remittances or a reliable local source of money) report selling a portion of the harvest of sweet potatoes, bananas, or beans, as a way of earning badly needed cash. As a result, despite the diversity of crops cultivated, these additional foods account for only a relatively small part of the diet.

The other commonly consumed energy-rich foods are cooking fats and refined sugar. Sugar, purchased from local shops, is usually consumed in liberally sweetened tea, which is often the sole breakfast item consumed. Tea is occasionally also taken as a late afternoon break. Midday and evening meals are usually composed of ugali, consumed in substantial quantities, with an accompanying side dish of stewed vegetables, meat, fish, or beans. These are usually cooked with onions and sometimes tomatoes that have been sauted in a little fat. Typically, local cooking recipes do not mix major ingredients together, so a meal might consist of ugali with vegetables or ugali with meat. Only on unusual days would a vegetable dish and a meat dish be consumed at the same meal or even on the same day.

While the staple starchy and energy foods provide the bulk of calories and assuage hunger, they are not sufficient sources of protein or other nutrients. With the exception of milk, protein-rich foods are consumed infrequently. Although there is some seasonal variation, servings of meat and fish are usually consumed only once or at most twice in a week, while beans are eaten on average less than twice a week. The infrequent consumption of protein-rich foods is partly explained by their high cost—beef was approximately Ksh 26 per kg

($1.60 in 1986) and tilapia, the preferred fish, costs Ksh 6–7 (about $0.35–0.45) per fish. Additionally, dependence on the market for access to grain reduces the income available for purchasing meat or milk, and increases the inclination to sell beans or milk to obtain grain.

Market cost not only reduces the frequency of protein consumption but also limits the size of portions that are served when protein is available. For example, the typical purchase of meat is only one kilogram shared by all members of the household. This results in an average of only 150-gm uncooked weight of meat/person/week. Most Hamisi residents prefer to eat tilapia, the most desired and expensive fish, but instead usually consume inexpensive dried *omena* (a small sardine-like fish). *Omena* is usually sold in the market piled in a small bowl costing Ksh 3.0 ($0.18). One bowl will be stewed and shared by all members of the household.

Milk consumption is more frequent, and is the most important source of protein in the Hamisi diet, but it is also consumed in very small quantities. On 43% of days surveyed, milk was the only protein-rich food consumed, but in quantities roughly the same as the amount of milk most of us would use to whiten two mugs of coffee. Milk is taken only in tea or is used to make a sauce when stewing vegetables, rather than consumed by the cupful. Milk intake, both from on-farm production and market purchase, averages 6.3 liters/household/week. With a mean household size of 6.6 people, this is the equivalent of only 136 ml per person per day (approximately 1/2 cup). Because on-farm milk production is low, most families are forced to spend considerable sums of money (average of Ksh 17.3 or $1.06/week) buying commercial milk sold in the local market, and the quantities consumed remain very low.

Discussion of Diet and Food Security

Of all the economic changes that have occurred in Hamisi, the decline in livestock keeping is especially significant for nutrition and food security. With population increase and the expansion of cultivation and the resulting decline in herd size, much less protein is available in the diet today. Earlier in the century, Hamisi farmers report that meat was not consumed frequently, as their livestock were primarily a wealth reserve to be used for brideprice payments, family emergencies, or ritual occasions. Nevertheless, large quantities of dairy protein and fat were originally part of the Hamisi diet. Milk offtake from cattle was used in a sauce for rich vegetable stews and as part of the cooking liquid for breakfast porridge (uji). Some of the cream was separated and clarified to produce ghee or butterfat that was used in cooking and for personal grooming. Animals were also occasionally bled to provide an additional protein-rich resource for the diet. The staple ugali and uji were frequently made with supplements of milk, ghee, and/or blood, and these additions significantly increased the caloric density of these staples, especially for uji which is the primary baby food used even today (Chaiken, 1988; Conelly, 1992b).

In contrast to the relative abundance of the past, today the dietary data for Hamisi present a portrait of people whose consumption of protein and calorically rich foods is very marginal and who are highly dependent on market purchases. This is true even for well-to-do households that have better quality houses, more land available, and larger herds of livestock. The more prosperous households in Hamisi are slightly less dependent on the market than is typical for the community as a whole, but prosperous households also have more mouths to feed—an average of 8.5 members vs. 5.0 and 6.2 for lower and middle status households. As a result, the data show no significant difference in frequency or quantity of nutritionally valuable foods consumed per person on the basis of income level. Similarly, cash crop production also does not correlate with nutritional adequacy in Hamisi. Households with coffee and tea holdings, for example, do not have noticeably better or worse levels of carbohydrate or protein consumption than the community as a whole (See Kennedy & Cowgill, 1988, and Fleuret & Fleuret, 1980, for some of the long debate on the impact of cash cropping on nutrition, and Howard & Millard 1997 for a comparable case).

Likewise, comparisons between households with a resident male head with those where the male is absent (typically looking for work or engaged in wage labor outside the community) failed to show any significant variation in food consumption patterns. Research reported by Onyango et al. (1994) for nearby Busia also found food consumption patterns did not vary significantly with the presence or absence of a male head of household. Female household heads in Busia produced less food than those with a resident male, but had more cash (in part from remittances) to purchase food than homes with a resident male, essentially balancing out the equation. Anthropometric studies of Luhya children in Busia did indicate, however, higher rates of stunting or chronic nutritional stress in female headed households (47.8%) than in male headed households (38.8%). The explanation for this anomaly, however, is not clear from the Onyango study.

Summary and Conclusion

This study of Hamisi agriculture supports the view that a high level of agro-diversity is a common strategy for coping with intense population pressure and extreme land scarcity in tropical farming systems. The data from Hamisi, as well as other studies from Africa discussed in this [selection], suggest that given a favorable environment with adequate rainfall, we might in fact expect to find the highest levels of crop diversity in the most intense agricultural systems. In these circumstances, diversity can be seen as a strategy that not only minimizes risk, but also maximizes the amounts of food and income that can be produced from very small parcels of land.

This suggests that cases such as Geertz' study of agricultural change in Indonesia, where intensification leads to specialization and the elimination of agro-diversity, are the exception rather than the rule. Specialization in the context of intensive agriculture might be expected to occur only in certain circumstances—such as where large-scale irrigation is practiced or where state control or market forces lead farmers to focus production on a single or narrow range of crops. Likewise, special technical circumstances, such as the introduction of plowing, especially if mechanized, would favor monocropping over diversity in intensive farming systems (Cohen, 1989). Even in these cases, farmers will often attempt to maintain as much diversity as possible, for example by developing very small, intensely cultivated, and highly diverse "homegardens" as a complement to the monocropping of irrigated rice.

A high level of diversity, however, does not guarantee a successful adaptation in terms of quality of life if population pressure is intense. At lower densities, the diversification, intensification, and commercialization of agriculture may permit a stable or even improving standard of living, despite population growth. At very high population density levels, we argue that farm size becomes so restricted that living standards are likely to decline and diet quality and food security are jeopardized. Under the stress of intense population pressure in Hamisi, it might be argued that the emphasis on diversity is more an indicator of desperation than adaptation, as farmers attempt to exploit their limited resources in every possible way. Their efforts have permitted them to survive under very difficult circumstances, but not to flourish.

As a result, despite a favorable environment for agriculture and a sophisticated system of intensive cultivation that fosters diversity, most Hamisi households have very poor nutrition and lack food security. Dietary intake is inadequate and most families are highly dependent on market purchases. Most people consume the equivalent of only a few tablespoons of protein-rich food each day, served largely as a relish to give taste to the meal rather than providing significant amounts of calories or nutrients, and an alarming number of children exhibit signs of chronic undernutrition. Published data on clinical observations of children support this claim of marginal nutritional status. A nationwide survey conducted with the support of UNICEF indicated that 41% of children in Kakamega District were stunted, a measure of long-term nutritional deprivation. This figure includes the less densely populated western and northern parts of the district where average farm size is larger, so rates of stunting may be higher i n Hamisi than for the district as a whole. In contrast, other high potential areas of the country, such as the Kikuyu highlands near Nairobi, have a stunting rate of only 34% (CBS, 1983). The poor nutrition in Hamisi reflects economic vulnerability caused by extreme land scarcity and a farming system that apportions almost 25% of landholdings to cash crops that cannot be eaten and that produce an unpredictable and marginal income for most families.

What changes might permit Hamisi farmers to improve their economic prospects and their nutritional status, even as population continues to increase? Though the issue remains controversial, increased production of non-edible cash crops for the export market does not seem to be the answer. A greater commitment to crops such as tea or coffee is likely to increase even further the vulnerability of households already heavily dependent on market purchases for their food (Howard & Millard, 1997). As an alternative, a successful strategy in some heavily populated areas of the tropics has been the development of high value food crops marketed to nearby urban centers, a process increasingly common in Kikuyu areas near Nairobi. Netting and Stone (1996) report this development trajectory for the Kofyar in Nigeria, who sell a wide range of grain and root crops on the market. This intensive agricultural system also maintains a high level of diversity, with the marketing of surpluses of crops that can be eaten in times of scarcity. Similarly, Eder (1982, 1989) describes an intensive farming system on Palawan Island in the Philippines where market gardening of vegetables to a nearby city has led to economic prosperity for many farmers, although this development has also contributed to increased socioeconomic inequality.

This type of development, however, may be difficult to implement for Hamisi. The marketing of French beans is a step in the right direction, but because of a lack of competition the private company that purchases the beans pays less than half the rate received by farmers in areas closer to Nairobi (Argwings-Kodhek, 1995). Market gardening for sale in regional urban centers will require a better infrastructure than currently exists, with more reliable and less expensive transportation to the cities of Kisumu and Kakamega. Perhaps a more serious obstacle would be competition from hundreds of similar communities, many of which are located closer to the cities than Hamisi, that would be able to market their produce more quickly and cheaply. In these constrained circumstances, despite the continued industriousness and ingenuity of the people, the future prospects for Hamisi, and many other communities in western Kenya, remain bleak.

NO    Michael Mortimore and Mary Tiffen

Population and Environment in Time Perspective: The Machakos Story

Introduction: Linkages Between Population Growth and the Environment

The linkages between population growth and environmental degradation are controversial. The view, widely held, that rapid population growth is incompatible with sustainable management of the environment is influenced, knowingly or not, by the neo-Malthusian belief that resources are limited. According to literature prepared for the United Nations Conference on Environment and Development (the Rio Earth Summit), 'the number of people an area can support without compromising its ability to do so in the future is known as its population carrying capacity'.

In agricultural terms, each agro-ecological zone is believed to have a carrying capacity which must not be exceeded if environmental equilibrium is to be maintained. In the words of Mustapha Tolba, formerly head of the United Nations Environment Programme, 'when that number is exceeded, the whole piece of land will quickly degenerate from overgrazing or overuse by human beings. Therefore, population pressure is definitely one of the major causes of desertification and the degradation of the land'. According to such a view, degradation threatens to diminish food production, and thereby the human carrying capacity, in a cumulative downward spiral.

More sophisticated estimations of population supporting capacities take account of technological alternatives to the low-input systems that are found in much of the tropical world. A recent study shows that if technology is varied (or levels of inputs increased), the limits rise accordingly. The critical constraint, for practical purposes, is access to technology. However, poverty inhibits investment, and the poor are said to be incapable of conserving their environment: rather, 'poverty forces them to exploit their limited stocks just to survive, leading to overcropping, overgrazing, and overcutting at unsustainable rates. A vicious circle of human need, environmental damage and more poverty ensues'.

A negative view of the effects of population pressure on the environment, which has been underwritten by several UN organizations, the Rio Earth

From Michael Mortimore and Mary Tiffen, "Population and Environment in Time Perspective: The Machakos Story," in Tony Binns, ed., *People and Environment in Africa* (John Wiley & Sons, 1995), pp. 69–70, 72, 75–78, 81–84, 86–87. Copyright © 1995 by John Wiley & Sons, Ltd. Reprinted by permission. Notes and references omitted.

Summit, and influential writers carries great weight in the environmental debate. Nevertheless it is not supported by some well-documented situations in Africa.

One of these, Machakos district of Kenya, is the subject of a recent study of resource management by African smallholders. The study covered the period 1930–90, which is long enough to control for rainfall variability, and for changes in the political economy. Profiles of change were constructed for all the major environmental and social variables. The linkages in what the World Bank has called the 'population, agriculture and environment nexus' were systematically investigated. The study shows positive, not negative influences of increasing population density on both environmental conservation and productivity.

Characteristics of Machakos District, Kenya

Machakos lies in south-east Kenya. Its northernmost point is about 50 km from the capital, Nairobi, from which it stretches some 300 km southwards. Since at least the eighteenth century it has been inhabited by agropastoralists known as the Akamba, who also populated the neighbouring Kitui district. Men looked after the livestock and cleared new land, while women cultivated a small plot for food crops. . . .

Setting the Scene for an Ecological Disaster

When British rule was imposed on Kenya, the Akamba people were confined in the Ukamba Reserve by the colonial government's Scheduled Areas (White Highlands) policy. . . . It was bounded by European settlers' farms and ranches on the north and west. To the east and south were uninhabited Crown lands, on which the government allowed only grazing, by permit. Thus encircled, the Akamba grew in numbers, and in livestock, while clearing extra land for shifting cultivation of maize and other crops, and chopping down trees for fuel burning and construction of their homes. Despite their protests, the government refused to relax its policy of containment.

During the period 1930–90, the population of Machakos district grew from 238 000 to 1 393 000, and an annual rate of increase of over 3% was maintained from the 1950s until after 1989. After 1962, the Akamba were allowed to settle on the semi-arid former Crown lands, and also took over some of the Europeans' farms in government schemes. Thus the land available to them effectively doubled. However, the growth of the population reduced the amount of land available to less than a hectare per person by 1989.

This conjunction of rapid population growth with unreliable rainfall, frequent moisture stress, low soil fertility and high erodibility, suggests the likelihood, on the premises outlined above, of population-induced degradation on a grand scale. This was indeed the diagnosis offered in assessments of the reserve in the 1930s. A disastrous series of droughts (in 1929, 1933, 1934, 1935 and 1939) caused major crop failures, losses of livestock, pest outbreaks, the deterioration of vegetal cover and accelerated erosion. In 1937, Colin Maher, the government's soil conservation officer, wrote despairingly:

> The Machakos Reserve is an appalling example of a large area of land which has been subjected to uncoordinated and practically uncontrolled development by natives whose multiplication and the increase of whose stock has been permitted, free from the checks of war and largely from those of disease, under benevolent British rule.
>
> Every phase of misuse of land is vividly and poignantly displayed in this Reserve, the inhabitants of which are rapidly drifting to a state of hopeless and miserable poverty and their land to a parching desert of rocks, stones and sand.

No less than eight official visits, reports and recommendations were commissioned between 1929 and 1939, strongly reflecting an official consensus view centred on overstocking, inappropriate cultivation, and deforestation in a reserve thought already to be overpopulated in relation to its carrying capacity. Did events bear out this gloomy prognosis? . . .

A Farming Revolution

Change in Machakos was multi-faceted. Technical innovation was not restricted to soil and water conservation. . . . An inventory of production technologies in Machakos identified 76 that were either introduced from outside the district or whose use was significantly extended during the period of the study. They included 35 field and horticultural crops, 5 tillage technologies and 6 methods of soil fertility management. The technical options available to farmers were thereby extended, adding flexibility to the farming system. Such flexibility is a great advantage in a risky environment.

Making Money From Farming

In the 1930s, capital was mostly locked up in livestock, and occasional sales provided needed cash. From cultivating maize, beans and pigeon peas for subsistence, farmers have since moved into marketing crops. The most successful of these, until its price fell in the 1980s, was coffee. Some Akamba learnt to grow it while employed on European coffee farms, but in 1938 the government expressly forbade 'native' coffee growing in order to protect the European producers' interests. After the overturning of this ban in 1954, strict rules were enforced in the growing, processing and marketing of coffee. African producers successfully achieved high grades. Coffee was an attractive component of rehabilitation programmes, since it was profitable and had to be grown on terraces. Coffee output increased spectacularly in the 'boom' of the later 1970s. It generated investment funds, and supported improved living standards, in the sub-humid zone. This had spill-over effects in the drier areas, through the demand of coffee-growing farmers for agricultural labour, for food and livestock products in which they might no longer be self-sufficient, and for a whole range of consumer goods, housing improvements and services. By 1982–83 over 40% of rural incomes in Machakos district were being generated by non-farm businesses and wages.

In contrast to coffee, cotton, which is recommended for the drier areas, was not a success in the long term. Its price was only rarely high enough to compensate for its tendency to compete with the food crops for labour and capital, and the profit margins were reduced by marketing inefficiencies. Three attempts to promote cotton—in the 1930s, 1960s and 1978–84—ran into the sand. Output limped along, and in 1991 the closed ginnery at Makueni offered silent testimony of failure.

Both coffee and cotton were sold to monopsonist parastatal marketing boards and required government support in extension and supervision, and in supervising officially sponsored co-operatives for input provision, processing and grading. By contrast, expanded growing of a great variety of perennial and annual fruit and vegetables was closely linked with the growth of Kenya's canning industry, the Nairobi and Mombasa retail markets, and exports of fresh vegetables by air. Itinerant Asian buyers, firms operating contract-buying and enterprising Akamba, as individual traders or in formal co-operatives or informal groups, have all played a role. A generally high value per hectare facilitated the skilful exploitation of wet micro-environments, even in the driest areas, and the development of technologies such as micro-irrigation and the cultivation of bananas in pits. Fruit production is attractive to women farmers, as trees do not compete with food crops (for which they often have the main responsibility).

Akamba farmers have adapted rather well to the opportunities provided by the market. No amount of promotion can succeed without incentives. But given these, innovation in both production and marketing aspects is commonplace. Meanwhile, livestock sales, on which they depended for market income in the 1930s, have steadily declined.

Achieving Food Sufficiency

The staple food of the Akamba is white maize. The shortness of the two growing seasons, and the high probability of drought, call for varieties that either resist drought or escape it by maturing quickly. In the 1960s, the government's local research station began a search for drought-escaping varieties that culminated in the release of Katumani Composite B (KCB) maize in 1968.

The new maize was promoted by the extension service, and it was steadily, if unspectacularly, adopted by the farmers. Various surveys suggest that from two-thirds to three-quarters use it, but it is not known how much of the maize area is planted to it, nor what proportion of output it contributes. Of 40 farmers interviewed in five locations in 1990, only a third said they used it exclusively, and another third used it together with local and hybrid varieties.

This was no 'green revolution'. The ambivalent response has, however, an explanation. Given the unpredictable rainfall, farmers need to keep their options open. Their local varieties, though slower to mature, are more resistant to drought, and hybrids do better in wetter sites. KCB is liked because, in combination with other varieties, it strengthens this flexibility rather than undermines it. Some farmers cross-pollinate it with their local varieties, further enhancing their adaptive choice.

With and without KCB, food crop production per person kept up with population growth from 1930 to 1987 although imported foods remained necessary after a series of bad seasons. The district's dependence on imported food in the period 1974–85 was less than in 1942–62 (8 kg per person annually compared with 17), notwithstanding major droughts in both periods. Food output per person in 1984 (after three seasonal droughts) was slightly higher than in 1960–61 (after two seasons with drought, and one with floods).

Faster Tillage

In view of the reputation then enjoyed by the Akamba for resistance to change, it is surprising that the ox-plough, introduced to the district as early as 1910, had spread to about 600 (or 3%) of the district's households by the 1930s. Farmers trained their own cattle, and ploughs were cheap (about equivalent to the price of a cow in 1940); furthermore the technology was being tested and developed on nearby European farms where some Akamba worked. Ownership greatly increased the area a farmer could cultivate, and enabled him to sell maize or cotton. Its adoption called for the cessation of shifting cultivation, and facilitated the adoption of row planting and better weeding, in place of broadcasting seed.

After the Second World War, ex-soldiers who had seen ploughs in India returned with the capital to buy their own. Proceeds from trade and employment outside the district were also invested in ploughs. The government made ox-ploughs the basis of a new farming system imposed on a supervised settlement at Makueni location. The government, and traders, provided some credit. Coffee (after 1954), horticulture and cotton (in some years) generated investment funds. Adoption accelerated in the 1960s and was more or less complete by the 1980s. Surveys found 62% or more of farmers owning a plough, the remainder being too poor, or having fields too small and steep for its use.

The plough proved to be both a durable and a flexible technology. The first ones in use were adapted to opening new land, with teams of six or eight oxen. Farmers later selected a lighter, two-oxen instrument suitable for work on small, terraced, permanent fields. The Victory mouldboard plough, though much criticized on technical grounds, is used everywhere, and for several operations—primary ploughing, seed-bed preparation and inter-row weeding—and attempts to promote its replacement by a more expensive tool-bar have failed. It saves labour, and is also used by women. The 'oxenization' of Akamba agriculture was, in a measure, a triumph of capitalization in a capital-poor, risk-prone and low productive farming system.

Fertilizing the Soil

Shifting cultivation used to rely on long fallows for replenishing the soil. In the 1930s, there was very little systematic manuring. But the fertility of arable land, as measured by yields, was low. The Agricultural Department favoured farmyard manure over inorganic fertilizers. It also, unsuccessfully, promoted composting.

It was not until the 1950s that manuring became widespread, in the northern sub-humid areas. By this time, arable fields were fixed, and cultivated every year. The silent spread of this practice can be judged from the fact that by the 1980s, 9 out of 10 farmers were doing it, in both wetter and drier areas. Now, most arable land is cultivated twice a year—in both rainy seasons—and composting is being adopted by small farmers with few livestock.

By contrast, the use made of inorganic fertilizers is minimal, the bulk of it on coffee. Manure is made in the *boma* (stall or pen) and supplemented with trash and waste. The amount applied depends on how many livestock there are, and how much labour and transport are available when needed. Every farmer knows that, under present technical and economic conditions, sustaining output depends on the use made of *boma* manure. Few can afford inorganic fertilizers in quantity.

Feeding the Livestock

At the beginning of our period (the 1930s), Akamba women cultivated food crops at home, while their men used to take the livestock away to common grazing lands for several months of the year. However, common grazing land vanished as it was transformed into new farms. After about 1960 settlement on Crown lands could no longer be restrained, and thousands of families moved into them. Each household must now keep its animals within the bounds of the family farm, or obtain permission to use another family's land, often in return for some rent or service. More than 60% of the cattle, sheep and goats are stall-fed or tethered for a part or all of the year. When in the *boma,* cut fodder and residues are brought to them, which requires additional labour. Fodder grass is grown on terrace banks. These changes are most advanced in the sub-humid zone. A third of the livestock are grazed all the time, mostly in the dry semi-arid zone. The effort required to maintain livestock is making grade or crossbred cattle popular (estimated to number about 9% of the total in 1983 and to have grown rapidly since), whose milk yields and value are superior to those of the native zebu, though their increased health risks call for frequent dipping.

Farming the Trees

From the 1920s, the Forest Department believed that reafforestation was necessary to arrest environmental desiccation, and supply the growing need for domestic fuel and construction timber. For several decades the department struggled, under-resourced, to reserve and replant hilltop forests. In 1984, however, estimates of household fuel requirements put the need for new plantations at 226 000 ha (15 times the area of gazetted forest reserves!). The destruction of surviving natural woodland seemed an imminent possibility.

But sites photographed in 1937 and 1991 showed little sign of woodland degradation. A fuel shortage has failed to develop on the expected scale and the district does not import wood or charcoal in significant quantities. Indeed it exports some. Part of the explanation for this expert miscalculation lay in ignoring the use made of dead wood, farm trash, branch wood from farm trees, and hedge cuttings, for domestic fires. The other part lay in failing to appreciate a major area of innovative practice: the planting, protection and systematic harvesting of trees. Forest policy in the 1980s was shifting towards farm forestry promotion, but in this it was following, not driving, farmers.

Tree densities on farmland in one location, Mbiuni, averaged over 34 per ha (14 when bananas are excluded) by 1982. Furthermore, the smaller the farm, the greater the density. The range of trees planted includes both exotic and indigenous, both fruit and timber species. Akamba women generally manage fruit trees, while the men look after the timber trees. Owners of grazing land manage the regeneration of woody vegetation, which is used for timber, fuel, browse, honey production, edible and medicinal products.

Producing More With Less

These were some of the features of a revolution in farming wrought in unpromising circumstances. What was the driving force behind these changes?

The growth of population had two important outcomes: the subdivision of a man's landholdings among all his sons, according to Akamba custom, and the increasing scarcity of land as former communal grazing and Crown land became new private farms.

As holdings shrank in size, the arable proportion rose, leaving less and less land for grazing while the cultivated area per person stagnated. [T]he percentage of arable land increased from the older settled areas to the new, and from the wetter to the drier areas.

These changes created the imperative for intensification. By intensification we mean the application of increasing amounts of labour and capital per ha to raise crop yields. Crop–livestock integration is intrinsic to this process in dryland farming systems in Africa. It was driven by the needs for draft energy on the farm, for fodder (the stalks of maize and haulms of beans, for example), for manure and milk.

Two changes—to intensive livestock feeding systems, and to permanent manured fields, often under plough cultivation—were pivotal in this transformation, whose outcome was an increasingly efficient system of nutrient cycling through plants, animals and soil. The changes could not have occurred without security of title. Akamba custom had already recognized individuals rights in land, including the right of sale, in the older settled areas by the 1930s. Security has been reinforced by statutory registration of title, a slow legal process which began in Machakos in 1968 and has still not covered all areas.

Equally important were sources of investment capital. This was not only required for terraces and ploughs. To clear and cultivate new land, build hedges or plant trees requires labour which often has to be hired, as well as tools and expertise. The off-farm incomes earned by Akamba men inside and outside the district have contributed for decades to agricultural investment. Such incomes are often high in households with small farms, and there is little evidence that investment per ha falls where farmers are poorly resourced in land.

The outcome of this process of intensification was an increase in the value of output per square kilometre (at constant prices) from 1930 to 1987. This was calculated by taking output data for the only three available years before 1974 (1930, 1957 and 1961), selecting two later years which were climatically average (1977 and 1987), and converting all the values into maize equivalent at 1957 prices. The year 1957 was an unusually good one and 1961, as already noted, unusually bad, hence the upward trend was interrupted. The trend continued despite the additions of large areas of the more arid types of land in 1962. Output per capita closely reflected this curve.

We conclude that, contrary to the expectations expressed in the 1930s, the Akamba of Machakos have put land degradation into reverse, conserved and improved their trees, invested in their farms, and sustained an improvement in overall productivity. . . .

No Miracle in Machakos?

What happened in Machakos did not contravene the laws of nature, as the Malthusian paradigm would express them, but rather grew logically from a conjunction of increasing population density, market growth and a generally supportive economic environment. The technological changes we have described, in conservation and production, cannot be adequately understood as exogenous, as mere accidents that gave breathing space on a remorseless progression towards irreversible environmental degradation and poverty. Rather, as argued long ago by Ester Boserup and more recently by Julian Simon, they were mothered by necessity. Technological change was an endogenous process, in which multiple sources and channels were employed, involving selection and adaptation by farmers.

Increasing population density is found, then, to have positive effects. The increasing scarcity (value) of land promoted investment, both in conservation and in yield-enhancing improvements. The integration of crop and livestock production improved the efficiency of nutrient cycling, and thereby the sustainability of the farming system.

The Machakos experience offers an alternative to the Malthusian models of the relations between population growth and environmental degradation. Elsewhere in Africa, there are more documented cases of positive associations, though it would be foolish to ignore the differences.

Successful intensification under rising densities has certain preconditions. These are peace and security, for trade and investment, and a marketing and tenure system in which economic benefits are shared by many, rather than monopolized by a few. Degradation may occur, as it did in Machakos, when a change from a long fallowing system is first needed, but when population densities or other conditions are not conducive. Normally, as population grows, so do the opportunities for specialization and trade. To the stick of necessity the market adds the carrot of incentives and resources for investment in new technologies.

In the past, development planners tried to transform farming systems that were seen as inefficient and technically conservative. In fact, they are changing themselves, as studying them in time perspective shows, and there is scope for supporting positive change with appropriate policies. The guiding principle must be to go with the grain of historical change. This means encouraging investment, by encouraging trade and by improving farm-gate prices (for example, by improving roads and by avoiding heavy taxation of agricultural products). There is also a need to protect investment when crises (e.g. famines) threaten to force households to sell their assets. Increasing the technical options available to local resource users in a risky environment is one of the most productive avenues to pursue, by encouraging endogenous experimentation; by increasing information through general education as well as agricultural extension, and by creating new avenues for technological development and transfer.

POSTSCRIPT

Is Food Production in Africa Incapable of Keeping Up With Population Growth?

Mortimore and Tiffen argue that resource practices of the Akamba farmers in the Machakos District of Kenya contradict the Malthusian theory that unchecked population tends to rise faster than food production capability. They state that, over a 60-year period, farmers were able to reverse land degradation, enhance their livestock, invest in their farms, and increase productivity despite increasing population density. An interesting aspect of the Machakos story was this rural community's ability to positively engage with the market. This stands in contrast to a large amount of Africanist scholarship that sees interaction with the global market as extremely problematic. Mortimore and Tiffen's analysis of the Machakos situation is not without its critics. One of the main concerns is that they did not address properly the problem of economic differentiation over time (that is, a growing gap between the rich and poor). For an excellent, and concise, articulation of this concern, see Diana Rocheleau's September 1995 piece in the journal *Environment* entitled "More on Machakos."

Unlike many hard-line neo-Malthusians, Conelly and Chaiken agree that some good may come of increasing population densities, such as agricultural intensification and a diversifying farm system. They, however, assert that these positive developments have not led to an improved standard of living. Conelly and Chaiken state that the food and nutritional security of Hamisi inhabitants is not ideal. It is important to note, however, that this same observation could be made for many areas of the world, including those with low population densities. What seems less clear in the Hamisi case is the degree to which food and nutritional conditions have changed over time. Conelly and Chaiken's conclusions are debatable if one questions the anecdotal evidence used to suggest that conditions were better in the past. If one does accept that circumstances have grown worse with increasing population growth, one possible conclusion is that Hamisi is further along in the population growth/agricultural intensification trajectory than Machakos. In other words, perhaps Hamisi has reached an agricultural intensification plateau (or point of diminishing returns) at which Machakos has yet to arrive.

For further reading on this topic, see a volume edited by B. L. Turner II, Goran Hyden, and Robert Kates entitled *Population Growth and Agricultural Change in Africa* (University Press of Florida, 1993) that includes case studies from around the continent. Or see an article by Lambin et al. in the December

2001 issue of *Global Environmental Change* entitled "The Causes of Land-Use and Land-Cover Change: Moving Beyond the Myths." For a more Malthusian perspective, see a book by Kevin M. Cleaver and Götz A. Schreiber entitled *Reversing the Spiral: The Population, Agriculture, and Environment Nexus in Sub-Saharan Africa* (World Bank, 1994).

ISSUE 8

Are Abundant Mineral and Energy Resources a Catalyst for African Development?

YES: Oliver Maponga and Philip Maxwell, from "The Fall and Rise of African Mining," *Minerals and Energy* (2001)

NO: Sunday Dare, from "A Continent in Crisis: Africa and Globalization," *Dollars and Sense* (July/August 2001)

ISSUE SUMMARY

YES: Oliver Maponga, chair of the Institute of Mining Research at the University of Zimbabwe, and Philip Maxwell, professor at the Western Australian School of Mines at Curtin University of Technology, describe a resurgence in the African mining industry in the 1990s after several lackluster decades. They assert that mineral and energy mining can make a positive contribution to economic development in Africa.

NO: Sunday Dare, a Nigerian journalist, describes how "much sorrow has flowed" from Africa's resource blessing. While Dare blames African leaders for corruption and resource mismanagement, he also implicates transnational corporations (TNCs) as key contributors to this problem. He states that TNCs have acted as economic predators that support repressive African leaders in order to garner uninterrupted access to resources. The result, Dare suggests, is that Africa's "raw materials are still being depleted without general development."

The authors in this issue are basically wrangling over the validity of the "resource curse thesis" in contemporary Africa. This thesis posits that countries with great natural resource wealth have a tendency to grow more slowly than resource-poor countries. There are at least two types of problems (that potentially inhibit economic growth) frequently linked to resource abundance in

African countries. First, is the notion that resource abundance may lead to a simplification or concentration of national economies, with increasing dependence on a single export mineral or fuel. The idea is that decision makers will tend to gravitate toward the easy money and shirk investments in other aspects of the economy. Nigeria is a potential example of this problem as productivity in a number of sectors (most notably food production and manufacturing) declined with the rise to prominence of the petroleum industry in the 1970s.

The second issue often linked to resource wealth is civil conflict. This problem may manifest itself in a couple of different ways. Rival groups may fight for control of key resources and then finance their military aggression through proceeds from the sale of these resources (such as diamond trading by rebel groups in Sierra Leone). Or, corrupt governments may sequester the proceeds from resource extraction to line their own pockets and to bankroll the suppression of opposition groups—a tactic that often leads to future opposition and unrest.

In their selection, Oliver Maponga and Philip Maxwell impugn the resource curse thesis in contemporary Africa. While acknowledging several lackluster decades in the mineral and energy sector, they describe a resurgence in the African mining industry in the 1990s. They maintain that mineral and energy extraction can make a positive contribution to economic development in Africa, ending their article on the optimistic assertion that "resource development will be seen as a blessing rather than a curse by the year 2010."

Sunday Dare counters that the resource curse thesis is alive and well in Africa. Dare blames African leaders for corruption and resource mismanagement, but he also implicates transnational corporations (TNCs) by asserting that TNCs have acted as economic predators that support repressive African leaders in order to gain access to resources.

In addition to the issues already raised, there are a few other points introduced by the authors in these texts that deserve some foregrounding. In the 1960s, when many African nations gained independence, there was a great deal of concern that transnational or multinational corporations acted as neo-imperialistic forces on the African continent. Dependency theorists suggested that major industries (including mines) either needed to be run by the government or indigenous entrepreneurs, or strictly regulated, in order to ensure that profits remained within a country's borders. This thinking led to a wave of nationalizations (or government takeovers) of foreign-owned enterprises in the 1960s and 1970s. Since the early 1980s these nationalized industries have gradually been re-privatized as part of the World Bank and International Monetary Fund's structural adjustment programs. The rationale for these privatizations is that governments are incapable of efficiently running enterprises and that foreign ownership needs to be permitted in order to encourage outside investment and technology transfer. Both of these policies, nationalization on the one hand and privatization on the other, are highly contested in the literature and, not surprisingly, appraised quite differently by the authors of the following selections.

Oliver Maponga and Philip Maxwell  **YES**

The Fall and Rise of African Mining

Introduction

Africa has twenty per cent of the earth's landmass and, in 2000, was home to about 800 million people. There is a highly prospective geology in much of the continent and minerals play an important role in the economic activities of many of its more than fifty sovereign nations. One reflection of this is the observation by [Graham] Davis that, in 1991, there were sixteen mineral dependent economies in Africa. They included Algeria, Angola, Botswana, Cameroon, Congo, Gabon, Guinea, Libya, Mauritania, Namibia, Niger, Nigeria, South Africa, Togo, Zaire (now the Democratic Republic of Congo) and Zambia.

Yet in a relative sense, the continent's minerals industry is still in its infancy. It has not achieved its full potential despite a long history of mining. This is in apparent contrast with the situation in Australia, Canada, the United States and Latin American nations such as Chile, where mineral discovery and exploitation has been an important source of subsequent economic development. Over the past decade there has been considerable debate about the role of natural resources in the economic development process. Authors such as Auty, Sachs and Warner and Gelb have argued that natural resources have been an economic curse for many nations. A contrary view appears in Davis.

One can classify almost every African nation as a developing economy. According to the African Development Bank, 39 African countries had GNP [gross national product] per capita levels of less than USD 1000 in 1998. The most affluent were Libya (USD 6160), Gabon (USD 4170), South Africa (USD 3310) and Botswana (USD 3070)—each major mineral producers. This compared with GNP per capita levels of USD 30600 in the United States, USD 20050 in Australia and USD 19320 in Canada in 1999.

Despite the resource curse argument, further expansion of the resources sector remains a major hope for the growth and development of many African nations. This is because of its potentially strong contribution to Gross Domestic Product [GDP] and to export income, as well as its role in generating forward and backward linkages to other parts of these economies.

There was widespread nationalisation of minerals in Africa early in the post-independence era after 1960. In response to this foreign mineral investors showed little interest in mineral exploration or investment in most African nations during the 1970s and 1980s. This led to a loss of earlier mineral competitiveness in countries such as Zambia, Zaire, Guinea, Nigeria and Tanzania. South Africa and Botswana stood almost alone as the continent's world-class hard rock mineral producers during this time. The continent's oil and gas also struggled, particularly after 1980 when oil prices began a long period of nominal and real price decline.

Since the early 1990s the situation in Africa and elsewhere has been changing. The demise of the Soviet Union and communism, and the continuing push of global capitalism has been an important part of this story. Over the past decade, there has been a major revision of mineral policy throughout the continent to encourage foreign investment in exploration and new mineral projects. In this new area, African nations have been attracting more mineral exploration spending. This has also led to a series of new mineral projects, which are likely to boost the economic fortunes of several nations, particularly if political decision makers choose to invest taxes and royalty payments wisely. . . .

The Performance of African Economies— 1970 to 1990

An important . . . issue . . . is whether the stagnation of the resources sector in Africa adversely affected the relative performance of its mineral dependent economies. In addressing this issue it is useful initially to compare GDP growth rates in African economies with those elsewhere. . . . African economic growth between 1970 and 1979 lagged behind that in Asia and South America. Yet it was higher than in OECD [Organization for Economic Cooperation and Development] member nations. During the 1980s, the situation changed, with African economic growth comparable to that in South America, but lagging considerably behind Asia and the OECD.

When one moves to consider the performance of mineral and non-mineral economies in Africa, the results seem consistent with those put forward by Sachs and Warner. Between 1975 and 1990, the real GDP at factor cost in Africa's mineral dependent economies grew at 0.9 per cent per annum less than in the non-mineral economies. Economic growth rates in the five African fuel-exporting nations were remarkably similar to those in the eleven hard rock mineral exporting economies.

It is useful also to consider the performance of some of the related quality of life indicators used by authors such as Davis. He considered movements in indicators such as life expectancy, infant mortality, daily calorie supply, primary school enrolment and the adult literacy rate. Focusing on the fortunes of 22 mineral economies and 57 never-minerals economies in developing nations between 1970 and 1991, Davis found that more favourable movement in these measures in the mineral dependent economies.

. . . Between 1975 and 1990, the average life expectancy in Africa's mineral dependent economies increased by about 4.8 years. The fuel exporting economies experienced an increase of 6.1 years while other mineral economies rose by 3.4 years. The corresponding change in non-mineral dependent economies was 3.3 years.

Daily calorie supply, which had been at similar levels in 1975 in mineral and non-mineral economies, increased by about 100 calories more in non-mineral economies than in mineral economies. Daily calorie intakes rose 260 calories in fuel exporting nations but fell by more than fifty calories in non-fuel mineral economies. Infant mortality rates began at somewhat higher levels in the non-mineral economies than in the mineral economies in 1975. They fell by somewhat more in the non-mineral economies over the next fifteen years but still remained at higher levels in those nations.

So the findings from this analysis of quality of life variables between 1975 and 1990 are mixed. A longer time period of analysis would seem necessary to reach more definitive conclusions.

The Past Decade—Is a Renaissance Underway?

The combination of a poor investment climate, political instability and the lack of supportive infrastructure including poor geological information contributed to the decline in the minerals industry in Africa generally up until 1990. In this environment it is hardly surprising that mineral economies did not perform well. But the economic geography of the world minerals industry has been changing in recent years. A reversal of the disinvestment trend between the 1960s and the 1980s has taken place since 1990 in many developing nations. Prospective nations in Latin America, Africa and Asia have received increased exploration spending. Inflows of foreign direct investment [FDI] have also been rising. An important early indicator of a broader change has been an increase in gold production from African nations outside of South Africa. The discussion of this section considers the extent of change and some of the reasons for it. It concludes by reflecting on whether the new operating environment has had any noticeable effect on the economic welfare of mineral dependent nations.

(a) Exploration Spending and Foreign Direct Investment in Mineral Development

The Metal Economics Group, which collects key mining data from large resource sector companies, reported a significant increase in the share of exploration spending in Africa during most of the 1990s. There was an apparent increase in the region's share of exploration spending from around six per cent in 1993 to more than thirteen per cent in 1998. It fell to 11.5 percent in 1999. Exploration spending in Africa increased from USD 165 million in 1993 to USD 670 million in 1997. Between 1997 and 1999 there was a decline in mineral exploration activity throughout the world. The fall in estimated mineral exploration spending in Africa to USD 323 million in 1999 reflects this. By the late 1990s Africa had overtaken the Asia-Pacific region as the second most favoured investment destination in the developing world after Latin America. . . .

Foreign direct investment (FDI) inflows to Africa increased generally during the 1990s, compared with the 1970s and 1980s. UNCTAD [United Nations Conference on Trade and Development] estimated that FDI inflows rose from USD 2 billion in 1990 to USD 10 billion in 1998. Despite the increase in the 1990s, investments by transnational corporations into Africa represented only 1.2 percent of global FDI flows and five percent of total foreign direct investment into developing nations. Although these data refer to direct investment in all industries, the authors of the UNCTAD report acknowledge that investment in natural resources in Africa has been the main focus of foreign direct investment.

The major presence of Canadian and Australian exploration and mining companies in Africa in the 1990s reflects this change. For Canadian-based resource companies, the growth of junior exploration companies, coupled with the success of Canadian stock exchanges, provided major stimulus for overseas investment. Campbell notes that Canadian companies responded aggressively to new liberalised operating environments in Africa in the 1990s. By 1996 there were over 170 Canadian-registered companies operating on the African continent. In the early 1990s hardly any Canadian companies had operations there. Campbell attributes the overseas involvement of Canadian mining companies particularly to competitive advantages as well as to the availability of risk capital from the Vancouver Stock Exchange.

Australian-based resource companies also became active in exploration and mine development in Africa during the 1990s. In 1985, Bridge Oil was apparently the only Australian resources company with interests in Africa. By 1992 fifteen Australian companies had African interests and this had increased to seventy-five by 1997. They operated in thirty African nations. . . . Australian companies had 134 projects in thirty African nations in 1999. This compared with 88 projects in 1995. Data collected by the Minerals Council of Australia also show increased offshore exploration spending by Australian mining companies, with Africa as one of the major destinations.

One reflection of the revival of Africa as a major mineral producer is the recent growth of the gold industry outside of South Africa. Despite declining prices, the gold sector has been the largest beneficiary of the new investment boom. Output increased from 40 tonnes in 1990 to more than 140 tonnes in 1998. Successful exploration has led to significant new gold mines coming on stream in Ghana, Mali, Tanzania, Guinea and Burkina Faso.

Tanzania's first mine in more than thirty years, Golden Pride, opened in 1998. The opening of Mali's Sadiola gold mine in 1997 propelled the country to one of the top twenty gold producers in the world. Mali now ranks as Africa's fourth largest gold producer behind South Africa, Ghana and Zimbabwe. Ghana's experience since 1988 illustrates the renaissance of the region's minerals industry more than any other country. Ghana's gold output increased from 12 tonnes in 1988 to over 70 tonnes in 1998. Australian companies opened two new gold mines at Damang (Ranger Minerals) and Obotan (Resolute) in the late 1990s.

Despite these positive developments . . . between 1989 and 1998 . . . Africa lost market share for ten of the eleven minerals reported. One might explain this apparently perverse trend by arguing that the new exploration and investment is taking several years to exert its full impact. Other forces may, however, be important. Before turning to this later issue we consider further some of the positive influences.

(b) Positive Influences on the Competitiveness of African Mining

Several factors have played a role in facilitating a re-emergence of African minerals and energy. They include:

- the modernisation of mining regimes through new legislation, formal mineral policy statements and observation of international agreements;
- availability of investment insurance;
- greater political stability in several prospective nations;
- a movement of privatisation of state mining companies; and
- changed conditions in home countries which have led to international expansion.

Otto reports that "Between 1985 and 1995 over 90 nations introduced, or commenced working on, new or major revisions to mining sector legislation." African nations were well represented in this group. This has created a more favourable environment for private sector investment. Campbell recognises the importance of these reforms to inflows of Canadian investment in Africa in the mid-1990s.

During the 1990s, fiscal regimes continued to evolve as competition for shrinking investment budgets increases. Countries like Botswana, one of the continent's most stable and attractive economies, introduced a new Mines and Minerals Act in 1999. This abolished free equity for government and simplified licensing systems. The Government of Namibia enacted a new Diamond Act in 1999, while Nigeria introduced its Mining and Minerals Decree No. 34. Recent changes to fiscal regimes in Benin, Burkina Faso, Central African Republic and Cote d'Ivoire have resulted in increased investor activity. Because of these changes the highly prospective Birimian greenstone belts in West and Central Africa have become major attractions for Canadian and South African mining companies.

The efforts of the World Bank to regenerate the African minerals industry are also playing a facilitating role. This occurs particularly through the activities of three of its associated organisations, the International Development Association (IDA), the Multilateral Investment Guarantee Agency (MIGA) and the International Finance Corporation (IFC).

Through its technical assistance program, the International Development Association currently makes credit available to countries wishing to reform their minerals sector. The International Finance Corporation (IFC) provides equity finance for projects in the developing world and helps mobilise private loans for private investors in the developing world. The Multilateral Investment Guarantee Agency provides investment risk guarantees to lower the risk profile of mining projects for foreign investors. Forty-two African nations were signatories to the MIGA convention by 1999. MIGA provides insurance against transfer restrictions, expropriation (direct and creeping), war and civil disturbance and against breach of contract by governments and helps mobilise insurance. The organisation also provides a mechanism for the settlement of investment disputes. Between the commencement of its guarantee scheme in Africa in 1991 and 1999, MIGA had issued 61 guarantees. Two examples of the work of IFC and MIGA relate to loans to finance the Sadiola gold mine in Mali and the Mozal Aluminium Smelter in Mozambique.

Institutions in Britain, France, Belgium, the United States and Canada have also been active in supporting investment risk protection measures to encourage movement of their companies into Africa. Two examples of these institutions are OPIC (for U.S. companies) and the Export Development Corporation (for Canadian companies).

Despite some notable exceptions, Africa was more politically stable in the 1990s, compared with the previous 30 years. The changes in political climate in countries such as Mozambique, Angola and the Democratic Republic of Congo, together with the post-apartheid government in South Africa, created a more favourable investment climate in the minerals sector. This had a particularly favourable impact for the major South African mining companies. They followed a program of greater diversification into the rest of the continent to replace aging and costly mines at home. Relative political stability in Mozambique and Angola led to legal mining resuming in these nations. Two diamond properties in the Luo and the Catoca in the kimberlite areas of Angola came into production in 1996. The aluminium smelter in Mozambique became a viable project after the UN supervised elections in the mid-1990s. Successes of foreign mining companies in Ghana, Tanzania and Mali and Mauritania have also enhanced the region's image.

Whereas privatisation of state-owned enterprises in most developed countries began in the mid-1980s, the process started much later in developing nations. Gathering momentum after 1990, it has been an important influence in attracting foreign direct investment. There have been several different privatisation approaches. They have included change in ownership status, complete divestiture or partial divestiture. Seventy percent of total FDI inflows into Africa in 1999 were in Angola, Egypt, Nigeria, South Africa and Morocco where major privatisations had occurred. UNCTAD reports that South Africa received USD 1.4 billion of foreign direct investment between 1990 and 1998. Ghana attracted USD 769 million, Nigeria, USD 500 million, Zambia USD 420 million, and Côte d'Ivoire USD 373 million.

Changing policy and legislative frameworks in the developed mining nations have also influenced the move of foreign mining companies into Africa. Factors such as more demanding environmental legislation and native title issues at home have been particularly important in encouraging European, North American and Australian resource companies to consider moving offshore. The activities of junior exploration companies have played a key role in promoting this change. Willing to take more risks, they have often been thwarted at home because mining and exploration tenements are more difficult to obtain.

(c) Mining Sector Reform and Economic Performance

Despite the above changes, the impact of reform in mineral rich African nations did not translate into economic or quality of life gains during most of the 1990s. . . . [B]etween 1990 and 1998, GDP at factor cost in Africa's mineral dependent economies grew at an average rate of 1.7 percent per annum. It compares with a growth rate of 3.1 percent in non-mineral economies. Furthermore, . . . a reduction in life expectancy in the mineral economies from 1990 to 1998 [occurred], as opposed to an increase in the non-mineral economies. Daily calorie supply increased at a faster rate in non-mineral economies and infant mortality rates declined more quickly than in their mineral dependent counterparts. . . . [F]uel exporting mineral economies fared considerably better than the non-fuel exporting mineral economies.

It is tempting to appeal directly to the resource curse thesis to provide an explanation for these trends. It would seem premature, however, to do so for at least two reasons. Firstly, the reform and renaissance of the African minerals sector is a process from which tangible benefits may take perhaps five or ten years more to flow into expected economic gains. Without consistent and continuing investment, a minerals industry loses its competitive strength. Rebuilding this in a consistent manner will take perhaps two decades, since it requires investment in infrastructure and human capital as well as exploration and mine development. The new policy and regulatory frameworks are establishing a good foundation for this to take place but the process is far from complete. Secondly, the poor performance of quality of life indicators in the non-fuel mineral economies appears to reflect the impact of the HIV/AIDS epidemic in Southern Africa.

An associated line of empirical investigation is to review the most recent rates of economic growth—say between 1996 and 1998—in African nations. In the 16 mineral dependent economies, real GDP at factor cost grew from an estimated USD 348.5 billion to USD 365.3 billion during this time—a growth rate of 2.3 percent per annum. In the 37 non-mineral economies estimated GDP at factor cost grew from USD 143.5 billion to USD 155 billion. This was an annual growth rate of 4.1 percent per annum. It is clear that a minerals-driven recovery is not yet under way.

(d) Mineral Endowment, Labour Productivity, Comparative and Competitive Advantage

. . . Africa's mineral endowment is substantial. The continent's favourable geology and the relative lack of recent mineral exploitation means that there are many high quality and potentially profitable deposits available, other things being equal. Recent mineral sector reforms and other legislative changes have apparently "levelled the playing field" for potential African mineral producers.

Despite this, developed nations such as the United States, Canada and Australia retained their percentage shares of world mineral production during the 1990s. This followed an increase between 1980 and 1990, despite a significant decline in ore grade at operating mines. By utilising new mineral processing

technologies, adopting modern mining methods and operating more efficiently in other ways (particularly through outsourcing), resource companies
in these nations improved their fortunes. In recent papers on the U.S. copper industry, Aydin and Tilton and Tilton and Landsberg find that labour productivity
tripled between 1975 and 1985. Between 1985 and 1995 this enabled the US industry to recover the share of world copper output lost in the preceding ten
years. New technology and innovation appeared either as important or more
important than mineral endowment in shaping comparative advantage in the
copper industry. Australia's re-emergence in the 1980s as a major world gold
producer and its holding of this position in the 1990s seem consistent with
these findings.

It is useful to reflect on a model such as the Porter "diamond" in assessing
the current position of the African mineral industry. The four attributes in this
model that determine competitive advantage are

- factor conditions. As well as mineral endowment, these include the
  quality of human and knowledge resources, the availability of capital
  and the quality of infrastructure.
- home demand conditions and the potential for their growth,
- the presence of and development of related and supporting industries, and
- firm strategy, structure and rivalry.

"Chance" events such as major technological or input cost discontinu-
ities, surges in world demand, shifts in world financial markets or exchange
rates, political decisions by foreign governments or wars may also affect these
main determinants.

The role of government policy stance is also important. As well as establishing a suitable mineral policy framework its activities in other areas such as
education, taxation, environmental protection, road, rail and telecommunica-
tions have an important complementary role. These will enhance factor conditions, encourage the growth of domestic demand and promote the growth of
supporting industries. If applied over an extended period of a decade or two it
will set the stage for movement through the factor-driven stage of national
competitive development towards the investment-driven and innovation-
driven stages. It is during these stages that nations move towards the "developed
economy" status. Nations such as Canada, Australia and the United States have
reached the innovation-driven stages with their mineral industries and this has
facilitated the broader development of the rest of their economy.

Porter notes that "to progress, the developing nations face the daunting
task of upgrading all four parts of the national "diamond" sufficiently to reach
the threshold necessary to compete in advanced industries." Yet successes of
Japan, Korea, Taiwan and Hong Kong in the second half of the twentieth century provide important case studies of this taking place. The recent strong economic performance of China and India is also encouraging. From a mineral and
energy sector perspective, the emergence of nations such as Saudi Arabia, Chile

and the United Arab Emirates suggests promise. The development experiences of countries such as South Africa, Brazil and Mexico are also worthy of mention.

An important conclusion from this discussion is that new mine development (or the expansion of existing mines) requires more than a favourable mineral endowment and the delineation of viable deposits with recent mineral exploration activity. As well as a favourable policy environment, technology and management issues are an important part of mineral sector development. Strong supporting infrastructure and a capable human capital base both are necessary to reap major benefits and to retain competitiveness over longer periods of time. Political stability is essential. For new mineral development to provide a platform for economic growth and development in African nations with major mineral endowments, as it has in Canada, Australia and the United States, the process will take perhaps a generation or more. It has recently taken each of the High Performing Asian Economies (Japan, China and the Asian Tigers) two or three decades to build their human and physical capital bases to a point where they have been able to compete favourably with the developed nations.

Concluding Remarks

With high geological prospectivity in many nations a safer and more attractive investment climate, Africa has recently shown the capacity to capture a greater share of the investment budgets of global mining companies. If this is continue further, African nations must continue to change images of a continent of civil unrest, starvation, deadly diseases and economic disorder.

The [selection] has shown that in general, FDI in Africa has increased in recent years due to improved regulatory frameworks, permitting profit repatriation, tax incentives provision and trade liberalisation. Progress on modernisation is uneven on the continent but countries have generally embraced the importance of private-sector led investment growth and realised that governments have to play a regulatory role. Sustaining a long-term renaissance will depend on many factors including the pace of reform, political stability, economic growth and developments in other investment destinations such as Asia and Latin America.

Ethnic disturbances and military coups remain major challenges to political stability. Political stability is also related to good governance, low levels of corruption and transparency. Weak institutional capacity remains a major constraint in many African countries.

The success of many current projects will be critical in ensuring a sustained flow of investors to Africa. Good macro-economic performance is a key for continued inflow of foreign investment and also for local investors. The change in the policy and regulatory environment, deregulation and privatiation programs are each important areas in reversing investment trends towards Africa. Programs supported by the United Nations and the World Bank to improve the operating environment in the region also appear to have an important place in this process.

Although the recent signs appear promising, it remains too early to appraise whether the "renaissance" of the African minerals sector will provide the impetus for the large number of mineral dependent nations to move forward significantly in their development. Given the difficult experiences of many African nations in their immediate post-colonial histories, an optimistic hope is that resource development will be seen as a blessing rather than a curse by the year 2010. Perhaps by that stage, economists will be writing of an African economic miracle under way.

References

African Development Bank (2000). *African Development Report 2000.*

Auty, R. (1986). "Multinational Resource Corporations, Nationalization and Diminished Viability: Caribbean Plantations, Mines and Oilfields in the Seventies," in C. Dixon et al. (Edss), *Multinational Corporations and the Third World,* Croom Helm, London, pp. 160–187.

Auty, R. (1993). "Determinants of state mining enterprise resilience in Latin America," *Natural Resources Forum,* pp. 3–11.

Aydin, H., and J. Tilton (2000). "Mineral Endowment, Labor Productivity, and Comparative Advantage in Mining," *Resource and Energy Economics,* Vol. 22, pp. 281–293.

Blainey, G. (1993). *The Rush that Never Ended: A History of Australian Mining,* Fourth edition, Melbourne: Melbourne University Press.

British Geological Survey (2000). *World Mineral Statistics.* Keyworth: Nottingham.

Campbell, B. (1998). "Liberalisation, Deregulation, State Promoted Investment—Canadian Mining Interests in Africa," *Journal of Mineral Policy, Business and Environment,* Vol. 13, No. 4. pp. 14–34.

Crowson, P. (1998). *Minerals Handbook 1998/99,* Mining Journal Books, London.

Davis, G. (1995). "Learning to Love the Dutch Disease: Evidence from the Mineral Economies," *World Development,* Vol. 23, No. 10, pp. 1765–99.

Ericsson, M. (1991). "African Mining: Light at the End of the Tunnel," *Review of African Political Economy,* Vol. 51, pp. 96–107.

Gelb, A.H. (1988). *Oil Windfalls: Blessing or Curse?* Oxford University Press, New York.

Minmet Australia (2001). Home Page, available at www.minmet.com.au.

Minerals Council of Australia (various years): *Mineral Industry Survey,* Canberra, available at www.minerals.org.au/.

Norrie, K. and D. Owram, (1996). *A History of the Canadian Economy,* 2nd edition, Toronto: Harcourt Brace Jovanovich.

O'Brien, J. (1994). *Undoing a Myth: Chile's debt to Copper and Mining,* Ottawa: International Council on Metals and the Environment.

O'Neill, D. (1992). "Australian Miners in Africa: An Australian Perspective," *Mining Review,* Vol. 16, No. 4, pp. 34–40.

Otto, J.M. (1997). "A National Mineral Policy as a Regulatory Tool," *Resources Policy,* Vol. 23, Nos. 1/2, pp. 1–7.

Otto, J.M. (1998). "Global Changes in Mining Laws, Agreements and Tax Systems," *Resources Policy,* Vol. 23, No. 4, pp. 79–86.

Porter, M. (1990). *The Competitive Advantage of Nations,* Macmillan, Basingstoke.

Premoli, C. (1998). "Mineral Exploration in Africa: The Long View," in S. Vearcombe and S.E. Ho (Eds.), *Africa: Geology and Mineral Exploration, AIG Bulletin,* No. 25, pp. 35–46.

Radetzki, M. (1985). *State Mineral Enterprises: An Investigation into Their Impact on International Mineral Markets,* Resources for the Future, Washington, DC.

Radetzki, M. (1990). *A Guide to Primary Commodities in the World Economy,* Basil Blackwell, Oxford.

Sachs, J., and A. Warner (1995). "Natural Resource Abundance and Economic Growth," Development Discussion Paper No. 517a, Harvard Institute for International Development, Cambridge, USA.

Sinclair, W.A. (1976). *The Process of Economic Development in Australia*, Melbourne: Longman Cheshire.

Slater, C.L. (1996). "The Investment Climate for Gold Mining: Comparing Australia with the USA," *Natural Resources Forum*, Vol. 20, No. 1, pp. 37–48.

South African Chamber of Mines (1998) Available at www.bullion.org.za/.

Tilton, J. (1989). "The New View of Minerals and Economic Growth," *The Economic Record*, Vol. 65, No. 190, pp. 265–278.

Tilton, J., and H. Landsberg (1999). Innovation, Productivity Growth and the Survival of the U.S. Copper Industry, in R.D. Simpson (Ed.), *Productivity in Natural Resource Industries: Improvement through Innovation*, Resources for the Future, Washington, DC.

UNCTAD (1999). *Foreign Direct Investment in Africa: Performance and Potential*, UNCTAD/ITE/IIT/Misc. 15, United Nations, Geneva.

UNCTAD (2000). *The World Investment Report 2000: Cross-border Mergers and Acquisitions and Development*, United Nations, Geneva.

United States Geological Survey (2000). *Mineral Industry Surveys,* available at minerals.usgs.gov/minerals/pubs/commodity/mis.html, Washington, DC.

World Bank (1992). *Strategy for African Mining,* Technical Paper No. 181, Mining Unit and Energy Division, Washington, DC.

A Continent in Crisis:
Africa and Globalization

From the oil fields of the Niger Delta in Nigeria, to the diamond and copper fields of Sierra Leone, Angola, and Liberia, to the rich mineral deposits of the Great Lakes region, to the mountain ranges, plains and tourist havens of the East African countries, the continent of Africa is undoubtedly blessed.

From these blessings, however, much sorrow has flowed. During the colonial era, most Africans did not benefit from the continent's resources. African economies were geared toward cultivating raw materials for export, and roads, health care, and other infrastructure were available only in areas where those materials were produced. The end of colonialism unleashed struggles for political control, social emancipation, and access to resources—struggles that, in turn, have degenerated into conflicts and internecine wars. Retarded in their development, unbridled in their lust for power, steeped in official corruption, chaotic in their political engineering, many African states are now sprinting toward total collapse.

Much of the blame for Africa's spiral of violence belongs to generations of opportunistic and venal African leaders, who have done little to develop their societies and emancipate their peoples. But the expansion of corporate dominance has accentuated the steady descent into near economic strangulation and political chaos. Many transnational corporations (TNCs) have acted as economic predators in Africa, gobbling up national resources, distorting national economic policies, exploiting and changing labor relations, committing environmental despoliation, violating sovereignties, and manipulating governments and the media. In order to ensure uninterrupted access to resources, TNCs have also supported repressive African leaders, warlords, and guerrilla fighters, thus serving as catalysts for lethal conflict and impeding prospects for development and peace.

TNCs and the African State

In the post-colonial period, many African leaders have exerted dictatorial control over their societies. Through their undemocratic policies, they have spread dissatisfaction among the people, which has manifested over time in nationalistic

From Sunday Dare, "A Continent in Crisis: Africa and Globalization," *Dollars and Sense* (July/August 2001). Copyright © 2001 by Sunday Dare. Reprinted by permission of the author.

feelings and even popular rebellions. These political tensions, in turn, have generated fierce conflicts over resource control.

In response to these conflicts, many African governments have embraced collaboration with TNCs and other foreign investors. Lacking the technological capacity to harness massive reserves of oil, gold, diamonds, and cobalt, these leaders grant licenses to foreign corporations to operate in their domain, and then appropriate the resulting revenue to maintain themselves in power. For example, both the late General Sani Abacha of Nigeria and the late Mobutu Sese Seko of Zaire looted hundreds of millions of dollars in government funds derived from corporate revenue, stashed the money in private foreign accounts, and used it for political patronage and to silence political opponents.

In turn, these authoritarian governments have stifled economic growth. Frequent changes in leadership through *coups d'etat* have made it impracticable to implement development plans. Each new government comes in with a new set of policies that often undermines earlier progress by previous governments. Also, many African leaders have used revenue to reward political pals with bogus contracts for white-elephant projects that contribute nothing to development. In the late 1990s, for example, President Daniel Arap Moi of Kenya built an airport—which handles almost no traffic—in his own hometown of Eldoret. And of course, money used as handouts to members of the ruling elite and to fight political opposition is money *not* spent on development.

After decades of economic mismanagement and political gangsterism, most African societies are in terrible shape. African unemployment rates are at crisis levels, with over 65% of college graduates out of jobs. Because manufacturing is at a low ebb, unskilled workers suffer a similar fate. Wages are also low. According to the *United Nations Development Report,* the average unskilled worker earns about 55 cents daily, while the average white-collar employee brings home a monthly check of between $50 and $120. Many African societies are characterized by minimal opportunities for education and self-development, collapsed infrastructure, and a debilitating debt burden.

These conditions have made the continent even more susceptible to international financial control. Typically, TNCs seek out societies with low production costs, poor working conditions, and abundant and easily exploitable resources, where profits can be maximized and repatriated without legal constraints. The icing, of course, is a political leadership that is weak, corrupt, and ready to cut deals. TNCs make huge investments in countries that meet those criteria, and many African countries fit the bill.

Economic Exploitation in Africa

The sheer growth of TNCs in recent decades has had profound consequences for Africa. The average growth rate of TNCs is three times that of the most advanced industrial countries. Of the 100 biggest economies in the world, more than half are corporations. TNCs hold enormous power to transform the world political economy, and Africa's vast resources, cheap labor, huge population, and expanding markets are crucial to their plans.

The globalization optimists maintain that global capital has served as a dynamic engine of growth, opening the window for diverse opportunities in terms of goods and services, creating employment, and boosting government revenues. This has been true in a few cases. In South Africa and Nigeria, for example, gold mining and oil companies respectively have brought new technology, attracted subsidiary industries, and made it possible for indigenous personnel to acquire skills.

However, any such benefits are far outweighed by activities that deplete local resources, stifle local or indigenous industry, and subvert the fragile democratic process. Africa is still confined to the role it played in the industrial revolution(s) that preceded globalization. Its raw materials are still being depleted without generating development.

In addition, the continent's increasing dependence on imported capital and consumer goods and services has left various sectors of the domestic economy comatose. African markets are specially targeted as dumping grounds for new and second-hand goods. Because of stiff competition from these products, infant manufacturing established earlier in Africa has quickly withered away. For example, imports of used clothing from the United States are threatening to destroy Kenya's domestic textile industry.

Finally, as soon as the TNCs have African economies firmly in their grip, they deploy funds and patronage to manipulate the media and influence government policies. Governments, in turn, grant them *carte blanche* to sidestep labor and environmental laws.

Other global institutions have contributed to the deregulation of African economies. The Generalized Agreement on Tariffs and Trade, the World Trade Organization, and the International Monetary Fund (IMF) all promote increased liberalization of international trade. The IMF's structural adjustment programs (SAPs) require African states to freeze wages, devalue currency, remove public subsidies, and impose other austerity measures, which have brought about even greater unemployment and under-utilization of productive capacity. These policies have caused considerable turmoil in Africa. In the early 1980s, Uganda was rocked by weeks of demonstrations, as industrial workers and students took to the streets to denounce President Milton Obote's IMF-imposed economic program. In 1990, Matthew Kerokou of the Benin Republic in West Africa was swept out of power in a wave of anti-SAP riots.

Corporations and Lethal Conflict

At the advent of the new millennium, Africa is hurting badly. Most parts of the continent are embroiled in independence wars, ethnic conflicts, violent wars for political and resource control, and cross-border conflicts. In a recent study of armed conflicts around the world, the University of Maryland's Center for International Development and Conflict Management found that 33 countries were at high risk for instability. Of these, 20 were in Africa.

In their quest to unravel the forces generating conflict in Africa, human-rights groups are closely scrutinizing TNCs. TNCs are not always responsible for the genesis of the crisis. But some of the deadliest conflicts that litter Africa's political landscape, at least in the last decade, can definitely be traced to the expansion and domination of TNCs.

This is especially true in extractive states where resources with global appeal, value, and markets are found. While the state's interest in generating revenue from these resources coincides with that of the TNCs, the latter's interest in maximizing profits conflicts with the welfare of the citizens. Thus, the state is caught between protecting a vital source of revenue, and defending the rights and privileges of its citizens. Too often, the state, in order to ensure an ongoing flow of revenue, sides with the TNCs against the citizens.

For example, the oil-producing Niger Delta region is perpetually at war with the government and oil corporations. For decades, successive Nigerian governments have been beholden to the TNCs that possess the technology, technical expertise, and capital to exploit the country's oil. The exploitation has resulted in serious environmental damage, developmental neglect, human-rights abuses, economic oppression, and inequitable resource allocation. These abuses, and the need for redress, are at the heart of the conflict. In recent months, calls for secession by the oil-yielding region have grown louder. As other parts of the country caught up in oil politics fight to defend their interests, the drums of war continue to beat. . . .

TNCs have played a major role in the Nigerian conflict. They gave their unalloyed support to the brutal military regimes of Generals Ibrahim Babangida and Sani Abacha. Under Abacha, Ken Saro-Wiwa and eight other activists were hanged for crusading against the government and the oil companies. Officials at Royal Dutch Shell, which dominates the lucrative Nigerian oil industry, admitted in a press statement that the company could have stopped the hangings if it had so desired. Shell executives also confessed publicly to purchasing arms for the Nigerian State Police, who have attacked community residents and picketers. Also, in 1998, oil giant Chevron used its own helicopter to carry Nigerian soldiers, who stormed Parambe, an oil-yielding community, and killed several protesters.

In Angola, meanwhile, the global trade in diamonds—widely known as "blood diamonds" or "conflict diamonds" because of their lethal consequences—has helped to perpetuate more than 20 years of civil war. With revenue from the illegal mining and sale of rough diamonds, the UNITA [Portuguese acronym for the National Union for the Total Independence of Angola] rebels, led by Jonas Savimbi, have been able to purchase and stockpile ammunition to prolong the war. Though fully aware of this, a number of corporations—such as American Mineral Fields (AMF), Oryx, and the world's leading diamond company, De Beers—continue to do business in the war-torn territory, where more than half a million citizens have been killed.

Also, Jean-Raymond Boulle, AMF's principal shareholder, is known to have invested millions of dollars in support of corrupt African governments and rebel leaders in order to secure juicy mineral contracts. The late Congolese leader Laurent Kabila used Boulle's jet and funds to prosecute his war against Mobutu of Zaire. When Kabila came to power in 1997, AMF secured exploration rights to 600 million pounds of cobalt and three billion pounds of copper, among other deals.

Similarly, the Foday Sankoh-led Revolutionary United Front (RUF) in Sierra Leone derives most of the funds it has used to unleash terror and mayhem on the country's people from the international trade in conflict diamonds. The RUF continues to fight a war that has claimed more than 80,000 lives, with no end in sight. According to the U.S. State Department, revenue from rough and uncut diamonds mined in conflict areas forms a large percentage of the commodity's over $50 billion in annual sales.

The conflicts that grip Africa can also be traced to a steady flow of arms. To safeguard their economic interests, Western corporations are procuring weapons and providing arms training in areas of conflict. Embattled African leaders, anxious to defend their own interests and protect their hold on power, readily grant contracts to private security armies run by TNCs. Since the early 1990s, the growth of the corporate private security sector in Africa has been phenomenal. TNCs have become direct parties to conflicts by recruiting or hiring private security companies to help protect their installations, operations and staff. In the process, they have connived with governments and sometimes with rebels, whichever is most expedient, thereby instigating further conflict and perpetuating civil war.

For example, in 1995, Executive Outcome (EO), a private security company, arrived in Sierra Leone. The Sierra Leonean government paid EO almost $40 million in cash, along with mining concessions, to assist in its campaign against the RUF. The peace it secured did not last; rather, the country was plunged into deeper crisis. EO was also in Angola, training the national army and helping to recapture lucrative mineral fields. In 1993, the Angolan state oil company, Sonangol, contracted with EO to provide security for its installations against UNITA attacks. The Angolan government also signed a three-year, $40 million contract with EO to supply military hardware and training.

In addition, J. & S. Franklin, a British supplier of military equipment, won a contract to train the Sierra Leonean military using the notorious U.S.-based Gurkha Security Guards. The Guards have carried out a series of military attacks in various mining areas to protect the activities of TNCs. J. & S. Franklin has also won supply contracts with several other African governments engaged in conflicts.

Clearly, where minerals abound, TNCs find the lure irresistible. They accrue substantial profits from diamonds, which are sold in wedding rings, bracelets, and necklaces all over Europe and the United States. TNCs also profit enormously from cobalt, a vital raw material for the manufacture of jet fighters. The unbridled lust for excessive gain has deadened the senses of corporate giants like De Beers, Royal Dutch Shell, Chevron, AMF, and others to the damaging impact of their activities.

Out of the Labyrinth

Because of the twin problems of rogue leadership and the exploitative tendencies of TNCs, Africa is caught between a rock and a hard place. As the history of conflicts in Africa shows, the extraction of mineral resources creates and reinforces government corruption, which easily begets repressive societies. As would be expected, poverty, unemployment, and insecurity spread, while social services decay. This leaves the citizens more prone to take up arms and fight in oil, diamond, and copper wars, as the conflicts in Sierra Leone, Angola, Nigeria, Sudan, Liberia, and the Great Lakes region all attest.

How can Africa wrench itself from itself and curb the rampaging TNCs? In response to corporations' bad behavior, and their brazen disregard for the political stability and economic viability of the states in which they operate, international human-rights organizations have tried to establish mechanisms of accountability.

The key, however, is action by Africans on their own behalf. Their options for ending the circle of violence and economic exploitation are few but practicable. Africa needs a new generation of leaders to define and pursue a dynamic political and economic agenda. The African states must renegotiate their terms of trade in the international marketplace. Diverse groups must achieve a sense of national pride and internal cohesiveness, in order to create an atmosphere conducive to implementing development programs. As long as the resources have not yet been depleted, there is still hope of rising again for a continent that has tarried for too long in the labyrinth.

POSTSCRIPT

Are Abundant Mineral and Energy Resources a Catalyst for African Development?

The issue of resource abundance and economic growth in Africa has received renewed attention in recent years due to at least three developments: increased interest by the United States in African energy resources leading up to and after the 2003 Iraq War, particularly brutal civil conflicts in Sierra Leone and Angola financed by the sale of mineral resources, and growing investments by South African entrepreneurs in the mining sectors of other African countries since the end of apartheid.

First, with escalating concerns about dependence on oil from the Middle East, the United States increasingly has become interested in oil exploration and oil field development in a number of African countries. In a January 13, 2003, *Los Angeles Times* article entitled "U.S. Quest for Oil in Africa Worries Analysts, Activists," Warren Vieth reports that "the Bush administration's search for more secure sources of oil is leading it to the doorsteps of some of the world's most troubled and repressive regimes: the petroleum-rich countries of West Africa." While the Bush administration states that it is committed to improving conditions in Africa, the concern among Africanist scholars and policy analysts is that production will be put ahead of balanced reform and development. The history of U.S. interaction with other oil-rich African nations does not portend well for good governance and broad-based development.

Second, in the last several years the links between brutal civil conflicts and the diamond trade in Sierra Leone and Angola have become increasing clear. This recognition led to the appellation of diamonds from these countries as blood, or conflict, diamonds. See a recent book by Greg Campbell entitled *Blood Diamonds: Tracing the Deadly Path of the World's Most Precious Stones* (Westview Press, 2002).

Finally, the mining sector in Africa has been transformed in recent years by the infusion (some would say invasion) of South African capital and expertise. This is related to the fact that South Africans were often prohibited from investing in other parts of the continent prior to the end of apartheid in 1994. Given the history of antagonism between South Africa and the frontline states, many countries are leery of a growing South African presence in their economies. Despite these concerns, some are grateful to have a new eager investor. See a February 17, 2002, *New York Times* article by Rachel Swarns entitled "Awe and Unease as South Africa Stretches Out."

145

ISSUE 9

Are Integrated Conservation and Development Programs a Potential Solution to Conflicts Between Parks and Local People?

YES: William D. Newmark and John L. Hough, from "Conserving Wildlife in Africa: Integrated Conservation and Development Projects and Beyond," *BioScience* (July 2000)

NO: Roderick P. Neumann, from "Primitive Ideas: Protected Area Buffer Zones and the Politics of Land in Africa," *Development and Change* (July 1997)

ISSUE SUMMARY

YES: William D. Newmark, research curator at the Utah Museum of Natural History, University of Utah, and John L. Hough, global environment facility coordinator for biodiversity and international waters for the United Nations Development Programme, acknowledge the limited success of integrated conservation and development programs to date in Africa, but see great promise for success in the future. They call for more adaptive management in which activities are monitored, evaluated, and reformulated in an interactive fashion.

NO: Roderick P. Neumann, associate professor and director of graduate studies in the Department of International Relations at Florida International University, argues that protected area buffer zone programs have not lived up to their initial intent of greater participation and benefit sharing. Rather, these programs duplicate more coercive forms of conservation practice associated with parks and facilitate the expansion of state authority into remote rural areas.

A number of national parks were established in African countries during the colonial era. While these parks ostensibly were established for the preservation of natural resources, they also served to sequester resources for the European population. In the postcolonial era, national park systems have persisted and have been expanded in many instances, particularly in East and Southern Africa where relatively larger numbers of charismatic megafauna still reside. African park systems have been bolstered by a global environmental movement as well as a burgeoning ecotourism industry. In countries such as Kenya, Tanzania, Zimbabwe, Botswana, and South Africa, Western tourists flock to see the "big five," a term used to refer to the biggest, rarest, or most cherished animals traditionally sought after by trophy hunters (elephant, lion, rhinoceros, leopard, and African or Cape buffalo). Nature tourism has become big business. In Kenya, for example, it is the leading source of foreign exchange. Here, Richard Leakey, the former head of the Kenyan Wildlife Service and famous archeologist, had established a shoot-to-kill policy to combat against suspected poachers found in national parks.

The poaching problem is symptomatic of a much deeper set of social issues (numbering at least three) surrounding parks and protected areas in Africa. First, local people often were evicted without compensation from areas where national parks were established. This led to feelings of resentment and compromised livelihoods in many instances; i.e., local people did not have the same resource base to rely on in their new location. Second, feelings of resentment and the inability of many national governments to effectively patrol park borders has led local people to encroach on parks in search of sustenance. Some local people may view parks as open-access resources, i.e., resources with no particular owner (as opposed to private property (belonging to an individual) or common property (controlled by a group)). In the absence of a legitimate controlling force or owner (government may be viewed as an illegitimate force), open access resources tend to be overexploited as everyone tries to maximize their personal gain with little to no regard for the overall ecological health of the resource. Finally, big game animals rarely respect park boundaries and may wander onto the lands of communities abutting national parks. This is problematic because large animals may pose safety risks and destroy field crops. Elephants in particular are known for their ability to inflict a substantial amount of crop damage.

In lieu of the aforementioned problems, a number of conservation programs have been initiated that take into consideration the needs of local people. Part of the incentive for these programs is a genuine concern about the welfare of local people, but there is also recognition that many conservation initiatives, including parks, are doomed to failure unless they enlist the support of local people. A key way of securing such support has been to share ecotourism revenues with local people in exchange for their participation in protecting wildlife resources. Such programs typically work with the communities neighboring national parks and may even involve the establishment of buffer zones (i.e., areas surrounding national parks where the activities of local people are restricted).

In the following selections, William D. Newmark and John L. Hough assert that they see great promise in integrated conservation and development programs in Africa. In contrast, Roderick P. Neumann states that these programs duplicate more forceful forms of conservation practice associated with parks and make possible the increase of state authority in remote rural areas.

William D. Newmark and
John L. Hough

 **YES**

Conserving Wildlife in Africa: Integrated Conservation and Development Projects and Beyond

Conservationists in Africa are struggling to develop new approaches to protect the continent's spectacular natural heritage. The challenge is to design strategies that not only will ensure the long-term viability of species and ecosystems but also will be politically and economically acceptable to local communities and governments. One approach that has gained considerable attention in recent years is the integrated conservation and development project (ICDP), which attempts to link the conservation of biological diversity within a protected area to social and economic development outside that protected area. In ICDPs, incentives are typically provided to local communities in the form of shared decision-making authority, employment, revenue sharing, limited harvesting of plant and animal species, or provision of community facilities, such as dispensaries, schools, bore holes, roads, and woodlots, in exchange for the community's support for conservation.

The ICDP approach to conservation in Africa began in earnest in the 1980s and 1990s, although efforts to link wildlife conservation with local development go back to the 1950s in a few protected areas in Africa, such as Ngorongoro Conservation Area in Tanzania. Currently, much of the funding by major bilateral and multilateral donors to protected areas in Africa is in the form of ICDPs. A recent review (Alpert 1996) suggests that there have been more than 50 such projects in 20 countries.

Given the popularity of ICDPs, it is discouraging that so many reviews (Kiss 1990, Hannah 1992, Wells et al. 1992, Kremen et al. 1994, Western et al. 1994, Barrett and Arcese 1995, Gibson and Marks 1995, Oates 1995, Alpert 1996) indicate that most ICDPs have had only limited success in achieving both conservation and development objectives. Thus, a lively and important debate about the appropriateness of the ICDP model is under way in the conservation and development community (Kramer et al. 1997). Recent critiques of ICDPs in Africa (Kiss 1990, Hannah 1992, Stocking and Perkin 1992, Wells et al. 1992, Kremen et al. 1994, Western et al. 1994, Barrett and Arcese 1995, Gibson and

Marks 1995, Oates 1995, Alpert 1996, Hofer et al. 1996) highlight many of the problems associated with these projects. Based on our own field observations in more than 15 African countries and the critiques of other workers, as well as a review of many project proposals, reports, and evaluations, we discuss these problems. In addition, we argue that the lack of success of many ICDPs is attributable in part to a series of erroneous assumptions made frequently by many designers of ICDPs. Finally, we suggest that ICDPs need to be viewed as just one of a variety of tools available to conservationists and development workers, and that both alternatives to ICDPs and tools and techniques that complement ICDPs need to be actively explored.

Rationale for the ICDP Approach

The ICDP approach to conservation in Africa has gained popularity for several reasons. One reason is the recognition that wildlife populations have declined dramatically throughout Africa over the last 30 years, primarily because of habitat loss. Surveys suggest that over 65% of the original wildlife habitat in Africa has been lost (Kiss 1990) as a result of agricultural expansion, deforestation, and overgrazing, which have been fueled by rapid human population growth and poverty. Given the underlying determinants of habitat loss, it has been argued that conservation activities in the field must be intimately linked with development (IUCN 1980).

A second reason for the popularity of ICDPs relates to the challenges of conserving biological diversity within existing protected areas. Throughout Africa, protected areas are becoming increasingly ecologically isolated as a result of agricultural development, deforestation, human settlement, and the active elimination of wildlife on adjacent lands. This phenomenon, in combination with the small size of most protected areas, indicates that in the absence of intensive management, most protected areas in Africa will not be large enough to conserve many species, as illustrated by recent patterns of extinction of large mammals in Tanzanian parks (Newmark 1996) as well as large carnivores in southern and East African protected areas (Woodroofe and Ginsberg 1998). Additionally, rural poverty and external markets will continue to encourage both subsistence and commercial poaching of many species within protected areas. Analysis in Zambia suggests that it costs $200 per km^2 per year to effectively control commercial poaching of species such as elephant and rhinoceros in protected areas (Leader-Williams and Albon 1988). Unfortunately, few, if any, African countries have such financial resources, and central governments are unlikely to allocate significantly more funds for wildlife management in the future, given the many other competing demands for governmental resources. Recognition of these problems has led many workers to argue that the only way to enlarge and link existing protected areas (Newmark 1985, 1996) and control commercial poaching (Owen-Smith 1993) is to develop cooperative relationships with adjacent communities.

A third reason for the popularity of ICDPs is that such programs are perceived as an effective mechanism for addressing problems of social injustice. Protected areas have adversely affected many indigenous people in Africa. For example, all of the large savanna parks of East Africa have been established on former Masai rangelands (Århen 1985, Parkipuny and Berger 1993). Many donors view ICDPs as a means to develop supportive relationships with the communities that must bear much of the social costs of protected areas.

Finally, ICDPs are attractive because of the recognition that past methods of management have been ineffective in curbing poaching and have frequently created confrontational relationships with local communities. The former "fences and fines" approach to conservation is viewed as anachronistic and counterproductive, and many conservationists view the ICDP approach as a valid alternative.

Critiques of ICDPs in Africa

A number of assessments of the effectiveness of ICDPs in Africa have been conducted. Two things are striking about these reviews: the consensus among workers that nearly all ICDPs have either not achieved their objectives or that progress has been modest, at best, and the multiple explanations given for the limited success of ICDPs. These explanations fall into three broad categories: assessment problems, internal constraints, and external forces.

Assessment problems   Project evaluators have identified two important constraints that have hindered the objective assessment and demonstration of success of many ICDPs. One is that many projects were at an early stage of implementation when they were assessed. The early evaluations (Kiss 1990, Hannah 1992, Wells et al. 1992) of ICDPs in Africa concluded that success was limited in meeting both conservation and development objectives, but also that most of these projects had not been under way long enough to be fairly evaluated. Reviewers noted that the normal 3–5 year project cycle may be inappropriate for ICDPs, as it was found to be during the 1970s for rural development projects, which required considerably longer project cycles to achieve project objectives. Given that a number of ICDPs in Africa have now been in operation for more than a decade, this issue should be less of a constraint; however, there is as yet little substantive evidence of improvement in success.

A second constraint on assessment is the absence of ecological monitoring. Kremen et al. (1994) examined 36 projects worldwide, 23 of them from Africa, and found that over half of the projects had no ecological monitoring and only two contained a comprehensive ecological monitoring component. The lack of ecological monitoring in most projects has prevented a rigorous evaluation of the impacts of development activities, particularly resource exploitation, on biological diversity. The lack of ecological monitoring has also meant that feedback useful for guiding the future course of project activities is frequently absent (Kremen et al. 1994). Wells et al. (1992) noted that few of the 18 ICDPs they studied in Africa, Asia, and Latin America were able to demonstrate—largely because of the absence of ecological monitoring—that the development activities occurring outside of the protected areas enhanced the conservation of biological diversity within the protected areas.

Internal constraints Project evaluations have also identified four internal constraints common to many ICDPs. First, public goods may not alter the behavior of individuals, as Gibson and Marks (1995) have suggested; they maintain that many ICDPs in Africa will fail in their goal of conservation because the incentives presented to communities are public goods and are insufficient to alter individual behavior. Furthermore, these incentives may have differential effects on different groups within the communities (Noss 1997). Gibson and Marks (1995) also argue that the economic incentives that many ICDPs offer are often ineffective because project designers frequently overlook the social importance of many activities, such as hunting. Metcalfe (1994) also highlights the difficulty of distributing public benefits to individuals as one of the key challenges facing the Communal Areas Management Programme for Indigenous Resources (CAMPFIRE) in Zimbabwe.

A second internal constraint is that the organizational structure of many ICDPs often mimics earlier ineffective colonial structures. Gibson and Marks (1995) suggest that many local people remain disenfranchised from most ICDPs in Africa because the ultimate authority for wildlife continues to reside with the state. They maintain that although a number of ICDPs have devolved authority over wildlife to local communities, that authority is limited and local communities should have greater control over the use of wildlife. Most wildlife departments accept the rhetoric of such a change in approach, but they can find it difficult to effect that change because doing so demands new sets of skills, a shift from competitive to collaborative relationships with other agencies and institutions, and changes in the internal institutional culture (Hough 1994a). These difficulties have been problematic for ICDPs in Madagascar; government and donor efforts to overcome them have resulted in a number of changes in institutional mandates and structure (Hough 1994a, McCoy and Razafindrainibe 1997).

A third internal constraint is that the offtake associated with many harvesting schemes may be unsustainable over the long term. Barrett and Arcese (1995) and Hofer et al. (1996), for example, have argued that the large mammal harvesting schemes associated with many ICDPs in savanna ecosystems in Africa may be unsustainable because wildlife populations in these ecosystems are inherently variable. They suggest that because managers are frequently under considerable political pressure to maintain a constant flow of benefits (in this case, meat, skins, or revenues) to local communities, they may find it extremely difficult to reduce the offtake when wildlife populations are declining. They also suggest that if wildlife managers do reduce offtake, the project could lose community support. Less work has been done on the sustainability of plant and animal harvesting in nonsavanna biomes in Africa, but some research on woodlands in southern Africa (Shackleton 1993) and forests in East and West Africa (Fa et al. 1995, FitzGibbon et al. 1995, Slade et al. 1998, Wilkie et al. 1998) indicates that the current offtake for many species in those areas is likewise unsustainable.

A fourth internal constraint is that development activities frequently conflict with conservation objectives. In many projects, such conflicts are a result of the inability of managers to effectively control resource exploitation by communities or individuals (Stocking and Perkin 1992), the nonsustainable use of resources, or the ecologically disruptive nature of the development activities. For example, one ICDP in Tanzania placed fish ponds in an important wetland habitat in the East Usambara Mountains. These ponds, although effective in providing additional protein to villagers, severely disrupted scarce riparian habitat (William D. Newmark, personal observation).

External forces   Finally, project evaluations have identified three external forces that adversely affect many ICDPs in Africa. First, sources of potential revenues for communities are usually unreliable and insufficient. Because exchange rate fluctuations and political turmoil often make tourist revenues unreliable, basing cash inducements to communities on tourism is unwise (Barrett and Arcese 1995). The dramatic decline in tourism in recent years in Uganda, Kenya, Comoro Islands, and Zimbabwe highlights the high vulnerability of this industry to political unrest and economic downturns. Additionally, as Barrett and Arcese (1995) noted, there are few protected areas in Africa where the revenues from gate receipts exceed the cost of management; thus, it is unlikely that many communities will ultimately benefit from such revenue-sharing practices. Furthermore, as Norton-Griffiths and Southey (1995) have pointed out, if opportunity costs are taken into account, protected areas and their buffer zones may impose economic penalties on their surrounding communities that far outweigh any potential financial advantages from revenue-sharing arrangements.

Second, external market forces are increasingly manipulating resource use patterns in Africa. The urbanization that is taking place in Africa has created a growing demand in many cities and towns for resources such as meat, timber, and firewood (Barrett and Arcese 1995). These urban markets will produce increasingly strong market incentives to exploit rural natural resources, which could circumvent or undermine ICDP activities. For example, regional urban market forces have encouraged the commercial poaching of large mammals in and around Serengeti National Park for meat (Hofer et al. 1996): Between 1970 and 1992, the population of Cape buffalo in Serengeti National Park declined between 50% and 90% over portions of their range (Campbell and Borner 1995, Hofer et al. 1996). Similarly, Hannah et al. (1998) found that distant market forces have had significant negative impacts on the success of ICDPs in Madagascar.

Third, ICDP development activities may induce migration into the project area (Wells et al. 1992, Barrett and Arcese 1995, Noss 1997). Evidence for such in-migration comes from other rural development projects in Africa. For example, a United Nations–supported irrigation project that was initiated in the early 1980s near Lake Manyara in Tanzania was largely responsible for the 40% growth in population in the area between 1978 and 1988 (Yanda and Mohamed 1990).

Why ICDPs' Success Has Been Limited

There are, in our opinion, several overarching factors responsible for the limited success of ICDPs in Africa. These include erroneous assumptions, unintended social relationships, and inadequate knowledge about the project environment.

Erroneous assumptions   That local communities are hostile to protected areas, that raising living standards will inevitably result in conservation, and that buffer zones are panaceas have proved to be erroneous assumptions that are detrimental to the success of ICDPs. Because protected areas in Africa have historically excluded local people and have a colonial legacy (Anderson and Grove 1987, Neumann 1998), it is generally assumed that these areas are surrounded by hostile communities and enjoy little, if any, support among local people (Lusigi 1981, Wells 1996). The attitudinal research that has been conducted in Africa indicates that this assumption is overly simplistic.

Surveys in South Africa (Infield 1988), Rwanda (Harcourt et al. 1986), Tanzania (Newmark and Leonard 1991, Newmark et al. 1993), and Nigeria (Ite 1996) have found that an overwhelming majority of people living adjacent to protected areas in these countries agreed on the need for the protected area or were opposed to abolishing the parks or making them available for agriculture. On the other hand, surveys showed that most people living adjacent to protected areas in South Africa (Infield 1988), Botswana (Parry and Campbell 1992), and Tanzania (Newmark et al. 1993) held negative or neutral attitudes toward managers of protected areas. Furthermore, surveys in South Africa, Botswana, and Tanzania found that local people's support or opposition to protected areas, managers of protected areas, and wildlife is based on utilitarian values (Infield 1988, Mordi 1991, Parry and Campbell 1992, Newmark et al. 1993). In these countries, local people expressed support for protected areas because national parks and related reserves protect important watersheds, generate foreign exchange, or maintain critical hydrological functions. Similarly, local people expressed support for wildlife primarily because wildlife is viewed as a source of food. However, those who held negative or neutral attitudes toward managers of protected areas did so because they felt that managers provided few services or benefits for their communities.

Thus, the documented instances of the unpopularity of ICDPs with local people (e.g., the Cross River National Park project in Nigeria; Ite 1996) and the overall lack of success of many ICDPs do not result from local people's opposition to conservation or protected areas per se. Rather, they are a result, in part, of the inherent limited capacity of ICDPs and—in the eyes of many local people—managers of protected areas to provide sufficient tangible incentives to alter the attitudes and behavior of local people toward the ICDPs (see, e.g., Ferraro and Kramer 1997, McCoy and Razafindrainibe 1997).

A second erroneous assumption of the ICDP model is that improving the living standards of people living adjacent to protected areas will necessarily enhance conservation within the protected area (Wells et al. 1992, Wells 1996). Studies of conservation attitudes of people in South Africa (Infield 1988) and Tanzania (Newmark and Leonard 1991, Newmark et al. 1993) have found a positive correlation between affluence and conservation attitudes, but it is unlikely that an improvement in the living standards of communities near protected areas will inevitably lead to enhanced long-term viability of many species within the protected areas. For example, although providing employment to local people in Zambia improved living standards and reduced hunting pressures on species in protected areas (Lewis et al. 1990), such correlations do not always hold. Ferraro and Kramer (1997) found that the hiring of poachers at Ranomafana National Park in Madagascar actually increased levels of poaching because these new employees used their earnings to hire more people to expand their poaching operations. It is also unclear whether species in protected areas that are threatened indirectly by habitat loss outside of these reserves, perhaps by agricultural intensification, would be helped by an improvement in the living standards of local communities. Thus, encouraging landscape-wide compatible land use adjacent to protected areas may be more important for conserving species in protected areas than simply stimulating local economic development.

A third erroneous assumption is that buffer zones are panaceas. These management zones are promoted frequently in many ICDPs as peripheral areas where living conditions of local communities are to be enhanced through selective resource use and where habitat degradation will be reduced through habitat restoration. However, it is unclear how those goals are to be achieved: None of the ICDPS that promote the use of buffer zones have explained how an already overexploited area can be used to increase productivity and provide additional habitat for wildlife (Little 1994).

Unintended social relationships   Aside from the problems caused by problematic assumptions underlying the ICDP approach are those that stem from ICDPs' creation of unintended social relationships with local communities. In the effort to win the support of local communities for conservation, ICDPs frequently share park revenues, provide employment, or permit access to plant and animal resources. However, most provide only nominal opportunities for community-wide participation and often fail to link development benefits directly to community conservation obligations. The result is that many ICDPs may unintentionally promote dependency rather than reciprocity and have often treated local communities as recipients of aid rather than partners in development.

Inadequate knowledge about the project environment   Finally, in our opinion, many ICDPs have had limited success because the social and ecological environment surrounding the project is often poorly understood and dynamic. This inadequate understanding of the project environment has contributed greatly to the difficulty in transferring seemingly successful components of ICDPs from one region to another. In most ICDPs, scientific input is normally limited to a "rapid" preproject ecological and social appraisal of the project area. However, these appraisals, by their very nature, have a limited capacity to capture the complex ecological and social (Gezon 1997) relationships that surround most projects. Moreover, they provide a tenuous baseline for subsequent project monitoring, assessment, and adaptation.

ICDP designers are often reluctant to incorporate a significant research component into these projects. Part of this reticence stems from the crisis nature of most conservation initiatives: Research is often viewed as a hindrance to action and an expensive luxury. Yet incorporating a significant research component into ICDPs is essential if the ecological and social dynamics encompassing each project are to be accurately defined and if conservation and development are to be truly integrated.

Lessons Learned

Several lessons can be drawn from our own and others' observations. One is that multiple ecological, social, political, economic, and institutional problems confront ICDPs. Not only are ICDPs themselves complex, but so is the environment in which they operate. However, ICDPs seem to rarely build in mechanisms for analyzing and adapting to these changes. Furthermore, project designers and scientists' understanding of the mechanisms governing this environment is generally inadequate. For example, little is known about the long-term primary and secondary impacts of resource harvesting on the structure and function of most tropical ecological communities. Similarly, many social scientists counsel that it is unwise to devolve total authority to local communities (West and Brechin 1991), but little is known about how much authority over the use of natural resources should be transferred to local communities or how to ensure that project benefits are equitably distributed and not captured by local elites (Lutz and Caldecott 1996). A second lesson, related to the first one, is that more thorough, and ongoing, ecological and social assessment and analysis is required, both during the design phase of ICDPs and during their implementation.

A third lesson is that project planners need to examine in more detail the effects of external factors such as markets, land tenure, and population growth on proposed project activities. ICDPs may need to include project components or explicit linkages to other initiatives, which address external constraints well beyond the limited geographic focus of the ICDP. For example, efforts to control commercial meat poaching in protected areas may require not only upgraded law enforcement within protected areas and improved grazing management in the buffer zones but also favorable pricing and marketing systems for domestic livestock in distant urban areas.

A fourth lesson is that although linking conservation with development may be desirable, the simultaneous achievement of these two objectives may be impossible because of inherent contradictions. In these cases, success may be enhanced by addressing each of these objectives separately but in parallel, tightly linked interventions, rather than within the same project. Decoupling these two objectives does not negate the importance of development to conservation and conservation to development; rather, it implies that protected-area organizations, which have been responsible for implementing most ICDP rural development activities so far, should delegate these activities to organizations with the appropriate mandate, expertise, and experience. The implications are that protected-area institutions should serve more as facilitators than as implementers of rural development activities, although they must work closely with local communities to attract assistance that addresses the needs of local communities without adversely affecting the protected areas. Some protected-area institutions, such as the Kenya Wildlife Service and Tanzania's Division of

Wildlife, appear to have already adopted this approach for revenue-sharing schemes between protected areas and local communities, but it needs to be extended to cover the full range of development activities. Another implication of decoupling conservation and development objectives is that an ongoing mechanism is needed within ICDPs for negotiating the compromises and seeking out the win–win solutions that meet both conservation and development needs.

Future Direction?

Several immediate challenges need to be addressed in designing future conservation initiatives in Africa. One is the need to develop mechanisms for ensuring that ICDPs respond to the real complexity of their ecological and social environments and that they effectively monitor, analyze, and adapt to this environment as it changes.

A second challenge is the need to assess, implement, and evaluate alternative and complementary approaches to ICDPs that address the external forces affecting ICDPs through actions such as economic and land-tenure policy reform, landscape-wide conservation planning, conflict resolution, community-based natural resources management, and enhanced management capacity of protected-area institutions. Although project experience with and evaluation of these approaches is insufficient for a rigorous assessment of their overall effectiveness in comparison to (or as complements of) ICDPs, some preliminary observations can be considered now.

- **Economic and land tenure policy reform.** Such reforms can greatly assist in reducing external environmental pressures on protected areas, particularly external market forces and in-migration. For example, the international ban on ivory trading has significantly reduced elephant poaching throughout Africa.
- **Landscape-wide conservation planning.** Given that most protected areas in Africa are small and that many are becoming ecologically isolated, it is important that land-use activities that are compatible with wildlife conservation be encouraged on a landscape-wide scale adjacent to protected areas, and activities that are incompatible must be actively discouraged. It is particularly important that land use be controlled within wildlife corridors linking existing protected areas as well as within wildlife dispersal zones. In most savanna ecosystems in Africa, pastoralism is considerably more compatible with wildlife conservation than agriculture; thus, efforts should be made to maintain existing pastoral systems adjacent to protected areas. Similarly, for use adjacent to protected areas in tropical forest ecosystems, native hardwood plantations and multilayer perennial agroforestry are better choices than agricultural monocultures and pastoralism (Thiollay 1995, Perfecto et al. 1996, Greenberg et al. 1997).
- **Conflict resolution.** Promoting dialogue between managers of protected areas and local communities, involving affected stakeholders in protected-area project planning and implementation, identifying areas

of common interest between protected areas and local communities, and including community representatives on advisory management boards for protected areas can greatly assist in reducing conflicts between parks and local people (Hough 1988, Lewis 1996). Such programs are attractive not only because they are relatively easy to implement but also because they are fairly inexpensive. Recent conflict resolution initiatives in areas adjacent to the Bwindi Impenetrable and Mgahinga Gorilla National Parks in Uganda indicate that such activities can greatly reduce tensions between local communities and park authorities (Wild and Mutebi 1996).

- **Community-Based Natural Resources Management (CBNRM).** Considerable success in generating compatible land-use regimes around protected areas has been claimed in Zambia, Zimbabwe, and Namibia through the use of CBNRM approaches, the most notable of which is the CAMPFIRE program (Murphree 1993, Metcalfe 1994). CBNRM differs from the normal ICDP approach in that, instead of offering development services in exchange for conservation, it devolves management responsibility for natural resources—wildlife—to local communities. Its success depends on communities seeing more value in managing their wildlife on a long-term sustainable basis than in pursuing short-term exploitation or alternative land uses. Yet a number of scientists associated with these projects believe that local communities will eventually be forced to forsake wildlife conservation for more intensive agriculture development because of demographic and social pressures (Hackel 1999). Therefore, complete devolution of authority to local communities may be unwise (West and Brechin 1991).
- **Enhancing the management capacity of protected-area institutions.** The capacity of most African protected-area institutions to address complex interactions between protected areas and local communities is limited (Hough 1994a, 1994b). The development of scholarships, courses, exchange programs, training manuals, and technical assistance that focus on ecological and social monitoring, conflict resolution, park planning, and modern law enforcement techniques would greatly enhance the capacity of protected-area institutions to address many of the protected-area–local community conflicts.

Although ICDPs in Africa have had only limited success, we feel that a refined ICDP approach may be appropriate in some circumstances, especially when protected areas and local communities are highly codependent; that is, when local communities control the habitat abutting protected areas—habitat that is vital to the long-term viability of protected-area species and ecological processes—and protected areas control resources used historically by local communities, the use of which could be managed to be both sustainable and ecologically nondisruptive. An argument for an ICDP is compelling in this case because, unlike most of the alternative approaches discussed above, an ICDP can simultaneously address issues of conservation and development on the ground.

However, the ICDP approach needs to be both refined and enhanced. Improvements to the ICDP model include increasing flexibility and enhancing the use of adaptive management, which is a process by which management activities in a complex biophysical and social environment are monitored, evaluated, and reformulated in an iterative fashion so as to evaluate alternative hypotheses, accumulate knowledge about the system, and reassess long-term objectives (Holling 1978, Walters 1986). Central to this approach is the formulation of well-articulated objectives and the rigorous testing of management activities, which typically entails incorporating adequate samples, replicates, and controls. Additional refinements of the ICDP model should include the incorporation of a comprehensive ecological and social monitoring component; a more rigorous assessment of resource-harvesting schemes; more use of ecological and social research as a basis for identifying and addressing the ecological, social, and economic links that affect ICDPs; and the recognition that protected-area institutions need to act as facilitators of development assistance.

Obviously, effective conservation of wildlife in Africa and elsewhere will depend on the willingness and capacity of both national institutions and donors to embrace a broad package of interventions. These interventions might well be applied in conjunction with improved ICDPs on a landscape-wide scale.

NO

Roderick P. Neumann

Primitive Ideas: Protected Area Buffer Zones and the Politics of Land in Africa

Introduction

The objective of this article is to critically evaluate the conceptualization and implementation of participatory, integrated conservation and development programmes in Africa. The focus of the paper is directed specifically at the interventions of international NGOs [nongovernmental organizations] into rural land use and access in communities bordering protected areas. These interventions are planned and implemented by conservation organizations such as the World Conservation Union (IUCN) and the World Wide Fund for Nature (WWF), headquartered in Europe and North America and operating on a global scale. The conservation programmes of these organizations are in turn increasingly funded by bi-lateral and multi-lateral donors like the World Bank, the European Community, and various national agencies from the First World. The geographic extent of protected areas alone would make an examination of international interventions crucial: nine African countries, including Namibia, Tanzania, the Central African Republic and Botswana, have 9 per cent or more of their land under strict protection in national parks and game reserves. Tanzania's total of nearly 130,000 km^2 exceeds the combined territories of Holland, Slovakia and Switzerland.

It is not merely the size of the land area under question, however, that makes an analysis of conservation interventions important. Efforts by conservation NGOs to include the lands surrounding protected areas as buffer zones under the jurisdiction of the state have major implications for the politics of land. In the cases of the international conservation interventions under examination, land politics can be viewed as operating at two geographical scales. The first is global: it raises questions about the relations of power between rural communities in Africa and international conservation NGOs, and about how power relations between local communities and the state are affected by global environmental agendas. In their conceptualization, global conservation strategies tend to gloss over the magnitude of political changes involved and invest international conservation groups and allied states with increased authority to

From Roderick P. Neumann, "Primitive Ideas: Protected Area Buffer Zones and the Politics of Land in Africa," *Development and Change*, vol. 28, no. 3 (July 1997). Copyright © 1997 by The Institute of Social Studies. Notes and references omitted.

monitor and investigate rural communities. Recent studies indicate that programmes attempting to integrate conservation with development serve to extend state power into remote and formerly neglected rural areas. The second scale at which land politics are affected is at the intra-community level. Many of the programmes and projects under review here emphasize land registration and tenure reform in general as key to stimulating the adoption of more resource-conserving land use in buffer zones. Research indicates that land conflict in rural Africa has often been heightened by land tenure reform and registration efforts. Conservation interventions will therefore undoubtedly engage with and influence ongoing negotiations and struggles over land ownership and access within communities. . . .

The 'New' Approach to Conservation in Africa

Calls to include 'local participation' and 'community development' as part of a comprehensive strategy for biodiversity protection in Africa are now ubiquitous, with organizations ranging from the World Bank to grassroots human rights activists offering endorsements. In outlining its lending policies, the World Bank emphasized that it would seek to integrate 'forest conservation projects with . . . macroeconomic goals' and involve 'local people in forestry and conservation management'. Writing for the World Conservation Union (IUCN), Oldfield asserted that 'new ideas are needed' in biodiversity conservation because '[l]ocal people all too often see parks as government-imposed restrictions on their traditional rights'. In short, the redistribution of the material benefits of conservation and the resolution of conflicts between conservationists and local communities are central elements in a purported 'new approach' to conservation in Africa.

The revamped conservation philosophy in Africa is manifested in the proliferation of integrated conservation-development projects (ICDPs). ICDPs take various forms, but all embody the idea that conservation and development are mutually interdependent and must be linked in conservation planning. An important rationalization for these initiatives is that 'conservation policies will work only if local communities receive sufficient benefits to change their behavior from taking wildlife to conserving it'. In other words, 'the basic notion of an exchange of access for material consideration is central to ICDPs'. 'Benefits' to local communities include those directly related to wildlife management (wages, income, meat), social services and infrastructure (clinics, schools, roads), and political empowerment through institutional development and legal strengthening of local land tenure. Additionally, ICDPs are often linked with cultural survival efforts and thus seek to incorporate indigenous knowledge and practices in conservation management. Indigenous peoples, so the argument goes, have been living sustainably in relatively undisturbed habitats for generations and can thus be active participants in implementing conservation policy.

The main features of ICDPs are embodied in protected area buffer zones, a particular land use designation that is gaining increasing currency within conservation circles in Africa. Government and non-government conservation officials support buffer zones as an ideal means to promote environmental protection while simultaneously improving socio-economic conditions on reserve boundaries. Buffer zones are now included in virtually all protected area plans and are viewed, along with other participatory ICDPs, as *the* key strategy for the future of biodiversity maintenance in Africa. The buffer zone idea is most directly traceable to UNESCO's 'Man and the Biosphere Programme' (MAB) biosphere reserve model, first proposed in 1968. There are now numerous published definitions for buffer zones. Generally they are lands adjacent to parks and reserves where human activities are restricted to those which will maintain the ecological security of the protected area while providing benefits to local communities. Though ecological and biological concerns have typically driven conservationists' designs for buffer zones and related strategies, they are increasingly presented as a means to strengthen local land and resource claims. The buffer zone idea originally entailed the legal demarcation of boundaries which would separate land uses in transitional stages, though sometimes authors use the term less discriminatingly.

Much of the writing on buffer zones has been light on analysis and evaluation, tending to be more 'philosophical and prescriptive'. At the foundation of this 'philosophy' is the notion that conservation will not succeed unless local communities participate in management of and receive material benefits from protected areas. Participation in buffer zones can best be accomplished by first securing local people's rights to land and resources. Writing in a World Bank Technical Paper, Cleaver argues that a 'key to success in better forest management [in Africa] will be local people's participation . . . This is best done through their ownership of land and of resources on the land . . . '. An additional rationale for supporting tenure reform as part of conservation planning is 'that private investment in environmental protection increases with security of tenure'. Thus, many new conservation proposals seek to integrate land surveying, titling, and registration efforts to improve land tenure security for buffer zone residents.

The issue of local land tenure in buffer zones is also seen to converge with cultural survival/indigenous rights efforts among conservationists and development experts. Writing for the IUCN, Oldfield suggests that where 'tribal and indigenous peoples' have customary land and resource rights, 'buffer zones should be established by vesting title to the lands with the local communities at the level of either the village or ethnic group'. Similarly, Cleaver recommends that '[w]here traditional authority still exists, group land titles or secure long-term user rights should be provided'. Rather than individual titling, most proposals suggest group titling to communities, so that '[l]and within the community can continue to be allocated according to customary practice'. In general, the policy rhetoric of institutions and organizations such as the IUCN and the World Bank presents indigenous land rights as complementing the goals of ICDPs.

A Kinder, Gentler Conservation?

Despite the sympathetic treatment of local land rights and emphasis on benefit sharing by buffer zone proponents, land alienations and local impoverishment seemingly continue apace. Many of the projects sound alarmingly similar to the fortress-style approach to protected areas which they supposedly replace. There are more reports of forced relocations, curtailment of resource access, abuses of power by conservation authorities, and increased government surveillance, than of successful integrations of local people into conservation management. Rather than representing a new approach, many buffer zone projects and other ICDPs more closely resemble colonial conservation practices in their socio-economic and political consequences. In actuality, many buffer zones constitute a geographical *expansion* of state authority beyond the boundaries of protected areas and into rural communities. Given the already substantial proportion of land placed in protected areas across Africa, the potential for spatially extending the reach of the state is tremendous. A few examples will illustrate.

In Madagascar, proposals to integrate conservation with rural development in buffer zones in fact involve new forms of state intervention and restrictions on land use. The Madagascar Environmental Action Plan, developed with the assistance of the World Bank, aims: 'to help farmers to sedentarize and to incite them to invest in the medium term in soil conservation, agroforestry and reforestation . . . To discourage shifting cultivation and other forms of deforestation, via integrated development in the zones surrounding protected areas'. The rationale of the Bank and the Madagascar government is virtually identical to ill-fated colonial efforts across Africa to convert shifting cultivators into 'progressive farmers'. Paradoxically, current conservation advocates, like their colonial predecessors, conceive of *tavy* (the local term for shifting cultivation in Madagascar) not as 'indigenous knowledge' in practice, but as a 'long-lived habit' which must be eliminated. Increased monitoring of land use activities by the state is required to implement conservation agendas in buffer zones. In the country's Mananara Biosphere project, the state has substantially increased the number of forest guards, engaged the support of local police, and placed forestry extension agents with surveillance duties in buffer zone communities. In general, Madagascar's present conservation policies 'stress the need to remove villagers from within protected areas [and] to create larger buffer zones'. At Montagne d'Ambre National Park, the government has recently added a buffer zone which has expanded the park authority's control over village lands and resources. In effect, park management has been 'encroaching upon local forest and land resources'.

In Tanzania, too, several buffer zone projects have been proposed or implemented with similar ramifications for local land and resource control. For instance, a buffer zone project is underway at the Selous Game Reserve, already the largest protected area on the continent at 50,000 km^2. In the 1980s, the Selous Conservation Programme was implemented under the aegis of the German organization Deutsche Gesellschaft Für Technische Zusammenarbeit (GTZ) in an attempt to address some of the conflicts between reserve authorities and local communities. A 1988 study produced for GTZ recommended that a buffer zone be established along the perimeter of the game reserve. The authors of the study recommended that within the buffer zone, '[t]he Game Authorities should have the final say. It should not be considered as part of village land'. The government subsequently established a buffer zone encompassing 3630 km^2 of adjacent forest, grazing pasture, and settlement under the jurisdiction of the reserve authorities. Similarly, a proposed buffer zone at Lake Manyara National Park, Tanzania, would be managed by park authorities who would oversee land use. In this case, restrictions on adjacent land uses are seen as essential '[t]o minimize conflicts across boundaries between the Park and adjacent villages'. As a final example, the Serengeti Regional Conservation strategy, on the boundaries of Serengeti National Park, was launched in 1985. The strategy includes three types of buffer zones including 'mandatory' buffer zones. In these areas, the ultimate resolution for land use conflicts is 'the removal of land uses that are incompatible with conservation'.

A final case comes from Cameroon. Korup National Park and its 'support zone' encompass 4500 km² of tropical rainforest in southwest Cameroon. The implementation of the Korup project, though formulated as a participatory ICDP, has meant an increase in the policing capacity of the state. Consequently, the buffer zone now has a much higher concentration of law enforcement officials than any other nearby government lands. Once again, when compliance with conservation objectives is not forthcoming, eviction and relocation are the ultimate solution. As Colchester points out, 'the same laws that made resettlement from [Korup National Park] necessary would also apply in the buffer zones to which the populations were relocated, making their presence there equally illegal'.

The above cases serve to reveal the relations of power between First World conservationists and rural African communities which are embodied within the new approach to conservation. As long as a 'tradition' of living 'in harmony with nature' is maintained in a manner suitable to buffer zone planners, local communities may remain on the land. However, it is the prerogative of First World conservationists (backed up by the power of the state) to determine whether land uses are compatible with their interests or suitable for the purposes of the buffer zones. A recent IUCN publication uses an example from Nigeria to describe how this works in practice. 'The SZDP [Support Zone Development Programme] and the Park Management Service will thus work closely together to monitor village behaviour, and to administer appropriate "rewards" and "punishments".' . . . In essence, these buffer zone management guidelines call for the geographical expansion of park authority to monitor and regulate the daily lives of local community members and to force compliance through systems of rewards and punishments.

In sum, though the documents of international conservation NGOs present ICDPs and buffer zones as participatory and locally empowering, the power to propose, design, and enforce buffer zones lies far distant from rural African communities. The concept of participation is severely limited and frequently based on an assumption that local indigenous communities live in harmony with their environment. In many proposals which suggest a place for people in buffer zones, the image of the Other as closer to nature is central. This image is best exemplified in an IUCN publication:

> Traditional lifestyles of indigenous people have often evolved in harmony with the local environmental conditions . . . Retaining the traditional lifestyles of indigenous people in buffer zones, where this is possible and

appropriate, will encourage the long-term conservation of tropical forest protected areas. Protecting the rights of local communities ensures that they remain as guardians of the land and prevents the incursion of immigrants with less understanding of the local environment.

'They' belong in buffer zones because they have co-evolved with the environment and will serve as protectors against the incursions of 'outsiders' who have lost that harmonious relationship with nature. Indigenous peoples thus bear a tremendous burden—to demonstrate to outsiders (i.e. Western conservationists) a conservative, even curative, relationship with nature while risking the loss of their land rights should they fail. As Stearman observes, there is a growing danger that indigenous peoples 'must demonstrate their stewardship qualities in order to "qualify" for land entitlements from their respective governments'. Their lifestyles must allow them to do what immigrants and, significantly, Westerners, cannot—produce and reproduce in an ecologically benign way. Conservationists' ideas for indigenous participation in buffer zones are structured by a long history of western notions of the non-western 'primitive'. . . .

Conclusion

. . . First, we need to recognize that past and present conservation policies are complicit in creating the climate of land tenure insecurity within which many rural African communities operate. The establishment of virtually every national park in sub-Saharan Africa required either the outright removal of rural communities or, at the very least, the curtailment of access to lands and resources. As a result buffer zones extend the authority of the park to monitor and restrict land and resource uses of populations already displaced by protected areas. Policies need to be reconceptualized as mechanisms for power-sharing between local communities and state and international institutions rather than as opportunities for extending state control. Research needs to be directed toward identifying and developing institutional mechanisms for controlling access and use of lands and resources that are seen as legitimate by affected communities and that have a detectable effect on conservation goals.

Second, research and policy needs to be directed toward identifying the lines of fracture in rural communities and how segments of the community are differentially and even adversely affected by conservation proposals. Specifically, we need to recognize that local communities are not homogeneous entities whose members share a common set of interests regarding land and resource rights and that conservation interventions, almost by definition, will produce winners and losers in struggles over access. Local politics in rural Africa often revolve around the competing land claims of men versus women or the poor versus more well-to-do peasants, within villages or even within households. Most importantly, we need to problematize the notion of traditional or customary land tenure as the product of years of intra-community struggle over rights, not a set of ancient laws frozen in time.

Finally, we need to understand how the development interventions in buffer zones relate to conservation. Many of the projects reviewed are designed not to improve livelihoods, but merely to defuse local opposition. This is a very short-sighted and short-lived 'solution' and simply 'buys' the support of (some segments of) local communities rather than integrating conservation with development. Whether the 'benefits' from conservation are reaching the people most directly involved in activities which threaten protected areas or, if they are, whether they have any marked effect on their land and resource decisions remains an open question. Research focused on the politics of land is needed to demonstrate the link between conservation and the improvement of local livelihoods. Moving in these directions will, I believe, lead us closer to a truly 'new approach' to biodiversity conservation.

POSTSCRIPT

Are Integrated Conservation and Development Programs a Potential Solution to Conflicts Between Parks and Local People?

This debate tugs at a couple of deeper philosophical issues, including the chasm between social and physical sciences in thinking about biodiversity protection and the ongoing debate within the environmental community regarding preservationist versus conservationist approaches.

While it may be an obvious point, the natural sciences have tended to prioritize wildlife in their design and evaluation of conservation programs, whereas the social sciences generally have emphasized the welfare of people. While some advocates of the sustainable development paradigm argue that their approach integrates the two concerns (development leading to better conservation, and sustainable use of resources leading to enhanced development), the reality is that sustainable development is conceptualized in very dissimilar ways by different constituencies, e.g., economists versus deep ecologists. Some African intellectuals complain that environmental concerns have been used to unfairly constrain their development prospects, a problem sometimes referred to as "green imperialism." In his 1997 article (January/February issue) in *The Ecologist*, entitled "The Authoritarian Biologist and the Arrogance of Anti-Humanism: Wildlife Conservation in the Third World," Bamchandra Guha, an Indian scholar, asserts that biologists have been overly influential in environmental policy making. According to Guha, "Biologists have a direct interest in species other than humans. . . . This interest in other species, however, sometimes blinds them to the legitimate interest of the less fortunate members of their own." In contrast, many environmentalists complain that the World Bank, for example, supports a view of sustainable development that is overly anthropocentric.

Within the environmental community itself, there is an ongoing debate about the effectiveness of preservationist versus conservationist approaches to biodiversity protection. Preservation calls for the total nonuse, or nonconsumptive use, of natural resources, an approach typically associated with parks. Conservation allows for the use of natural resources to meet human needs within certain biological limits. This debate is directly relevant to discussions in Africa regarding protected areas. Some countries, such as Kenya, have opted for a more park-based approach, whereas others, such as Namibia, rely on more of an integrated conservation and development model. The preservation-conservation debate has also influenced views toward trade in ivory. Until recently, there was a total ban on ivory trade under the Convention on International Trade in Endangered Species (CITES). Kenya, with a declining elephant population, argued that this was necessary to eliminate elephant poaching and the associated illegal trade in ivory. It was further suggested that limited legal trade in ivory would be unenforceable. In contrast, countries like Zimbabwe, Botswana, and South Africa, with stable or growing elephant populations, argued that the sanctioned culling of elephants would help fund conservation efforts and maintain elephant populations at sustainable levels. Against the wishes of many environmental organizations, CITES allowed for the limited sale of legal ivory beginning in the late 1990s.

ISSUE 10

Is Sub-Saharan Africa Experiencing a Deforestation Crisis?

YES: Kevin M. Cleaver and Götz A. Schreiber, from *Reversing the Spiral: The Population, Agriculture, and Environment Nexus in Sub-Saharan Africa* (The World Bank, 1994)

NO: Thomas J. Bassett and Koli Bi Zuéli, from "Environmental Discourses and the Ivorian Savanna," *Annals of the Association of American Geographers* (March 2000)

ISSUE SUMMARY

YES: World Bank economists Kevin M. Cleaver and Götz A. Schreiber argue that Africa is engaged in a downward spiral of population growth, poor agricultural performance, and environmental degradation.

NO: Academic geographers Thomas J. Bassett and Koli Bi Zuéli counter that it is dominant perceptions of environmental change, rather than concrete evidence, that lie behind the widely held belief that Africa is engaged in an "environmental crisis of staggering proportions."

It has long been assumed that Africa is experiencing high levels of deforestation and that drastic measures must be undertaken to curb this disturbing trend. While this characterization still pervades much of the literature on environmental change in Africa, this view has been challenged in recent years by a series of books and articles questioning the degree and nature of deforestation in Africa.

One may gain perspective on debates concerning the nature and extent of environmental degradation in Africa by considering a number of broader, theoretical viewpoints. Contemporary scholars who believe that population growth is the main cause of environmental degradation in Africa are often described as neo-Malthusians. The neo-Malthusian label stems from the work of the original Malthusian, Thomas Malthus, and his 1798 treatise entitled *Essay on the Principle of Population*, as well as the subsequent and related work of neo-Malthusians such as Paul Ehrlich and his book entitled *The Population Bomb* (Simon and Schuster, 1968). Malthus suggested that human population will tend to expand until it has outstripped the capacity of the land to support it. Neo-Malthusians have updated this idea by accounting for improvements in agricultural productivity and family planning technologies, while still maintaining that high rates of population growth lie behind most forms of environmental degradation. During the colonial era in Africa, many European administrators, foresters, and academics adopted Malthusian-leaning viewpoints, portraying Africa as a scene of environmental destruction driven by excessive population expansion and poor management practices. Since independence, the neo-Malthusian perspective has tended to flourish following natural disasters (such as the Sahelian droughts of 1968–1972 and 1984–1985) and during periods of heightened international attention to environmental issues (such as that surrounding the 1992 United Nations Conference on Environment and Development in Rio de Janeiro).

There have been a number of scholarly traditions countering the neo-Malthusian perspective. In Africa, cultural ecologists (an interdisciplinary field examining human-environment interactions in developing-country settings) have conducted field studies demonstrating that many traditional natural resource management practices are ecologically sound. Another closely related line of scholarship was developed in the wake of the pioneering work of Ester Boserup, who, in her 1965 book, *The Conditions of Agricultural Growth* (G. Allen and Unwin), argues that population increase leads to sustainable increases in agricultural production via a process of intensification over the medium to long term. In Africa, scholars such as Michael Mortimore have documented agricultural intensification in northern Nigeria and the highlands of East Africa.

More recently, scholars working in the tradition of political ecology have focused on the effect of broad-scale political and economic processes on local-level ecological dynamics (both real and perceived). Political ecologists have often drawn on Marxist-inspired discourse theory to highlight the role of political and economic power, rather than "neutral" empirical observation, in shaping our understanding and diagnosis of environmental problems.

In this issue, Kevin M. Cleaver and Götz A. Schreiber present deforestation as a major problem in sub-Saharan Africa and in West Africa in particular. While acknowledging that data on forest resources and rates of extraction are imperfect, the authors argue that this information is reliable enough to describe the scale and trend of the problem. Cleaver and Schreiber suggest that the causes of deforestation include a number of activities driven by population growth and international demand, especially forest conversion to farmland, road construction in environmentally fragile areas, timber extraction, and commercial fuelwood harvesting.

Thomas J. Bassett and Koli Bi Zuéli contend that there is only shaky evidence to support the perception of Africa as physically disintegrating due to the destructive practices of its inhabitants. They are particularly critical of National Environmental Action Plans (NEAPs) that the World Bank has sponsored in a number of African countries. NEAPs that misdiagnose the nature and causes of environmental change may, in fact, lead to policy prescriptions that acerbate, rather than assuage, real environmental problems.

Kevin M. Cleaver and
Götz A. Schreiber

 **YES**

Reversing the Spiral: The Population, Agriculture, and Environment Nexus in Sub-Saharan Africa

Introduction

Over the past thirty years, most of Sub-Saharan Africa (SSA) has experienced very rapid population growth, sluggish agricultural growth, and severe environmental degradation. Increasing concern over these vexing problems and the apparent failure of past efforts to reverse these trends led the authors to take a fresh look at the available research findings and operational experience. The objective was not to compile and address all of the agricultural, environmental, and demographic issues facing Africa or simply to juxtapose these three sets of problems. It was to gain a better understanding of the underlying causes and to test the hypothesis that these three phenomena are interlinked in a strongly synergistic and mutually reinforcing manner.

The need to survive—individually and as a species—affects human fertility decisions. It also determines people's interactions with their environment, because they derive their livelihood and ensure their survival from the natural resources available and accessible to them. Rural livelihood systems in SSA are essentially agricultural, and agriculture is the main link between people and their environment. Through agricultural activities people seek to husband the available soil, water, and biological resources so as to "harvest" a livelihood for themselves. Such harvesting should be limited to the yield sustainable from the available stock of resources in perpetuity so as to ensure human survival over successive generations. Improvements in technology can increase the sustainable yields or reduce the resource stock required. Population growth should thus be matched or surpassed by productivity increases so as to safeguard the dynamic equilibrium between the stock of resources and the human population depending on it for survival. Over the past thirty years, this has not been the case in most of Sub-Saharan Africa.

This study's findings confirm the hypothesis of strong synergies and causality chains linking rapid population growth, degradation of the environmental resource base, and poor agricultural production performance. Traditional African crop and livestock production methods, traditional methods of obtaining woodfuels and building materials, traditional land tenure systems and land use arrangements, and traditional gender roles in rural production and household maintenance systems were well suited to survival needs on a fragile environmental resource endowment when population densities were low and populations growing slowly. But the persistence of these traditional arrangements and practices, under severe stress from rapid population growth in the past thirty to forty years, is causing severe degradation of natural resources which, in turn, contributes further to agricultural stagnation.

Rapid population growth is the principal factor that has triggered and continues to stimulate the downward spiral in environmental resource degradation, contributing to agricultural stagnation and, in turn, impeding the onset of the demographic transition. The traditional land use, agricultural production, wood harvesting, and gender-specific labor allocation practices have not evolved and adapted rapidly enough on most of the continent to the dramatically intensifying pressure of more people on finite stocks of natural resources.

Many other factors also have a detrimental impact on agriculture and the environment. These include civil wars, poor rural infrastructure, lack of private investment in agricultural marketing and processing, and ineffective agricultural support services. Inappropriate price, exchange rate, and fiscal policies pursued by many governments have reduced the profitability and increased the risk of market-oriented agriculture, prevented significant gains in agricultural productivity, and contributed to the persistence of rural poverty.

A necessary condition for overcoming the problems of agricultural stagnation and environmental degradation will be, therefore, appropriate policy improvements along the lines suggested in the 1989 World Bank report on Sub-Saharan Africa's longer-term development prospects (World Bank 1989d). These policy changes will be instrumental in making intensive and market-oriented agriculture profitable—thus facilitating the economic growth in rural areas necessary to create an economic surplus usable for environmental resource conservation and to provide the economic basis for the demographic transition to lower population fertility rates. That this can occur has been demonstrated in a few places in Africa that pursued good economic and agricultural policy, invested in agriculture and natural resource conservation, and provided complementary supporting services to the rural population. This study provides evidence for both the causes of the problem and its solution.

The Three Basic Concerns

Population Growth

Sub-Saharan Africa lags behind other regions in its demographic transition. The total fertility rate (TFR)—the total number of children the average woman has in a lifetime—for SSA as a whole has remained at about 6.5 for the past twenty-five years, while it has declined to about 4 in all developing countries taken together. As life expectancy in Sub-Saharan Africa has risen from an average of forty-three years in 1965 to fifty-one years at present, population growth has accelerated from an average of 2.7 percent per annum for 1965–1980 to about

3.0 percent per year at present. Recent surveys appear to signal, however, that several countries—notably, Botswana, Kenya, and Zimbabwe—are at a critical demographic turning point. This study discusses the factors that have contributed to the beginning of the demographic transition in these countries.

Agricultural Performance

Agricultural production in Sub-Saharan Africa increased at about 2.0 percent per annum between 1965 and 1980 and at about 1.8 percent annually during the 1980s. Average per capita food production has declined in many countries, per capita calorie consumption has stagnated at very low levels, and roughly 100 million people in Sub-Saharan Africa are food insecure. Food imports increased by about 185 percent between 1974 and 1990, food aid by 295 percent. But the food gap (requirements minus production)—filled by food imports, or by many people going with less than what they need—has been widening. The average African consumes only about 87 percent of the calories needed for a healthy and productive life. But as with population growth, a few African countries are doing much better, with agricultural growth rates in the 3.0 to 4.5 percent per annum range in recent years (Nigeria, Botswana, Kenya, Tanzania, Burkina Faso, and Benin). The policies of these countries help show the way forward.

Environmental Degradation

Sub-Saharan Africa's forest cover, estimated at 679 million ha in 1980, has been diminishing at a rate of about 2.9 million ha per annum, and the rate of deforestation has been increasing. As much as half of SSA's farmland is affected by soil degradation and erosion, and up to 80 percent of its pasture and range areas show signs of degradation. Degraded soils lose their fertility and water absorption and retention capacity, with adverse effects on vegetative growth. Deforestation has significant negative effects on local and regional rainfall and hydrological systems. The widespread destruction of vegetative cover has been a major factor in prolonging the period of below long-term average rainfall in the Sahel in the 1970s and 1980s. It also is a major cause of the rapid increase in the accumulation of carbon dioxide (CO_2) and nitrous oxide (N_2O), two greenhouse gases, in the atmosphere. Massive biomass burning in Sub-Saharan Africa (savanna burning and slash-and-burn farming) contributes vast quantities of CO_2 and other trace gases to the global atmosphere. Acid deposition is higher in the Congo Basin and in Côte d'Ivoire than in the Amazon or in the eastern United States and is largely caused by direct emissions from biomass burning and by subsequent photochemical reactions in the resulting smoke and gas plumes. Tropical forests are considerably more sensitive than temperate forests to foliar damage from acid rain. Soil fertility is reduced through progressive acidification. Acid deposition also poses a serious risk to amphibians and insects that have aquatic life cycle stages; the risk extends further to plants that depend on such insects for pollination.

Unlike the situation of population growth and agriculture, there are few environmental success stories in Africa, although there remain large parts of

Central Africa that are little touched. In looking closely, however, places can be found, such as Machakos District in Kenya, where environmental improvements have occurred along with rapidly expanding population. Good agricultural and economic policy, and investment in social services and infrastructure, are found to be the critical ingredients to such success (English and others 1993; Tiffen and others 1994). These positive experiences form the empirical basis for an action program to overcome the downward spiral elsewhere. . . .

Agricultural Stagnation and Environmental Degradation

The Deteriorating Natural Resource Base and Ecological Environment

Much of Sub-Saharan Africa's natural resource base and ecological environment is deteriorating. If present trends continue, this deterioration will accelerate. The most pressing problem is the high rate of loss of vegetative cover—mainly the result of deforestation and the conversion of savanna to cropland—which in turn leads to loss of soil fertility and soil erosion. Global and regional climatic changes and deviations from longer-term average conditions are also causal factors—but human impact on the environment in Sub-Saharan Africa may itself be an important element contributing to these climatic changes.

Deforestation

In much of Sub-Saharan Africa, deforestation is a major problem—with significant local, national, and global consequences. Forests provide a multitude of products and serve many functions, including essential environmental ones. With deforestation, these are lost. Forests and woodlands are cleared for farming and logged for fuelwood, logs, and pulp wood. Data on forest resources and rates of extraction and clearing are imperfect, as are data on most of Africa's environmental resources, but information is continually improving and reliable enough to suggest the scale of the problem. In 1980, there were about 646 million hectares of forests and woodlands in Sub-Saharan Africa. A 1980 FAO/UNEP [Food and Agriculture Organization/United Nations Environment Programme] study estimated that 3.7 million hectares of tropical Africa's forests and open woodlands were being cleared each year by farmers and loggers (Lanly 1982). More recent estimates suggest that close to 2.9 million hectares were lost each year during the 1980s, mainly through conversion to farm land, but the rate of deforestation may be accelerating as the aggregate area still under forests continues to shrink. Reforestation during the 1980s amounted to 133,000 hectares per year, only about 5 percent of the area lost each year to deforestation.

Aggregate data obviously obscure important differences among regions and countries. Deforestation has been particularly rapid in West Africa, with East Africa and southern Africa also suffering substantial losses in forest cover. Large tracts of tropical forest still remain, especially in Zaire, Gabon, Congo, the Central African Republic, and Cameroon. It would take many years for Central

Africa's forests to be completely destroyed, but the process has started. In most of East Africa and southern Africa, as well as in the West African coastal countries, the process is far advanced. . . .

Managing the Natural Resource Base

Forests

About 30 percent of Sub-Saharan Africa's land area is classified as forests or woodlands. But only about 28 percent of this area is closed forest—compared with about two-thirds in Latin America and in Asia. About 34 percent is shrubland and 38 percent is savanna woodland; both are multiple-use resource systems, utilized for meeting local requirements for fuelwood and other tree and forest products as well as for farming and forage.

. . . The most important causes of deforestation are conversion to farmland, infrastructure development in environmentally delicate areas, timber extraction, and commercial fuelwood harvesting. Growing and migrating human populations as well as international demand for tropical timber drive these processes. Timber exports from Sub-Saharan Africa amount to about US$700 million per year at present. Cropland is expanding at a rate of 1 million ha annually—to a large extent at the expense of forest areas and woodlands. A number of agricultural development projects supported by external aid donors, including the World Bank, have facilitated the conversion of forest and rangelands into cropland.

The most important areas for action to stop the degradation of Sub-Saharan Africa's forest resources lie outside the immediate purview of forestry sector policy. They are: (a) reducing population growth, and (b) intensifying agricultural production at a rate which exceeds population growth, in order to encourage sedentary agriculture and livestock raising and to discourage further invasion of the remaining forests. Rapidly growing numbers of people, barely surviving in land-extensive agricultural systems, have no option than to continue to invade and destroy forests. This points again to the complex mutual dependency of agricultural and nonagricultural activities.

For the forests that remain, improved management for multiple uses will be vital. These uses range from the provision of critical environmental services to the supply of timber and nontimber products, and from tourism and recreational uses to mineral extraction. It is unrealistic to expect that all forests can be conserved in their present state. For almost all of Africa's forests the issue is not whether to use them or not to use them—but how to use them. If people (and governments) feel that there is little benefit from forests, they will continue to be mined for urgently needed export revenue or converted into agricultural land.

To address these problems effectively, there is no alternative to planning, orchestrated by governments. This can be done within Tropical Forestry Actions Plans (TFAPs), National Environmental Action Plans (NEAPs), or simply forestry master plans. Each will involve some form of land use and natural resource planning. Land use plans for forest areas should identify conservation areas, parks, areas designated for sustainable logging, mining areas, farming and grazing areas, and areas designated for infrastructure development. . . .

National Environmental Action Plans

The development of national environmental resource management strategies must be a national affair. The main instrument for this process is the National Environmental Action Plan (NEAP). NEAPs are currently being prepared or implemented with World Bank support by most African countries. They should contain strategies for addressing all of the issues of the nexus. The NEAP concept is multisectoral in approach, and oriented to bottom-up participatory planning and implementation. It provides a framework for integrating environmental concerns with social and economic planning within a country. The objective is to identify priority areas and actions, develop institutional awareness and processes, and mobilize popular participation through an intensive consultation process with NGOs and community representatives. Donor collaboration can also be effectively mobilized in this manner.

A successful national approach to environmental concerns involves several important steps:

- Establishing policies and legislation for resource conservation and environmental protection that are integrated into the macroeconomic framework and, if possible, assessing the costs of. degradation. These were, for example, estimated to be between 5 and 15 percent of GNP [gross national product] in Madagascar and more than 5 percent of GDP [gross domestic product] in Ghana.

- Setting up the institutional framework, usually involving a ministerial or higher-level environmental policy body, developing mechanisms for coordination between agencies, building concern in these agencies, balancing private and public sector concerns, decentralizing environmental management, and assuring continuous contact with local people. The preparation of regional land use plans could be an important component. The basic framework needed to guide the implementation of land tenure reform, forest policy reform, and other elements discussed above can also be included in NEAPs.

- Strengthening national capacity to carry out environmental assessments and establishing environmental information systems. This can be done to some extent by restructuring existing data and making them available to users. Pilot demand-driven information systems should also be initiated to strengthen national capacity to monitor and manage environmental resources. Local and regional research capacity will be crucial to the development of plant varieties and technologies which are truly adapted to local conditions.

- Developing human resources through formal and on-the-job training; introducing environmental concerns into educational curricula and agricultural extension messages; and increasing public awareness through media coverage, general awareness campaigns, and extension services.

- Establishing Geographical Information Systems (GISs) that incorporate adequate environmental information. Lack of operationally meaningful and reliable environmental data is a major problem. It tends to result in misconceptions about natural resource problems and the consequent risk that policy measures will be misdirected. Urgent needs include assessments of forest cover, soil erosion and soil capability, desertification risks, and the distribution of human and livestock populations. This is clearly an area in which donors can provide support and expertise and governments need to act. It is important to develop national capacity to gather and analyze information in-country: properly designed and operated Geographical Information Systems can be extremely helpful in this regard. GISs make use of aerial photography, remote sensing, and actual ground inspections and data collection. GISs will be particularly useful not only to monitor the progress of natural resource degradation and destruction, but—more importantly—to assess land capability for various uses and, thus, to provide the basis for sound land use planning.

NEAPs are intended to be evolutionary—developing policies through field experience as well as national-level analysis. They should lead to the empowerment of the nongovernmental sector, not just by providing funds for small-scale community activities through national environmental funds, but also by drawing large numbers of village and district representatives into consultative forums. A nongovernmental advisory body was part of the institutional arrangements set up, for example, under the Lesotho NEAP.

Considerable external support has been provided for the NEAP process, from bilateral and multilateral agencies and NGOs [nongovernmental organizations] (such as the World Wildlife Fund, the World Resources Institute, and the International Institute for Environment and Development). External expertise is made available to the countries undertaking NEAP preparation, and aid agency policies are coordinated in the process, with the NEAP forming the basis for coordination. Where NEAPs have led to the preparation of national environmental investment plans (as in Madagascar and Mauritius), donors have substantially oversubscribed the programs. A National Environmental Action Plan can therefore become the major preparatory instrument for addressing the issues discussed [here].

Environmental Discourses and the Ivorian Savanna

The image of an entire continent physically disintegrating due to the destructive land-use practices of its inhabitants conveys the magnitude of Africa's environmental problems—at least in the eyes of the World Bank. Yet there remains considerable uncertainty about the very processes generating this assumed degradation of the environment. As the Bank notes for Côte d'Ivoire..., "little data exist to confirm this fact." This is also the case for other parts of Africa (Watts 1987; Stocking 1987, 1996). Despite a lack of reliable evidence, the World Bank considers environmental degradation to be so widespread, that "the business" of environmental planning and regulation is now seen as a global affair, and one that falls within its own purview (Falloux and Talbot 1993: xiii–xiv). Indeed, since the late 1980s, the Bank has required low-income countries that receive International Development Association (IDA) funding to draw up national environmental-action plans (NEAPs).

As demonstrated in World Bank policy papers and country reports, the link between environmental planning and development assistance is articulated within the discourse of sustainable development (Williams 1995). The fuzzy green notion of sustainable development is readily adopted by the Bank because of its compatibility with its basic "technocratic, managerial, capitalistic, and modernist ideology" (Adams 1995: 93). Under the banner of sustainable development, the Bank now promotes a "win-win" strategy of combining economic growth with environmental conservation (World Bank 1992: 2–3; Biot et al. 1995). Never considering that its past policies and interventions are in any way implicated in the so-called environmental crisis, the Bank presents itself as an impartial observer and promoter of good stewardship. It is currently assisting dozens of African governments to develop NEAPs which, in assembly-line fashion, are being produced according to a blueprint (Greve et al. 1995). NEAPS are heralded as modern vehicles that will lead its member countries down the road to rational and orderly sustainable development.

In this [selection], we examine the contents of the NEAP for Côte d'Ivoire with special attention given to how environmental problems are constructed for the northern savanna region. Our objectives are [two]fold. The first is to high-

From Thomas J. Bassett and Koli Bi Zuéli, "Environmental Discourses and the Ivorian Savanna," *Annals of the Association of American Geographers,* vol. 90, no. 1 (March 2000). Copyright © 2000 by The Association of American Geographers. Notes and references omitted.

light the problem of data gaps in the environmental-planning and sustainable-development discourses. Determining if an environmental problem exists would appear to be a critical first step in the planning process. Yet what is striking about the NEAP process is how quickly environmental problems are identified and prioritized on the basis of so little data. We confront this data problem by contrasting the image of a highly degraded savanna environment found in Côte d'Ivoire NEAP reports with the actual landscape as experienced by farmers and herders and confirmed by our analysis of aerial photographs and land-cover changes.

Our second objective is to address some of the policy implications related to the disjuncture between local and regional patterns of environmental change and national and global environmental discourses. One result of this disjointed scale problem is that the actual dynamics of environmental change are being overlooked. We consider, in turn, the ecological and human consequences of ignoring ongoing biophysical changes while planners are busy addressing imaginary environmental problems. . . .

Research Area and Methods

. . . We have collected information on land-cover and land-use patterns using a variety of field research and analytical techniques. Survey-research methods were used to administer a questionnaire on environmental perceptions to a sample of 38 Senufo and Jula households in Katiali and to 42 Senufo households in the Tagbanga region. Group interviews were also held with Fulbe pastoralists in the Katiali area. To assess whether local perceptions of environmental change were congruent with scientific findings, we reviewed the specialist literature on human-induced modifications of savanna vegetation. We then examined aerial photographs located in Côte d'Ivoire and France for different time periods to compare land use/cover patterns for the two research sites. Geographic information systems (GIS) techniques were utilized to quantify land-cover trends. To gain a clearer understanding of vegetation change dynamics on the ground, we inventoried species along 50-m transects and in 10×10-meter plots following the contact-point method adapted from Daget and Poissonnet (1971) by César and Zoumana (1995). Finally, we collected environmental policy and planning documents and interviewed individuals involved in the NEAP process in government ministries and at the World Bank's regional headquarters in Abidjan. . . .

Discourses on Environmental Change

Before examining the contents of the Ivorian NEAP, it is useful to provide some information on its origins.

The NEAP Process

According to the World Bank, the NEAP process involves four stages: the identification of environmental problems and their underlying causes; setting priorities; establishing goals and objectives; and proposing new policies, institutional and legal reforms, and priority actions. The Bank considers this process to be straightforward. "It is relatively easy to identify problems and formulate appropriate responses to them" (Greve et al. 1995: 8–16). The more difficult phase, it argues, centers around the implementation of reforms and other policy actions.

The elaboration of a National Environmental Action Plan for Côte d'Ivoire began on the eve of the U.N.-sponsored conference on the environment and development held in Rio de Janeiro, June 3–4, 1992. The Ministry of the Environment and Tourism organized a national conference (May 19–21, 1992) during which a rationale for a NEAP was presented. The World Bank worked closely with and helped to train the staff members who organized this meeting and who subsequently carried out the first "civilian phase" in the preparation of the NEAP (World Bank 1994: 4). This stage involved holding regional meetings at which local civic and political leaders, government officials, and selected farmers and herders were invited to present their views on regional environmental issues. This form of "participatory planning" did not involve consultations with ordinary men and women living in rural areas about what they considered to be the most important environmental issues. At the Korhogo regional meeting, a small number of peasants were invited to participate in a setting that was dominated by civil servants representing the agricultural services, planning, and rural development. If "participation" means "the ability of people to share, influence, or control design, decision-making, and authority in development projects and programs which affect their lives and resources" (Peters 1996: 22), then this so-called civilian phase of the Ivorian NEAP process involved very limited participation. As the Korhogo region NEAP report reveals (see below), the voices that were heard and the stories that were ultimately accepted suggests that "participatory" planning in the NEAP process is more rhetoric than reality.

A second national conference took place on November 28–30, 1994, where the NEAP coordinating committee presented its *White Paper on the Environment* that summarized the results of these regional meetings (RCI 1994a). This document became the basis of the Côte d'Ivoire National Environmental Action Plan [NEAP-CI] whose appearance in June 1995 capped the second phase of the NEAP process. At the same meeting, the World Bank presented its own report on Ivorian environmental problems and policy recommendations, *Côte d'Ivoire: Towards Sustainable Development,* which it later revised and distributed in December 1994 (World Bank 1994). A comparison of the outlines and content of the two reports suggests that both the Bank and NEAP coordinating committee worked closely together. The Ivorian government approved the NEAP in 1996, thus bringing the formal planning phases to completion. The execution of the plan, however, quickly ran aground when the Ministry of Housing and the Environment established its own environmental agency to coordinate environmental projects estimated to cost $112 million. Aid donors, led by the World Bank, were critical of the structure of the agency, especially the lack of inter-Ministry coordination and the exclusion of NGOs [Nongovernmental Organizations] from its operations. They also expressed serious concerns about the competence of the individuals appointed by the Minister to head the agency (World Bank 1998). The NEAP proposed the establishment of a National Environmental

Agency that would be placed under the control of a Management Council, comprising twelve representatives from government and the private sector, including a representative of Ivorian NGOs. The NEAP process also envisioned the creation of regional commissions and departmental committees in which different social groups would participate in various environmental planning activities. In contrast to this partially decentralized planning apparatus, the Minister of the Environment "viewed environmental planning as a *business* and could only see millions of dollars falling from the sky," according to a well-placed member of the World Bank (World Bank 1998).

Discourse 1: The NEAP Report

What image of environmental change in the savanna emerged from this NEAP process? The most detailed picture is contained in the report summarizing the findings of the northern regional meeting held in Korhogo and published in March 1994. The Korhogo report presents a grim scene of environmental degradation in which peasants and pastoralists are blamed for the deforestation of wooded savannas. According to the report's authors, "vegetative cover is declining due to the practice of shifting cultivation, bush fires, and the anarchic exploitation of forests and overgrazing" (RCI 1994b: 14). This change in plant cover is specifically characterized by "a replacement of the tree savanna by the grass savanna." The report goes on to paint a portentous picture of this process:

> More generally, the progressive widening towards the south of (grassy) plant formations has climatic repercussions (temperature, rainfall. . .) which in turn affects vegetation cycles. To compensate for the corresponding lowering of productivity, the one recourse has been to use chemical fertilizers which engenders certain problems such as soil acidification, and the contamination of surface and subsurface waters (RCI 1994b: 14).

This image of a southerly advancing boundary of a vegetation type associated with more arid climates is similar to the "marching desert" view found in the desertification literature. That is, in the absence of field studies, it is assumed that the forms of environmental transformation purportedly taking place in the Sahel are also occurring in humid savannas. Bush fires and indigenous land-rights systems are signaled out as major forces in this assumed environmental devastation:

> We are increasingly witnessing a decline in vegetation cover, essentially due to bush fires used in an intensive and excessive manner leading to the disappearance of certain varieties . . . these assaults on the environment are essentially linked to the abuses of customary rights, the exaggerated interpretation of the declarations of [the former] President Félix Houphouet-Boigny such as *"Land belongs to the person who improves it"* and *"that which has been planted by the hand of man must not be destroyed, no matter where;"* and especially the absence of a rural land code (RCI 1994b: 12).

According to this analysis, and that of the World Bank (Cleaver and Schreiber 1994: 8–10), "traditional" land-rights systems are inadequate for the task of modern environmental planning. They may have worked in the past, but contemporary land conflicts and insecurity have prevented farmers from investing in land improvements that might increase agricultural output and conservation practices. The assumption is that only when customary rights give way to modern (i.e., freehold) tenure systems will the incentive to conserve natural resources exist. Indeed, the transformation of land-holding systems to freehold arrangements is considered in the Ivorian NEAP to be an important step towards addressing all sorts of environmental problems. [T]his modernization model of tenure change, environmental conservation, and agricultural growth, which informs the Ivorian NEAP and which is at the heart of World Bank rural-development policies in Africa. The model points to the extent to which environmental crises are integral to neoliberal development discourses. These policies only have meaning with reference to agricultural stagnation and/or environmental decline. The desertification narrative serves such a need. This is not to suggest that environmental degradation is not occurring in sub-Saharan Africa. There is ample evidence of it (Batterbury and Bebbington 1998), whether it be soil erosion on the Borana plateau of southern Ethiopia (Coppock 1993), soil-fertility decline in southwestern Burkina Faso (Gray 1997), or rangeland degradation in southern Botswana (Abel and Blaikie 1989). Yet all of these examples are situated in local-level dynamics of resource access, control, and management. They emphasize, as do the case studies of environmental conservation in the Kano Close-Settled Zone of northern Nigeria (Mortimore 1998) and in Machakos District, Kenya (Tiffen et al. 1994), the importance of approaching natural resource management at multiple and nested scales (both biophysical and social). In the policy arena, these examples point to the need to provide "locationally and culturally appropriate technical and economic options" to different groups of land users and the necessity of moving away from "regulation and intrusive administration" (Mortimore 1998: 190–93). In contrast, blueprint development and cookie-cutter planning models like NEAP tend to be highly regulatory and intrusive. The case of Côte d'Ivoire's NEAP is indicative.

With reference to preserving the country's biodiversity, the Ivorian NEAP makes a number of forestry policy recommendations. These include outlawing logging (*l'exploitation forestière*) above the eighth parallel, intensifying village tree planting, controlling bush fires, and creating a Forestry Police to enforce these new regulations (RCI 1996). The first recommendation is strikingly reminiscent of E. P. Stebbing's shelterbelt scheme, proposed more than sixty years ago (Stebbing 1935). The link between bush fires and desertification is commonly made in Côte d'Ivoire newspapers. An editorial in the ruling party's newspaper warned that dry-season bush fires "each year contribute to the desert's advance in our country" (Gooré Bi 1999). A National Commitee for Forest Protection and Bush Fire Control was organized in 1996 to raise public awareness about the assumed social and environmental costs of bush fires. The committee organizes a national awareness day each year in which statements are made linking bush fires with desertification. A national arbor day is also held each year, during which public officials and billboards urge citizens to plant trees to stop the purportedly advancing desert.

The image of an increasingly degraded wooded savanna giving way to a grass savanna and, ultimately, desert-like conditions not only persists in the minds of environmental planners, journalists, and public officials but is also firmly implanted in the perception of Ivorian environmental nongovernmental organizations (ENGOs). For example, Côte d'Ivoire's leading ENGO, the Green Cross of Côte d'Ivoire, devoted a special issue of its monthly information magazine to bush fires and forests. In his editorial, Green Cross President Gomé Gnohité Hilaire emphasized the importance of educating the citizens of Côte d'Ivoire about the urgency of protecting the nation's remaining forests. He exclaimed that the paucity of funds going to environmental education was having disastrous consequences:

> We have forgotten that raising awareness, that environmental education, is a daily and unending task. We have continued more than ever to utilize the forest as the green gold of our development without concerning ourselves with the consequences. Our forest ecosystem has continued to degrade, forests are savannized. The savannas are desertified (Gomé Gnohité 1998: 4).

How accurate is this image of environmental change in the northern savanna of Côte d'Ivoire found in national environmental planning documents and in the pages of NGO publications? Has the expansion of livestock raising and the area under cultivation led to an increase in grass savannas and a decline in tree cover? Is fire the great destructive force that environmental planners and NGOs believe it to be? . . .

The Lessons of the Katiali and Tagbanga Case Studies

A major finding of this comparative research on land-cover changes in the Katiali and Tagbanga areas is that, contrary to received wisdom, the savanna has become more wooded over the past thirty years. This finding runs counter to the dominant narrative, which assumes that the savanna has become less wooded and increasingly dominated by grass savannas. It also extends, both geographically and analytically, the findings of Fairhead and Leach on the expansion of wooded landscapes in the forest-savanna transition zone of Guinée (Fairhead and Leach 1996) to the sudanian savanna.

A second finding points to the diversity of savanna vegetation communities in the Korhogo region. The similarities and differences in the transformation of the Tagbanga and Katiali savanna areas underscore the importance of temporal and spatial variations in environmental change. This finding conforms to the scientific literature on savanna ecology that points to a wide range of plant communities, which are commonly distributed in mosaic form across the landscape. The most important factors influencing the nature and direction of vegetation change are farming systems, grazing pressure, population density, and changing fire regimes. These factors, which are themselves linked to changing political and economic processes extending beyond the region (e.g., cotton-development policies, immigration of Fulbe herders, or farmer-herder conflicts), interact with a host of biophysical factors such as soil type, slope, and rainfall to create temporally and locationally specific outcomes.

A third finding of this research is its relevance to environmental planning. Despite its problematic scientific status, the desertification narrative currently guides environmental policy. For example, NEAP-CI recommendations to combat the assumed reduction in tree cover include the regulation of bush fires through a range of increasingly coercive measures, restrictions on wood cutting, and the promotion of village-level tree planting (RCI 1994a). In light of the findings of this case study, such policy recommendations can be seen as misconceived and a waste of limited resources. The disjuncture between national and global environmental discourses and actual vegetation-change patterns is alarming. Our findings show that although desertification is not taking place, heavy grazing and early fires have significantly reduced the quality of the savanna for livestock raising. Tree and shrub invasion and a highly degraded herbaceous layer were evident in both the Katiali and Tagbanga study areas. Since livestock development is a priority of the Ministry of Agriculture, one would think that rangeland rehabilitation would be a centerpiece of the Côte d'Ivoire NEAP. Not surprisingly, it is nowhere to be found in NEAP documents which are more concerned with reforestation than range condition. While environmental analysts and planners are occupied with an imaginary environmental problem, tree and bush encroachment continues unabated. This disjointed scale problem also produces its contradictions. For example, the Ivorian NEAP's recommended regulation, permitting only *early*-dry-season fires, would result in further bush invasion. To improve range conditions, degraded areas will have to be protected from grazing for at least two or three years, and woody growth must be controlled by extremely hot (i.e., *late*) bush fires (César 1994).

Conclusion

Given the extraordinary amount of environmental planning currently underway in Africa and its far-reaching implications on land use, access and management, one obvious conclusion of this study is that further research on environmental-change dynamics is of utmost importance. Indeed, the World Bank places the identification of environmental problems and their underlying causes as the first step in the NEAP process. Yet, from all indications, the Bank does not consider this to be a particularly challenging phase. Despite the glaring gaps in our knowledge, the Bank believes that most environmental issues are easy to identify and can be classified along a simple color scheme. A conclusion of this [selection] is that identifying environmental problems and their causes is one of the most difficult and time-consuming stages in environmental planning and policy making.

One of the challenges in confronting the environmental-data problem is that so little data exist with any meaningful time depth. Even where data like aerial photographs do exist, their relatively small scale rarely permits one to make little more than very general statements. This situation demands that multiple approaches be pursued to determine the spatial and temporal dynamics of environmental change that are not apparent in aerial photos. In this study, we have combined household-survey research focused on farming systems and environmental perceptions with aerial-photo interpretation and vegetation transects to identify the general trends in vegetation change. This multiscale, multimethod approach yielded different results from the so-called "participatory approach" followed in the NEAP process, in which the opinions of selected individuals from different social strata were solicited in public meetings. Not surprisingly, peasants and herders were reticent in such fora.

A third point centers on competing discourses on environmental change. In contrasting the environmental narratives expressed in World Bank and NEAP documents with those of rural land users, it would be misleading to suggest that a homogeneous view prevails on either side. For example, peasant farmers downplayed their own role in transforming savanna vegetation through their agricultural activities by pointing their finger at Fulbe herders as the primary agents of environmental change. This tendency to "blame" the Fulbe must be contextualized in the often bitter conflicts that exist between farmers and herders in northern Côte d'Ivoire. Similarly, the tendency of the Fulbe to deny their use of fire as a range management tool and to "blame" farmers and hunters for bush fires must be seen in light of these land-use conflicts.

Fourth, it is also clear that the environmental-crisis lexicon is widespread. Despite the disparate goals of the World Bank, the Ministry of the Environment, and environmental nongovernmental organizations, they share an environmental-crisis imaginary that gives meaning and an immediacy to their missions. The visual imagery of an expanding desert is a powerful framing device that demands equally dramatic solutions, such as the establishment of green belts along the eighth parallel and imprisonment as punishment for lighting bush fires.

Finally, there are striking historical parallels between colonial-era writings on land degradation and control and contemporary environmental planning. Both blame farmers and herders for recklessly destroying the land and altering local climates, programmatic statements abound while good data are hard to come by, and proposed conservation measures invariably involve increased state intervention in the countryside with an emphasis on transforming land-rights systems, specifically the exclusion of local people from protected areas. These recurring themes, principal players, and silences regarding the goals and resource management strategies of farmers and herders suggest that we are operating within a regional discursive formation. What is different between the 1930s and the 1990s is the number of contestants involved in environmental management. In addition to the state, there are NGOs and development-aid organizations seeking to establish their authority and legitimacy as environmental advocates and stewards. The desertification narrative persists in part because it serves to mobilize support for these groups' varied agendas. This [selection] has privileged the voices and experiences of farmers and herders whose understanding of environmental change is more nuanced and sophisticated than the dominant narrative. From all indications, this local understanding of the nature and direction of environmental change is not reflected in the Côte d'Ivoire National Environmental Action Plan. As in the past and despite the rhetoric of decentralization and participatory planning (Little 1994; Ribot 1999; Schroeder 1999), this case study shows that the perceptions of ordinary men and women are marginalized in contemporary environmental planning in sub-Saharan Africa.

POSTSCRIPT

Is Sub-Saharan Africa Experiencing a Deforestation Crisis?

It is important to note that while Bassett and Zuéli question the accuracy of perceptions of deforestation in northern Côte d'Ivoire, they acknowledge that grassland degradation is occurring. The distinction they make between deforestation and degradation is critical because it is often assumed that deforestation equates with degradation, and that afforestation is a positive environmental trend. In the Ivorian savanna, the increase in woody species may actually be undermining the health of grasslands. As Bassett and Zuéli suggest, misperceived forest cover trends in the Ivorian savanna may, ironically, lead to policy prescriptions that exacerbate grassland degradation.

Another significant issue to keep in mind when evaluating these two selections is the question of scale. Cleaver and Schreiber are discussing deforestation at the scale of the African continent, whereas Bassett and Zuéli are examining vegetation trends at the scale of the savanna ecosystem in northern Ivory Coast. Is it possible that deforestation is occurring at the broad, continental scale while quite different, and even contrary, trends are operating in specific bioregions and localities? As Bassett and Zuéli suggest, it is critical not to assume that macro-scale trends are always reflected at the local level.

The findings of Bassett and Zuéli are not inconsistent with others who have come out in recent years with books and articles reassessing landscape change in Africa. Examples of these works include those by James Fairhead and Melissa Leach, *Misreading the African Landscape* (Cambridge University Press, 1996); Michael Mortimore, *Roots in the African Dust* (Cambridge University Press, 1998); and Jeremy Swift, "Desertification: Narratives, Winners and Losers," in Melissa Leach and Robin Mearns, eds., *The Lie of the Land: Challenging Received Wisdom on the African Environment* (Oxford & Heinemann, 1996). Newer work on African savannas suggests that, in addition to misrepresented biomass trends, the character of these ecosystems has been misunderstood. For example, a volume edited by Roy H. Behnke, Ian Scoones, and Carol Kervan, entitled *Range Ecology at Disequilibrium: New Models of Natural Variability and Pastoral Adaptation in African Savannas* (Westview Press, 1993), argues that African savannas may actually be more resilient than originally perceived. The book further asserts that transhuman pastoral livelihoods, rather than European modes of livestock rearing, are better suited to handle the inherent variability of African savannas.

Despite emerging new evidence to the contrary, Cleaver and Schreiber are in the majority when they sound the alarm about deforestation in Africa. Environmental organizations and the World Bank have been the most outspoken about their deforestation concerns. Examples of recent books and articles expressing apprehension about deforestation trends in Africa include Uma Lele et al., *The World Bank Forest Strategy: Striking the Right Balance* (World Bank, 2000); K. Boahene, "The Challenge of Deforestation in Tropical Africa" (1998); Claude R. Heimo et al., *Strategy for the Forest Sector in Sub-Saharan Africa* (World Bank, 1994); and Alan Durning, *Saving the Forest: What Will It Take* (Worldwatch Institute, 1993).

Wherever the truth may lie, the contrasting perspectives presented in this set of selections suggest that it is important not to make assumptions about forest cover change in Africa. Among other issues, students may more thoughtfully read assessments of environmental change in Africa by paying attention to the scale of analysis and suppositions regarding appearance of degradation on the landscape.

Ethnologue Country Index: Languages of Africa

The Ethnologue Country Index: Languages of Africa is a comprehensive source on the languages of Africa, including linguistic heritage and geographical distribution of the listed languages.

`http://www.ethnologue.com/country_index.asp?place=Africa`

Population Council: Africa

The Population Council's Africa page explains population and family planning, reproductive health, HIV/AIDS, and other issues by country and for the African continent as a whole.

`http://www.popcouncil.org/africa/africa.html`

Washington Post: AIDS in Africa

The *Washington Post* has an ongoing "Special Report" on the HIV/AIDS issue that contains news on the HIV/AIDS epidemic from African countries, debate and information on United States actions to fight the epidemic, and world HIV/AIDS information.

`http://www.washingtonpost.com/wp-dyn/world/`
`issues/aidsinafrica/index.html`

World Health Organization

The World Health Organization offers information on diseases and epidemiological facts for African countries as well as other areas of the world.

`http://www.who.int/en/`

PART 3

Social Issues

*P*erhaps more than any other set of contested African issues, those pertaining to the social sphere tend to provoke deep-seated emotional responses. This is also an area where differences in perspective between
Africanists and non-Africanists tend to be more apparent. Such a degree of
contestation is not surprising as these are, after all, deeply personal and
culturally specific issues dealing with sexuality, reproduction, gender
roles, intrahousehold dynamics, customs, and language. Despite the
highly private nature of some of the topics, this is also a realm that has
come under incredible public scrutiny given concern about the global
AIDS pandemic and the increasingly global nature of the human rights
and feminist movements.

- Should Female Genital Cutting Be Accepted as a Cultural Practice?

- Should International Drug Companies Provide HIV/AIDS Drugs
  to Africa Free of Charge?

- Is "Overpopulation" a Major Cause of Poverty in Africa?

- Is Sexual Promiscuity a Major Reason for the HIV/AIDS Epidemic
  in Africa?

- Is the Use of European Languages as the Medium of Instruction in
  African Educational Institutions More Negative Than Positive?

- Are Women in a Position to Challenge Male Power Structures in
  Africa?

ISSUE 11

Should Female Genital Cutting Be Accepted as a Cultural Practice?

YES: Richard A. Shweder, from "What About 'Female Genital Mutilation'? And Why Understanding Culture Matters in the First Place," *Daedalus* (Fall 2000)

NO: Liz Creel et al., from "Abandoning Female Genital Cutting: Prevalence, Attitudes, and Efforts to End the Practice," A Report of the Population Reference Bureau (August 2001)

ISSUE SUMMARY

YES: Richard A. Shweder, professor of human development at the University of Chicago, acknowledges the adverse reaction that most Westerners have to female genital cutting (FGC), but he also notes that women from certain African countries are repulsed by the idea of unmodified female genitals. He suggests, "We should be slow to judge the unfamiliar practice of female genital alterations, in part because the horrifying assertions by . . . activists concerning the consequences of the practice . . . are not well supported with credible scientific evidence."

NO: Liz Creel, senior policy analyst at the Population Reference Bureau, and her colleagues argue that female genital cutting (FGC), while it must be dealt with in a culturally sensitive manner, is a practice that is detrimental to the health of girls and women, as well as a violation of human rights in most instances. Creel et al. recommend that African governments pass anti-FGC laws, and that programs be expanded to educate communities about FGC and human rights.

When examining the issue of female genital cutting (FGC) in Africa (also known as female circumcision, female genital mutilation, or female genital alteration), it is difficult for many Westerners not to have an emotional reaction. In order to carefully evaluate this topic, the reader should try to keep as open a mind as possible.

This issue tugs at a deeper debate between those who believe that there are certain universal rights and wrongs, and that female genital cutting is simply wrong irrespective of the cultural context, and those who believe that a practice needs to be evaluated within its own cultural context. Advocates of the universality of certain norms often depict female genital cutting as a violation of basic human rights. They may further disparage defenders of female genital cutting as cultural relativists. Cultural relativism is often cast as problematic because it may be used as an excuse to say that anything goes. For example, some individuals have argued that slavery is appropriate in some cultural contexts.

Others would argue that, despite one's personal objections to the practice, female genital cutting must be viewed within the context of cultural pluralism. Cultural pluralists assert that there are separate and valid cultural and moral systems that may involve social mores that are not easily reconcilable with one another. In contrast to cultural relativists, cultural pluralists would maintain that everything is *not* always acceptable, and that there are certain universal norms (e.g., murder is wrong). The challenge for cultural pluralists is to determine if a practice violates a universal norm when it is viewed in its proper cultural context (rather than in the cultural context of another). The result of this deep philosophical divide is that we often see Western feminists pitted against multiculturalists (two groups that frequently function as intellectual allies in the North American context) over this controversial African issue.

In the following selections, Richard A. Shweder acknowledges that women from certain African countries have an adverse reaction to unmodified female genitals, just as many Westerners have an adverse reaction to female genital cutting. Shweder advises against judging unfamiliar practices too quickly. In order to reduce complications, he suggests that the practice be medicalized. In contrast, Liz Creel et al. recommend anti-FGC laws and education about FGC and human rights. They also believe that the use of medical professionals to perform the procedure should be discouraged.

Richard A. Shweder  **YES**

What About "Female Genital Mutilation"? And Why Understanding Culture Matters in the First Place

By Rites a Woman: Listening to the Multicultural Voices of Feminism

On November 18, 1999, Fuambai Ahmadu, a young African scholar who grew up in the United States, delivered a paper at the American Anthropological Association meeting in Chicago that should be deeply troubling to all liberal free-thinking people who value democratic pluralism and the toleration of "differences" and who care about the accuracy of cultural representations in our public-policy debates.

Ahmadu began her paper with these words:

> I also share with feminist scholars and activists campaigning against the practice [of female circumcision (FGM)] a concern for women's physical, psychological and sexual well-being, as well as for the implications of these traditional rituals for women's status and power in society. Coming from an ethnic group [the Kono of Eastern Sierra Leone] in which female (and male) initiation and "circumcision" are institutionalized and a central feature of culture and society and having myself undergone this traditional process of becoming a "woman," I find it increasingly challenging to reconcile my own experiences with prevailing global discourses on female "circumcision."

Coming-of-age ceremonies and gender-identity ceremonies involving genital alterations are embraced by, and deeply embedded in the lives of, many African women, not only in Africa but in Europe and the United States as well. Estimates of the number of contemporary African women who participate in these practices vary widely and wildly between eighty million and two hundred million. In general, these women keep their secrets secret. They have not been inclined to expose the most intimate parts of their bodies to public examination and they have not been in the habit of making their case on the op-ed pages of

American newspapers, in the halls of Congress, or at academic meetings. So it was an extraordinary event to witness Fuambai Ahmadu, an initiate and an anthropologist, stand up and state that the oft-repeated claims "regarding adverse effects [of female circumcision] on women's sexuality do not tally with the experiences of most Kono women," including her own. Ahmadu was twenty-two years old and sexually experienced when she returned to Sierra Leone to be circumcised, so at least in her own case she knows what she is talking about. Most Kono women uphold the practice of female (and male) circumcision and positively evaluate its consequences for their psychological, social, spiritual, and physical well-being. Ahmadu went on to suggest that Kono girls and women feel empowered by the initiation ceremony (see quotation, above) and she described some of the reasons why.

Ahmadu's ethnographic observations and personal testimony may seem astonishing. . . . In the social and intellectual circles in which most Americans travel it has been so "politically correct" to deplore female circumcision that the alarming claims and representations of anti-"FGM" advocacy groups (images of African parents routinely and for hundreds of years disfiguring, maiming, and murdering their female children and depriving them of their capacity for a sexual response) have not been carefully scrutinized with regard to reliable evidence. Nor have they been cross-examined by freethinking minds through a process of systematic rebuttal. Quite the contrary; the facts on the ground and the correct moral attitude for "good guys" have been taken to be so self-evident that merely posing the rhetorical question "what about FGM?" is presumed to function as an obvious counterargument to cultural pluralism and to define a clear limit to any feelings of tolerance for alternative ways of life. This is unfortunate, because in this case there is good reason to believe that the case is far less one-sided than supposed, that the "bad guys" are not really all that bad, that the values of pluralism should be upheld, and that the "good guys" may have rushed to judgment and gotten an awful lot rather wrong.

Six months before Fuambai Ahmadu publicly expressed her doubts about the prevailing global discourse on female circumcision, readers of the *Medical Anthropology Quarterly* observed an extraordinary event of a similar yet (methodologically) different sort. Carla Obermeyer, a medical anthropologist and epidemiologist at Harvard University, published a comprehensive review of the existing medical literature on female genital surgeries in Africa, in which she concluded that the claims of the anti-"FGM" movement are highly exaggerated and may not match reality.

Obermeyer began her essay by pointing out that "The exhaustive review of the literature on which this article is based was motivated by what appeared as a potential disparity between the mobilization of resources toward activism and the research base that ought to support such efforts." When she took a closer look at that "research base" . . . she discovered that in most publications in which statements were made about the devastating effects of female circumcision no evidence was presented at all. When she examined research reports actually containing original evidence she discovered numerous methodological flaws (e.g., small or unrepresentative samples, no control groups) and quality-control problems (e.g., vague descriptions of medical complications) in some of the most widely cited documents. She remarks: "Despite their deficiencies, some of the published reports have come to acquire an aura of dependability through repeated and uncritical citations."

In order to draw some realistic, even if tentative, conclusions about the health consequences of female circumcision in Africa, Obermeyer then introduced some standard epidemiological quality-control criteria for evaluating evidence. For example, a research study would be excluded if its sampling methods were not described or if its claims were based on a single case rather than a population sample. On the basis of the relatively small number of available studies that actually passed minimum scientific standards (for example, eight studies on the topic of medical complications), Obermeyer reported that the widely publicized medical complications of African genital operations are the exception, not the rule; that female genital alterations are not incompatible with sexual enjoyment; and that the claim that untold numbers of girls and women have been killed as a result of this "traditional practice" is not well supported by the evidence.

Many anthropologists and other researchers who work on this topic in various field settings in Africa have been aware of discrepancies between the global discourse on female circumcision (with its images of maiming, murder, sexual dysfunction, mutilation, coercion, and oppression) and their own ethnographic experiences with indigenous discourses and physical realities.

Perhaps the first anthropological protest against the global discourse came in 1938 from Jomo Kenyatta, who, prior to becoming the first president of postcolonial Kenya, wrote a Ph.D. thesis in anthropology at the London School of Economics. His thesis was published as a book entitled *Facing Mount Kenya: The Tribal Life of the Gikuyu,* in which he described both the customary premarital sexual practices of the Gikuyu (lots of fondling and rather liberal attitudes toward adolescent petting and sexual arousal) and the practice of female (and male) circumcision.

Kenyatta's words, published in 1938, have an uncanny contemporary ring and relevance. First he informs us that "In 1931 a conference on African children was held in Geneva under the auspices of the Save the Children Fund. In this conference several European delegates urged that the time was ripe when this 'barbarous custom' should be abolished, and that, like all other 'heathen' customs, it should be abolished at once by law."

He goes on to argue that among the Gikuyu a genital alteration, "like Jewish circumcision," is a bodily sign that is regarded "as the *conditio sine qua non* of the whole teaching of tribal law, religion and morality," that no proper Gikuyu man or woman would have sex with or marry someone who was not circumcised, that the practice is an essential step into responsible adulthood for many African girls and boys, and that "there is a strong community of educated Gikuyu opinion in defense of this custom."

Nearly sixty years later echoes of Jomo Kenyatta's message can be found in the writings of Corinne Kratz, who has written a detailed account of female initiation in another ethnic group in Kenya, the Okiek. The Okiek, she tells us, do not talk about circumcision in terms of the dampening of sexual pleasure or desire, but rather speak of it "in terms of cleanliness, beauty and adulthood."

According to Kratz, Okiek women and men view "genital modification and the bravery and self-control displayed during the operation as constitutive experiences of Okiek personhood."

Many other examples could be cited of discrepancies between the global discourse and the experience of many field researchers in Africa. With regard to the issue of sexual enjoyment, for example, Robert Edgerton remarks that "Kikuyu men and women, like those of several other East African societies that practice female circumcision, assured me in 1961–62 that circumcised women continue to be orgasmic," and similar remarks appear in other field reports.

. . . [E]thnographic reports are noteworthy because they suggest that instead of assuming that our own perceptions of beauty and disfigurement are universal and must be transcendental we might want to consider the possibility that there is a real and astonishing cultural divide around the world in moral, emotional, and aesthetic reactions to female genital surgeries. There is, of course, no doubt that our own personal feelings of disgust and anxiety about this topic are powerful and can be easily aroused and rhetorically manipulated either with pictures (for example, of Third World surgical implements) or with words (for example, labeling the activity "torture" or "mutilation"). But if we want to understand the true character of this cultural divide in sensibilities it may make good sense to bracket our own initial (and automatic) emotional/visceral reactions and to save any powerful conclusive feelings for the end of the argument, rather than have them color or short-circuit all objective analysis. Perhaps, instead of simply deploring the "savages," we might develop a better understanding of the subject by constructing a synoptic account of the inside point of view, from the perspective of those many African women for whom such practices seem both normal and desirable.

Moral Pluralism and the "Mutual Yuck Response"

People recoil at each other's practices and say "yuck" at each other all over the world. When it comes to female genital alterations, however, the "mutual yuck" response is particularly intense and may even approach a sense of mutual outrage or horror. From a purely descriptive point of view, that particular type of modification of the "natural" body is routine and normal in many ethnic groups. For example, national prevalence rates of 80–98 percent have been reported for Egypt, Ethiopia, the Gambia, Mali, Sierra Leone, Somalia, and the Sudan. In African nations where the overall prevalence rate is lower—for example, 50 percent in Kenya, 43 percent in Côte d'Ivoire, 30 percent in Ghana—this is typically because some ethnic groups in those countries have a tradition of female circumcision while other ethnic groups do not. For example, within Ghana the ethnic groups in the north and the east circumcise girls (and boys), while the ethnic groups in the south have no tradition of female circumcision. In general, for both boys and girls the best predictor of circumcision (versus the absence of it) is ethnicity or cultural group affiliation. For example, circumcision is customary for the Kono of Sierra Leone, but for the Wolof of Senegal it is not. For women within these groups, one key factor—their cultural affiliation—trumps other predictors of behavior, such as educational level or socioeconomic

status. Among the Kono, even women with a secondary-school or college education are circumcised, while Senegalese Wolof women—including the illiterate
and unschooled—are not.

There are other notable facts about this cultural practice. For one thing,
most African women do not think about circumcision in human-rights terms.
Women who endorse female circumcision typically argue that it is an important part of their cultural heritage or their religion, while women who do not
endorse the practice typically argue that it is not permitted by their cultural heritage or their religion.

Second, among members of ethnic groups for whom female circumcision
is part of their cultural heritage approval ratings for the custom are generally
rather high. According to the Sudan Demographic and Health Survey of
1989–1990, which was conducted in northern and central Sudan, out of 3,805
women interviewed 89 percent were circumcised. Of the women who were circumcised, 96 percent said they had circumcised or would circumcise their
daughters. When asked whether they favored continuation of the practice, 90
percent of circumcised women said they favored its continuation.

In Sierra Leone the picture is much the same, and the vast majority of
women are sympathetic to the practice. Even Olayinka Koso-Thomas, an anti-
"FGM" activist, makes note of the high degree of support for genital operations,
although she expresses herself with a rather patronizing voice and in imperial
tones. "Most African women," Koso-Thomas observes, "still have not developed
the sensitivity to feel deprived or to see in many cultural practices a violation of
their human rights. The consequence of this is that, in the mid-80s, when most
women in Africa have voting rights and can influence political decisions
against practices harmful to their health, they continue to uphold the dictates
and mores of the communities in which they live; they seem in fact to regard
traditional beliefs as inviolate." When it comes to maintaining their coming-of-
age and gender-identity ceremonies, Koso-Thomas does not like the way many
African women vote. She thinks she is enlightened about human rights and
health and that they remain in the dark. But she does recognize that, despite her
censure, most women in Sierra Leone endorse the practice of circumcision.

Third, although ethnic group affiliation is the best predictor of who cir-
cumcises and who does not, the timing and form of the operation are not consistent across groups. Thus, there is enormous variability in the age at which the
surgery is normally performed (any time from birth to the late teenage years).
There is also enormous variability in the traditional style and degree of surgery
(from a cut in the prepuce covering the clitoris to the complete "smoothing
out" of the genital area by removing all visible parts of the clitoris and most if
not all of the labia). In some ethnic groups (for example, in Somalia and the Sudan) the "smoothing out" operation is concluded by stitching closed the vaginal opening, with the aim of enhancing fertility and protecting the womb. The
latter procedure, often referred to as "infibulation" or Pharaonic circumcision,
is not typical in most circumcising ethnic groups, although it has received a
good deal of attention in the anti-"FGM" literature. It is estimated that it occurs
in about 15 percent of all African cases.

In places where the practice of female circumcision is popular, including Somalia and the Sudan, it is widely believed by women that these genital alterations improve their bodies and make them more beautiful, more feminine, more civilized, more honorable. . . .

So What About FGM?

So what about "FGM"? I shall treat this as a real question deserving a considered response rather than as a rhetorical query intended to terminate all debate. For starters, the practice of genital alteration is a rather poor example of gender inequality or of society picking on women. Surveying the world, one finds very few cultures, if any, in which genital surgeries are performed on girls but not boys, although there are many cultures in which they are performed only on boys or on both sexes. The male genital alterations often take place in adolescence and they can involve major modifications (including subincision, in which the penis is split along the line of the urethra). Considering the prevalence, timing, and intensity of the relevant initiation rites, and viewing genital alteration on a worldwide scale, one is hard pressed to argue that it is an obvious instance of a gender inequity disfavoring girls. Quite the contrary; social recognition of the ritual transformation of both boys and girls into a more mature status as empowered men and women is not infrequently a major point of the ceremony. In other words, female circumcision, when and where it occurs in Africa, is much more a case of society treating boys and girls equally before the common law and inducting them into responsible adulthood in parallel ways.

The practice is also a rather poor example of patriarchal domination. Many patriarchal cultures in Europe and Asia do not engage in genital alterations at all or (as in the case of Jews, many non-African Muslims, and many African ethnic groups) exclude girls from participation in this valued practice and do it only to boys. Moreover, the African ethnic groups that circumcise females (and males) are very different from each other in kinship, religion, economy, family life, ceremonial practice, and so forth. Some are Islamic, some are not. Some are patriarchal, some (such as the Kono, a matrilineal society) are not. Some have formal initiations into well-established women's organizations, some do not. Some care a lot about female purity, sexual restraint outside of marriage, and the social regulation of desire, but others (such as the Gikuyu) are more relaxed about premarital sexual play and are not puritanical. And when it comes to female initiation and genital alterations the practice is almost always controlled, performed, and most strongly upheld by women, although male kin often do provide material and moral support. Typically, however, men have rather little to do with these female operations, may not know very much about them, and may feel it is not really their business to interfere or to try to tell their wives, mothers, aunts, and grandmothers what to do. It is the women of the society who are the cultural experts in this intimate feminine domain, and they are not particularly inclined to give up their powers or share their secrets.

In those cases of female genital alteration with which I am most familiar (I have lived and taught in Kenya, where the practice is routine for some ethnic groups), the adolescent girls who undergo the ritual initiation look forward to it. It is an ordeal and it can be painful (especially if done "naturally" without anesthesia), but it is viewed as a test of courage. It is an event organized and controlled by women, who have their own view of the aesthetics of the body—a different view from ours about what is civilized, dignified, and beautiful. The girl's parents are not trying to be cruel to their daughter—African parents love their children too. No one is raped or tortured. There is a celebration surrounding the event.

What about the devastating negative effects on health and sexuality that are vividly portrayed in the anti-"FGM" literature? When it comes to hard-nosed scientific investigations of the consequences of female genital surgeries on sexuality and health, there are relatively few methodologically sound studies. As Obermeyer discovered in her medical review, most of the published literature is "data-free" or else relies on sensational testimonials, secondhand reports, or inadequate samples. Judged against basic epidemiological research standards, much of the published empirical evidence, including some of the most widely cited publications in the anti-"FGM" advocacy literature (including the influential *Hosken Report*), are fatally flawed. Nevertheless, there is some science worth considering in thinking about female circumcision, which leads Obermeyer to conclude that the global discourse about the health and sexual consequences of the practice is not sufficiently supplied with credible evidence.

The anti-"FGM" advocacy literature typically features long lists of short-term and long-term medical complications of circumcision, including blood loss, shock, acute infection, menstrual problems, childbearing difficulties, incontinence, sterility, and death. These lists read like the warning pamphlets that accompany many prescription drugs, which enumerate every claimed negative side effect of the medicine that has ever been reported (no matter how infrequently). They are very scary to read, and they are very misleading. Scary-looking, stomach-churning, anxiety-provoking lists of possible medical complications aside, Obermeyer's comprehensive review of the literature on the actual frequency and risk of medical complications following genital surgery in Africa suggests that medical complications are the exception, not the rule; that African children do not die because they have been circumcised (they die from malnutrition, war, and disease, not because of coming-of-age ceremonies); and that the experience of sexual pleasure is compatible with the genital aesthetics and related practices of circumcising groups.

Her findings are basically consistent with Robert Edgerton's comments about female circumcision among the Gikuyu in the Kenya of the 1920s and 1930s, when Western missionaries first launched their own version of "FGM eradication programs." As Edgerton remarks, the operation was performed without anesthesia and hence was very painful, "yet most girls bore it bravely and few suffered serious infection or injury as a result. Circumcised women did not lose their ability to enjoy sexual relations, nor was their child-bearing capacity diminished. Nevertheless the practice offended Christian sensibilities."

In other words, the alarmist claims that are a standard feature of the anti-"FGM" advocacy literature that African traditions of circumcision have "maimed or killed untold numbers of women and girls" and deprived them of their sexuality may not be true. Given the most reliable, even if limited, scientific evidence at hand, those claims should be viewed with skepticism and not accepted as fact, no matter how many times they are uncritically recapitulated on the editorial pages of the *New York Times* or poignantly invoked in a journalistic essay on PBS. . . .

This is not to say that we should not worry about the documented 4–16 percent urinary infection rate associated with these surgeries, or the 7–13 percent of cases in which there is excessive bleeding, or the 1 percent rate of septicemia. The reaction of many people to unsafe abortions, however, is not to get rid of abortions. Perhaps some antiabortion groups might be tempted by the argument that because some abortions are unsafe, there should be no abortions at all. However, a far more reasonable reaction to unsafe abortions is to make them safe. Why not the same reaction in the case of female genital alterations? Infections and other medical complications that arise from unsanitary surgical procedures or malpractice can be corrected without depriving "others" of a rite of passage and system of meaning central to their cultural and personal identities and their overall sense of well-being. What I do want to suggest, however, is that the current sense of shock, horror, and righteous "Western" indignation directed against the mothers of Mali, Somalia, Egypt, Sierra Leone, Ethiopia, the Gambia, and the Sudan is misguided, and rather disturbingly misinformed.

Liz Creel et al.

 **NO**

Abandoning Female Genital Cutting: Prevalence, Attitudes, and Efforts to End the Practice

Introduction

More than 130 million girls and women worldwide have undergone female genital cutting [FGC]—also known as female circumcision and female genital mutilation—and nearly 2 million more girls are at risk each year. The practice often serves as a rite of passage to womanhood or defines a girl or woman within the social norms of her ethnic group or tribe. The tradition may have originated 2,000 years ago in southern Egypt or northern Sudan, but in many parts of West Africa, the practice began in the 19th or 20th century. No definitive evidence exists to document exactly when or why FGC began. FGC is an ancient practice but has also been recently adopted, for example, among adolescents in Chad.

FGC is generally performed on girls between ages 4 and 12, although it is practiced in some cultures as early as a few days after birth or as late as just prior to marriage, during pregnancy, or after the first birth. Girls may be circumcised alone or with a group of peers from their community or village. Typically, traditional elders (male barbers and female circumcisers) carry out the procedure, sometimes for pay. In some cases, it is not remuneration but the prestige and power of the position that compels practitioners to continue. The practitioner may or may not have health training, use anesthesia, or sterilize the circumcision instruments. Instruments used for the procedure include razor blades, glass, kitchen knives, sharp rocks, scissors, and scalpels. A discouraging trend is the use of medical professionals (physicians, nurses, and midwives) in some countries (e.g., Egypt, Kenya, Mali, and Sudan) to perform the procedure due to growing recognition of the health risks associated with FGC and heightened concern regarding the possible role of FGC in HIV transmission. WHO [World Health Organization] has strongly advised that FGC, in any of its forms, should not be practiced by any health professional in any setting—including hospitals and other health centers.

FGC has health risks, most notably for women who have undergone more extreme forms of the procedure (see Box 1). Immediate potential side effects include severe pain, hemorrhage, injury to the adjacent tissue and organs, shock, infection, urinary retention and tetanus—some of these side effects can lead to death. Long-term effects may include cysts and abscesses, urinary incontinence, psychological and sexual problems, and difficulty with childbirth. Obstructed labor may occur if a woman has been infibulated. This involves cutting off the external genitalia and sewing together the two sides of the vulva, leaving a small hole for urination and menstruation. If the woman's genitalia is not cut open (defibulated) during delivery, labor may be obstructed and cause life-threatening complications for both the mother and the child, including perineal lacerations, bleeding and infection, possible brain damage to infants, and fistula formation.

TYPES OF FEMALE GENITAL CUTTING

Female genital cutting (FGC) refers to a variety of operations involving partial or total removal of female external genitalia. The female external genital organ consists of the vulva, which is comprised of the labia majora, labia minora, and the clitoris covered by its hood in front of the urinary and vaginal openings. In 1995, the World Health Organization classified FGC operations into four broad categories described below:

Type 1 or **Clitoridectomy:** Excision (removal) of the clitoral hood with or without removal of the clitoris.

Type 2 or **Excision:** Removal of the clitoris together with part or all of the labia minora.

Type 3 or **Infibulation:** Removal of part or all of the external genitalia (clitoris, labia minora, and labia majora) and stitching and/or narrowing of the vaginal opening, leaving a small hole for urine and menstrual flow.

Type 4 or **Unclassified:** All other operations on the female genitalia including

- pricking, piercing, stretching, or incising of the clitoris and/or labia;
- cauterization by burning the clitoris and surrounding tissues;
- incisions to the vaginal wall; scraping or cutting of the vagina and surrounding tissues; and introduction of corrosive substances or herbs into the vagina.

Note:
1. World Health Organization, *Female Genital Mutilation: Report of a Technical Working Group* (Geneva: WHO, 1996): 9.

All of these possible side effects may damage a girl's lifetime health, although the type and severity of consequences depend on the type of procedure performed (see Box 1). Infibulation or Type 3 is the most invasive and damaging type of FGC. Operations research studies conducted in Burkina Faso and Mali have shown that women who were infibulated were nearly two and a half times more likely to have a gynecological complication than those with a Type 2 or Type 1 cut. Risks during childbirth also increased according to the severity of the procedure. For instance, in Burkina Faso, women with Types 2 or 3 cutting had a higher likelihood of experiencing hemorrhaging or perineal tearing during delivery.

While it is difficult to determine both the number of women who have undergone FGC and how many have undergone each type of circumcision, WHO has estimated that clitoridectomy, which accounts for up to 80 percent of all cases, is the most common procedure. Fifteen percent of all circumcised women have been infibulated—the most severe form of circumcision.

FGC is practiced in at least 28 countries in sub-Saharan and north-eastern Africa but not in southern Africa or in the Arabic-speaking nations of North Africa, with the exception of Egypt. It is practiced at all educational levels and in all social classes and occurs among many religious groups (Muslims, Christians, animists, and one Jewish sect), although no religion mandates it. For countries presented here with DHS [Demographic and Health Surveys] data, prevalence varies from 18 percent in Tanzania to nearly 90 percent or more in Egypt, Eritrea, Mali, and Sudan. According to WHO estimates, 18 African countries have prevalence rates of 50 percent or more. Through migration, the practice has also spread to Europe, North and South America, Australia, and New Zealand. Although doctors, colonial administrators, and social scientists have documented the adverse effects of FGC for many years, governments and funding donors have become increasingly interested in the practice because of the public health and human rights implications.

Global efforts to end FGC have used legislation to provide legitimacy for project activities, to protect women, and to discourage circumcisers and families who fear prosecution. In the 1960s, WHO was the first United Nations (UN) specialized agency to take a position against female genital cutting. It began efforts to promote the abandonment of harmful traditional practices like FGC in the 1970s, focusing largely on gathering information about FGC's epidemiology and health consequences and speaking out about FGC at international, regional, and national levels. In 1982, WHO issued a formal statement to the UN Commission on Human Rights and recommended several actions:

- Governments should adopt clear national policies to end FGC, and educate and inform the public about its harmful aspects.
- Anti-FGC programs must consider the practice's association with difficult social and economic conditions and respond to women's needs and problems.
- Women's organizations at the local level should be encouraged to take action.

In 1988, WHO began to integrate FGC into the development context of primary health care. Over the intervening years, WHO shifted its position on

FGC from addressing the practice only in terms of health to acknowledging it as both a health and human rights issue. In the 1990s, FGC gained recognition as a health and human rights issue among African governments, the international community, women's organizations, and professional associations. The 1993 Vienna Human Rights Convention, the 1994 International Conference on Population and Development, and the 1995 Fourth World Conference on Women called for an end to the practice. When performed on girls and nonconsenting women, FGC violates a number of recognized human rights protected in international conventions and conferences, such as the Convention on Children's Rights, the Convention on the Elimination of All Forms of Discrimination against Women, and recommendations of the Committee on the Elimination of Discrimination against Women (CEDAW). These conventions explicitly recognize harmful traditional practices such as FGC as violations of human rights, including the right to nondiscrimination, the right to life and physical integrity, the right to health, and the right of the child to special protections.

Respect for international human rights law does not require that every culture use an identical approach to abandoning FGC. One Muslim scholar suggested that respecting different cultures means accepting "the right of all people to choose among alternatives equally respectful of human rights," and that human rights must include life, liberty, and dignity for every person or group of people.

In Africa, 10 countries—Burkina Faso, the Central African Republic (CAR), Côte d'Ivoire, Djibouti, Ghana, Guinea, Niger, Senegal, Tanzania, and Togo— have enacted laws that criminalize the practice of FGC. The penalties range from a minimum of six months to a maximum of life in prison. In Nigeria, three of 36 states (as of 2000) had also enacted legislation regarding FGC. In Burkina Faso, Ghana, and Senegal, these laws are enforced and circumcisers are imprisoned. In these countries, various groups educate the public about the law, use a variety of strategies (e.g., public service announcements and watchdog committees) to denounce FGC, and stop circumcisers by going to the police. Several countries also impose fines. In Egypt, the Ministry of Health issued a decree declaring FGC unlawful and punishable under the Penal Code. There have been several prosecutions under this law, which include jail time and fines. In addition, seven more developed countries that receive immigrants from countries where FGC is practiced—Australia, Canada, New Zealand, Norway, Sweden, the United Kingdom, and the United States—have passed laws outlawing the practice. Enforcement of these laws, however, is extremely uneven. France, on the other hand, consistently enforces general penal code provisions against providers of FGC but has not adopted specific legislation regarding FGC.

Why is FGC Performed?

The traditions surrounding FGC vary from one society to another. In some communities, FGC is a rite of passage to womanhood and is performed at puberty or at the time of marriage. In other communities, it may be performed on girls at a younger age for other reasons such as a celebration of womanhood, preservation of custom or tradition, or as a symbol of ethnic identity. The ritual cutting is often an integral part of ceremonies, which may occur over several weeks, in which girls are feted and showered with presents and their families are honored. It is described as a joyous time with many visitors, feasting, dancing, good food, and an atmosphere of freedom for the girls. The ritual serves as an act of socialization into cultural values and an important connection to family, community, and earlier generations. The ceremonies often involve three interrelated aspects:

- **Educational** A girl learns her place in society and her role as woman, wife, and mother.
- **Physical** A girl must undergo physical pain to prove she is capable of assuming her new role courageously without showing suffering or pain; the pain is experienced both through the actual cutting and through punishment received by girls in complete submission throughout weeks of initiation.
- **Vow of silence** Each girl must make a solemn pledge not to speak about her experience during the ceremony.

The reasons for performing FGC differ, but many practicing communities believe that it preserves the girl's virginity and protects marital fidelity because it diminishes her sexual desire. Practicing communities cite reasons such as giving pleasure to the husband, religious mandate, cleanliness, identity, maintaining good health, and achieving good social standing. At the heart of all this is rendering a woman marriageable, which is important in societies where women get their support from male family members, especially husbands. A circumcised woman will also attract a favorable bride price, thus benefiting her family. The practice is perceived as an act of love for daughters. Parents want to provide a stable life for their daughters and ensure their full participation in the community. Many girls and women receive little formal education and are valued primarily for their role as sources of labor and future producers of children. For many girls and women, being uncircumcised means that they have no access to status or a voice in their community. Because of strong adherence to these traditions, many women who say they disapprove of FGC still submit themselves and their daughters to the practice.

Understanding Why the Practice Continues

FGC is a cultural practice. Efforts to end it require understanding and changing the beliefs and perceptions that have sustained the practice over the centuries. Irrespective of how, where, and when the practice began, those who practice it share similar beliefs—a "mental map"—that present compelling reasons why the clitoris and other external genitalia should be removed. The details of these mental maps vary across countries, and there are distinctive features to each culture that providers, community workers, and others involved with anti-FGC campaigns need to take into consideration.

Figure 1 provides a conceptual framework for understanding the role of FGC in society. This mental map shows the psychological and social reasons, and the religious, societal, and personal (hygienic and aesthetic) beliefs that contribute to the practice. These beliefs involve continuing long-standing custom and tradition; maintaining cleanliness, chastity, and virginity; upholding family honor (and sometimes perceived religious dictates); and controlling women's sexuality in order to protect the entire community. In the countries surveyed by Demographic and Health Surveys, good custom/tradition is the most frequently cited reason for approving of FGC. Bad custom/tradition is also mentioned as one of the primary reasons for discontinuing the practice.

To encourage abandonment of FGC, health care providers, community workers, and others involved with anti-FGC programming need to understand the mental map in the communities where they are working. Communities have a range of enforcement mechanisms to ensure that the majority of women comply with FGC. These include fear of punishment from God, men's unwillingness to marry uncircumcised women, insistence that women from other tribes get circumcised when they marry into the group, as well as local poems

Figure 1

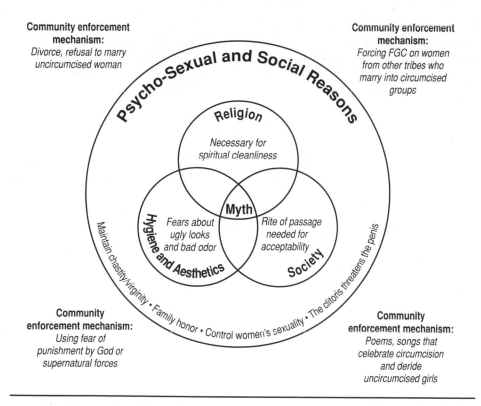

Why the Practice of FGC Continues: A Mental Map

Source: Asha Mohamud, Nancy Ali, Nancy Yinger, World Health Organization and Program for Appropriate Technology in Health (WHO/PATH), *FGM Programs to Date: What Works and What Doesn't* (Geneva: WHO, 1999): 7.

and songs that reinforce the importance of the ritual. In some cases, women who are not circumcised may face immediate divorce or forced excision. Girls who do undergo FGC sometime receive rewards, including public recognition and celebrations, gifts, potential for marriage, respect and the opportunity to engage in adult social functions. In other instances, girls and women are cut without an accompanying ceremony; thus, importance is attached to being circumcised rather than to having gone through a ritual.

The desire to conform to peer norms leads many girls to undergo circumcision voluntarily, yet frequently girls (and sometimes infants) have no choice in whether they are circumcised. A girl's family—typically her mother, father, or elder female relatives—often decides whether she will undergo FGC. Due to the influence of tradition, many girls accept, and even perpetuate, the practice. In Eritrea, men are more likely than women to favor ending the practice.

FGC could continue indefinitely unless effective interventions convince millions of men and women to abandon the practice. Many African activists, development and health workers, and people following traditional ways of life recognize the need for change but have not yet achieved such an extensive social transformation. . . .

Recommended Actions to End FGC

Data on attitudes, practices, and prevalence can provide important background information on opportunities for intervention. In addition, lessons from program experiences provide an important context for formulating abandonment campaigns. PATH [Program for Appropriate Technology in Health] and WHO developed the following recommendations for policymakers and program managers.

Recommendations for Policymakers

Policymakers are those who are in a position to influence policies and provide funding related to FGC.

> *1. Governments and donors need to support the groundswell of agencies involved in FGC abandonment with financial and technical assistance.*

An increasing number of agencies, especially NGOs [nongovernmental organizations], are involved in efforts to end FGC. However, programs tend to be small, rely heavily on volunteers and funds from foreign donors, and reach a small proportion of the people in need. Additional support is needed to make the growing network of agencies more effective and expand their reach.

In Egypt, 15 NGOs, including the Egyptian Fertility Care Society, the Task Force Against FGM, the Cairo Institute for Human Rights Studies, and CEOSS [Coptic Evangelical Organization for Social Services] have been instrumental in advocating the abandonment of FGC. In order to become more effective, these groups need to collaborate with one another, enhance training in advocacy and communications skill building, and evaluate the impact of their programs.

2. Governments must enact and use anti-FGC laws to protect girls and educate communities about FGC and human rights.

Passing anti-FGC legislation is one of the most controversial aspects of the FGC abandonment movement. It is extremely difficult to enforce anti-FGC laws. There is fear that heavy-handed enforcement may drive the practice underground. In fact, this has occurred in some countries. Still, most program planners and activists agree that anti-FGC legislation can demarcate right from wrong, provide official legal support for project activities, offer legal protection for women, and ultimately, discourage circumcisers and families for fear of prosecution. The key is to use the law in a positive fashion—as a vehicle for public education about and community action against FGC.

3. National governments need to be active both in setting policy and in expanding existing programs.

A key role for governments is to "scale up" successful community-based FGC abandonment activities. To date, most governments have provided support in the form of in-kind contributions to NGOs working in the communities. The excellent program models that NGOs have carried out on a pilot basis need to be expanded, either by direct government interventions or by increased support for the NGO networks.

4. To sustain programs, governments need to institutionalize FGC abandonment efforts in all relevant ministries.

Currently, none of the anti-FGC efforts underway are sustainable over the long run, in part because they have failed to change the social norms underlying FGC. The integration of FGC issues into government programs, however, has met with some success. In Burkina Faso, the National Committee to Fight Against the Practice of FGC has effectively promoted FGC abandonment through participation in national events such as the international day of population, by integrating FGC abandonment into all of the relevant ministries, and through training and awareness raising activities.

Efforts in other countries, such as Mali, have encountered more difficulties. While various ministries have expressed their support for anti-FGC activities, they have not been integrated into the relevant ministries, particularly in programs carried out by Mali's Ministry of Health. The primary nursing and medical schools in Mali do not include FGC as an adverse health practice in their curricula. Presently, PRIME II, a partnership of U.S.-based organizations, is working with the Ministry of Health to develop a national curriculum integrating FGC.

Governments have a responsibility to make political decisions and place FGC abandonment in the mainstream of reproductive health and development programs. Limited success has been achieved through increased fundraising,

greater integration of anti-FGC activities into government and civil society programs, and through solicitation of community support.

*5. Health providers at all levels need to receive training and financial support
to treat FGC complications and to prevent FGC.*

A key foundation for FGC abandonment is to make health providers aware of the
extent and severity of FGC-related complications and to give them the skills and
resources to treat these problems. Health providers often encounter women and
girls suffering from FGC-related complications, yet they are often not prepared to
treat and counsel women, or to prevent recurrence of the circumcision practice.
Because there is limited training on clinical treatments for circumcised women
or counseling women suffering from psychological or sexual problems, women
lack access to high quality, relevant services in most countries.

A 1998 operations research study in Mali sought to assess the use of health
personnel to address FGC. The study, which was conducted by an NGO, Association de Soutien au Developpement des Activités de Population (ASDAPO), and
the Ministry of Health, evaluated the effectiveness of a three-day training course
on identifying and treating medical complications related to FGC and counseling patients about the problem. The study focused on 14 urban and rural health
centers in Bamako and the Ségou region and included 107 health providers from
experimental and control sites. Results indicated that the course was highly effective in changing provider attitudes toward FGC. After receiving training,
three in four trained providers knew at least three immediate and long-term
complications of FGC. The study also indicated that providers felt they had limited competence in caring for FGC complications (even after receiving training)
and needed further training in how to discuss FGC with their clients.

*6. Governments, donors, and NGOs working on FGC abandonment should
continue to coordinate their efforts.*

Findings from field assessments reveal an impressive array of cooperative efforts and exchanges of information and resources among NGOs, government
institutions, and donors. Agencies typically invite each other to meetings and
training activities and coordinate at program sites to avoid duplication of efforts. Although occasional conflicts arise over funding and strategies, they
should not discourage agencies from continuing to coordinate and build on
each other's strengths.

*7. International agencies should assist staff of NGOs and government to de-
velop their advocacy skills.*

Advocacy is essential to ensuring that FGC abandonment programs are established and maintained until the practice of FGC ceases. Agencies involved in
abandonment efforts increasingly use advocacy for public education and to influence legislation, but they need to improve their skills.

POSTSCRIPT

Should Female Genital Cutting Be Accepted as a Cultural Practice?

After reading the two selections, there are at least five different points that one may want to consider before taking an informed stance on this issue. First, female genital cutting covers a very broad range of practices in the African context. In some instances, apparently contradictory statements from either author may have as much to do with the specific practices that they have in mind as they do with the fact that the authors have differing views on the general topic.

Second, the whole notion of choice, the prospect of giving girls and women the option of undergoing or not undergoing the procedure, implies that they are aware that there *is* a choice. Many girls and women may assume that FGC is simply what is done or what is normal.

Third, if a girl or young woman opts not to undergo female genital cutting, there could be serious social consequences in some settings. Within the context of the educational programs advocated by Creel et al., is there an obligation to make sure individuals are made aware of the medical dangers of the procedure as well as the social consequences of not being initiated?

Fourth, some African countries have a mix of ethnic groups that may or may not practice female genital cutting (e.g., Kenya). This level of homogeneity or heterogeneity could have implications for people's exposure to different practices (particularly in urban areas where ethnic groups tend to mix), as well as the chance of success or failure of education programs in this domain.

Finally, the different authors have very different views on medicalizing FGC in Africa. Shweder contends that medicalizing FGC will minimize health problems, whereas Creel et al. assert that it may help perpetuate the practice. How likely is making the procedure safer going to contribute to its spread?

For more information on this debate, see a volume edited by Bettina Shell-Duncan and Ylva Hernlund entitled *Female "Circumcision" in Africa: Culture, Controversy, and Change* (Lynne Rienner Publishers, 2000) that includes voices on both sides of the issue. The book includes a chapter by Fuambai Ahmadu ("Rites and Wrongs: Excision and Power Among Kono Women of Sierra Leone"), the African anthropologist mentioned at the start of Shweder's article. For a publication questioning the medical evidence against female genital surgeries, see an article by Carla M. Obermeyer entitled "Female Genital Surgeries: The Known, the Unknown, and the Unknowable," *Medical Anthropology Quarterly* (no. 13, 1999). Finally, for more of an anti-FGM perspective, see Susan Rich and Stephanie Joyce, *Eradicating Female Genital Mutilation: Lessons for Donors* (Wallace Global Fund for a Sustainable Future, 1990).

ISSUE 12

Should International Drug Companies Provide HIV/AIDS Drugs to Africa Free of Charge?

YES: Akin Jimoh, from "'Raise the Alarm Loudly': Africa Confronts the AIDS Pandemic," *Dollars and Sense* (May/June 2001)

NO: Siddhartha Mukherjee, from "Take Your Medicine," *The New Republic* (July 24, 2000)

ISSUE SUMMARY

YES: Akin Jimoh, program director of Development Communications, a non-governmental organization (NGO) based in Lagos, Nigeria, argues that the AIDS epidemic in Africa is linked to a number of factors, including the high cost of drugs. He describes how some of the big drug companies, in the face of international protests, begrudgingly agreed to lower the price of anti-HIV medications in Africa. "The companies, however, remain steadfast about keeping their patent rights, which would leave ultimate control over prices and availability in their hands."

NO: Siddhartha Mukherjee, a resident in internal medicine at Massachusetts General Hospital and a clinical fellow in medicine at Harvard Medical School, asserts that the availability of cheap anti-HIV drugs in Africa, without adequate health care networks to monitor their distribution and use, is dangerous. If such medications are not taken consistently and over the prescribed length of time, new strains of HIV are likely to develop more quickly that are resistant to these drugs. He states that investment in health care infrastructure must accompany any distribution of cheap anti-HIV medications.

Patents provide the inventor of a new product with exclusive rights to its sale for a specified time period before competitors may produce the same product under a generic label. Patents are seen as a critical element of the free market system in the United States as they provide an incentive to invent new products. As patent protection is not equally viewed across the globe, the United States has consistently sought to expand and enforce its view of this concept through international trade agreements, most notably via the World Trade Organization (WTO). The Agreement on Trade-Related Aspects of Intellectual Property (TRIPS), ratified in 1995, required countries to implement strict U.S.-style patent rules within a five to ten year period for a variety of products, including pharmaceuticals.

An important provision in this agreement was an exception for countries in a state of national emergency to be able to resort to compulsory licensing and parallel imports. In the case of pharmaceuticals, compulsory licensing allows a country to require a patent holder to grant authority to another company to produce the drug in exchange for a reasonable royalty. Parallel importation involves importing a product from another country, without permission of the original seller, for resale in another. This practice is officially forbidden under normal conditions because drug companies charge different prices (depending on market conditions) for the same product in different countries.

South Africa angered the United States in 1997 when it adopted a Medicines Act that allowed for both compulsory licensing and parallel importing (although the government's initial intent was to allow parallel importing only). Fearing that similar legislation might be passed in other countries, the Pharmaceutical Researchers and Manufacturers Association, a U.S. trade association, filed suit in South African courts against certain provisions of the legislation. In the face of considerable pressure from the pharmaceutical industry as well as the U.S. government, the South Africans were unyielding, insisting that their citizens should not have to pay more for anti-HIV drugs than, for example, Australians. A series of well-publicized protests from 1999 to 2001 in South Africa, and by AIDS activists in the United States, actually led the pharmaceutical industry and the U.S. government to back down.

In the following selections, Akin Jimoh contends that the AIDS epidemic in Africa can be linked to the high costs of drugs. Big drug companies agreed to lower the price of anti-HIV medications in Africa, however, they have kept their patent rights. Therefore, availability and pricing are still under the control of the drug companies. Siddhartha Mukherjee states that adequate health care networks are needed to monitor the distribution and use of anti-HIV drugs in Africa. Such medications must be taken consistently and over the prescribed length of time so that new strains of HIV will not develop that are resistant to these drugs. The health care infrastructure must be invested in rather than simply providing free anti-HIV medications.

Akin Jimoh  **YES**

"Raise the Alarm Loudly": Africa Confronts the AIDS Pandemic

W e were both standing on the sidewalk, watching the convoy of returning soldiers on their way to the military hospital in Victoria Island, Lagos, Nigeria. Amid the noise from the heavy-duty military vehicles and downtown traffic, my companion, Mohammed Farouk Auwalu, a former soldier in the Nigerian army, shook his head and muttered, "Many of them will most likely die soon or be out of the army like me with little or nothing to show for it. A lot of people don't know that many have died, others are dying, and many are walking in the shadow of death."

The convoy was returning from one of Nigeria's many peacekeeping missions elsewhere on the continent, but African wars were far from Auwalu's mind. He was talking about the specter of AIDS. In his mid thirties and married, Auwalu is now retired, not because he cannot perform his assigned duties, but because he is living with HIV. He currently heads the Nigeria AIDS Alliance, an awareness group formed by people living with HIV/AIDS.

The Pandemic

So far, AIDS has killed 17 million Africans. It has orphaned about 12 million children. And about 25.3 million Africans (about 9% of the continent's total population) now live, like Auwalu, with HIV. According to the World Bank, the HIV infection rate in pregnant women in Blantyre, Malawi, increased from less than 5% in 1985 to over 30% in 1997. In Francistown, Botswana, the rate climbed from less than 10% in 1991 to 43% in 1997. New figures from the United Nations Joint AIDS Program (UNAIDS) show that 3.8 million people in sub-Saharan Africa became infected with HIV during 2000. Meanwhile, 2.4 million Africans died of AIDS that year.

From the Horn of Africa to the Cape of Good Hope, HIV/AIDS is crippling national economies. Many African countries now face the enormous costs of fighting the epidemic and caring for the millions orphaned by AIDS, even as the most productive generation is decimated by the disease. A study published in the *South African Journal of Economics* in July 2000 concluded that, as a result of HIV/AIDS, South Africa's national income would be 17% lower in 2010 than it would have been otherwise. Overall, the World Bank estimates that HIV/AIDS has cut economic growth in Africa by about two thirds.

"The AIDS situation in Africa is catastrophic and sub-Saharan Africa continues to head the list as the world's most affected region," says Dr. Peter Piot, executive director of UNAIDS. "One of the greatest causes for concern is that over the next few years, the epidemic is bound to get worse before it gets better." AIDS has struck virtually all sectors of society. Families have been devastated; husbands, wives, brothers, and sisters are dead or dying. Women, young people, and children are among the hardest hit.

How did it get this bad?

- *Migrant labor.* The prevalence of migrant labor in Southern Africa has greatly contributed to the high infection rates in Botswana, South Africa, Malawi, Namibia, Zambia, and Zimbabwe. As migrant laborers move from one work site to another, leaving their families behind, many engage in multiple sexual relationships.
- *Low social status of women.* Women account for half of Africa's HIV-positive population, according to the UN, and the infection rate for women is on the rise. Data from several African countries show infection rates for teenage girls five to six times the rates for teenage boys. Poverty forces many girls and women to trade their bodies for money. Meanwhile, the low social and economic status of women, argues UN Secretary General Kofi Annan, results in a "weaker ability to negotiate safe sex."
- *Lack of open discussion.* Cultural and religious inhibitions on the discussion of sex-related issues hindered AIDS prevention at an early stage. Repression against the media also inhibited the flow of information. At an HIV/AIDS meeting in Mexico in 1988, U.S. journalist and science writer Laurie Garrett saluted by name a Kenyan journalist who had broadcast AIDS information over an independent radio station. He was arrested within hours. The Zimbabwean and South African governments have also routinely targeted journalists disseminating information about AIDS.
- *Lack of quick government action.* Olikoye Ransome-Kuti, a pediatrician and former health minister of Nigeria, says that, even in the mid 1990s, the Nigerian military regime allocated a mere $3,000 annually to AIDS control programs. Now, 5.4% of Nigerians between the ages of 15 and 49—about 2.6 million people—live with HIV/AIDS. In many African countries, political turmoil and war contributed to a delayed government response.
- *Weak health-care systems.* In the mid 1980s, most African countries achieved child-immunization rates, to take just one indicator of basic public-health provision, of over 80%. In the following decade, rates fell below 20% in many African countries. Lack of access to basic health services has increased the rate of non-sexual (mother-to-child) HIV transmission.

- *Economic austerity programs.* The AIDS epidemic began its full onslaught in the mid-to-late 1980s, when the International Monetary Fund imposed structural adjustment programs (SAPs) on many African countries. Under the SAPs, national currencies were devalued and subsidies to critical sectors of the economy discontinued. With minimal funds available to governments, social infrastructure and services, including health services, suffered. Keith Hansen, deputy head of the World Bank's AIDS Campaign Team for Africa, admitted that SAPs had weakened African economies. Austerity has deprived African countries of the means to fight the epidemic.
- *The high cost of drugs.* Pharmaceutical companies like Bristol-Myers Squibb of the United States, Glaxo-SmithKline of Great Britain, and Boerhinger Ingelheim GMBH of Germany sell their patented AIDS drugs for $10,000-15,000 per patient per year, three to five times the per capita income of South Africa (the highest in Africa).

Uganda, the place where AIDS first struck in Africa, now offers a model for combating the epidemic. The Ugandan government has helped bring about a mini-sexual revolution. In the mid 1980s, it began prevention campaigns on HIV/AIDS and other sexually transmitted diseases, and started promoting sex education generally. President Yoweri Museveni personally championed the AIDS-control program. Meanwhile, some debt relief and the creation of an anti-poverty program has resulted in a revival of the health system.

"When a lion comes to your village you must raise the alarm loudly," Museveni says. "This is what we did in Uganda; we took it seriously and achieved good results. AIDS . . . is not like small pox or Ebola. AIDS can be prevented as it is transmitted through a few known ways. If we raise awareness sufficiently, it will stop." Between 1997 and 2000, while the HIV infection rate climbed from about 13% to nearly 20% in South Africa and from about 25% to over 35% in Botswana, it has actually decreased in Uganda, from 9.5% to 8.3%. Since there is no cure for AIDS, lower infection rates reflect the deaths of some people who already had AIDS—but also a lower rate of new HIV infections.

The Patents War

In Pretoria, South Africa, this past March [2001], thousands of AIDS activists and HIV-positive youths descended on the country's High Court and the U.S. Embassy. Wearing "HIV-positive" T-shirts and baseball caps, hands locked together in solidarity, they marched in angry protest against the high cost of AIDS drugs. Their placards expressed their rage: "Lives Before Profits" and "AIDS Profiteer Deadlier Than The Virus." The battle over AIDS-drug patents had begun.

A new cocktail of generic AIDS drugs developed by the Indian drug company CIPLA threatens the big drug companies' lucrative monopolies. CIPLA has offered the drug at a cost of $350 per year per patient to the humanitarian organization Doctors Without Borders, and $600 per year per patient to African governments. In March, thirty-nine of the big pharmaceutical companies went to court to challenge the South African government's go-ahead on the sale of generic AIDS drugs, provoking the March protests.

A few weeks after the court battle began, Doctors Without Borders approached Yale University to convince it to release its patent on the AIDS drug dT4. Two Yale professors had developed dT4, which the University then licensed to Bristol-Myers Squibb. Professor William Prusoff, one of the developers, wrote during the height of the controversy that the drug should be either free or very inexpensive in sub-Saharan Africa, and expressed disappointment that it was not reaching the millions of people who desperately needed it. Not long after, Bristol-Myers announced that it would reduce the cost of d4T by 15% in the United States and 85% in the rest of the world, and that it would offer the drug for 15 cents per daily dose in the most afflicted areas of Africa. The other two pharmaceutical giants, GlaxoSmithKline and Boerhinger Ingelheim GMBH, are also expected to cut their AIDS-drug prices. The companies, however, remain steadfast about keeping their patent rights, which would leave ultimate control over prices and availability in their hands.

In response, the AIDS-devastated countries of Africa may resort to "compulsory licensing," ignoring the patents and proceeding with generic drugs. International convention recognizes the right of countries in states of national emergency to obtain or manufacture generic drugs, even in breach of drug-company patents. So far, President Thabo Mbeki of South Africa has resisted an official declaration of national emergency, though he promises to go forward with generic drugs. The U.S. government, under both former President Bill Clinton and current President George W. Bush, has promised not to challenge laws passed by African countries to improve access to AIDS drugs, even if U.S. patent laws are broken. It has not, however, pressed U.S. pharmaceutical firms to renounce their patent rights—which is why protestors targeted the U.S. embassy.

The battle is far from over. Even at 15 cents per day, or about $55 per year, AIDS drugs will remain beyond the means of most Africans. At the 8th Conference on Retroviruses and Opportunistic Infections in February 2001, doctors, scientists, and policymakers proposed that rich nations pay for drugs and other means to combat AIDS in Africa, with the United States paying $3 billion. Harvard economist Jeffrey Sachs explained that $3 billion would only cost the United States about $10 per person, the cost of a movie ticket and a bag of popcorn. Dr. Peter Piot of UNAIDS believes that this additional $3 billion would go a long way towards coping with the epidemic in sub-Saharan Africa—with half going to basic care for those already infected, the other half to prevention efforts.

Donors cannot, however, dictate how the battle against AIDS will be fought. A recent report issued by the Africa-America Institute, which champions a greater U.S. commitment to the fight against AIDS in Africa, concludes that donors need to support national priorities set by Africans themselves. Local circumstances vary greatly from country to country, the AAI argues, so international donors need to learn more about Africa and adapt their programs to the needs of each country. "If the U.S. and other donors want to make a difference in the fight against HIV/AIDS in Africa," AAI President Mora McLean says, "they need to listen to Africans and involve them as full partners in the global battle against the epidemic."

Take Your Medicine

Last week's [July 2000] International AIDS Conference in Durban, South Africa, was a spectacularly glum affair. While angry protesters outside the conference railed against greedy pharmaceutical companies, delegates inside recited dismal statistics about the plague, each more alarming than the last. In South Africa, approximately one in ten adults is HIV-positive; in Africa as a whole, AIDS now takes three times as many lives as the next most common cause of death. Of all the depressing numbers, there was only one that health officials felt confident about changing any time soon: the $15,000 it currently costs to treat just one person with anti-HIV drugs for a year.

The reason is something called "tiered pricing" or "equity pricing," a concept that UNAIDS, the United Nations agency dealing with AIDS, began promoting recently and that elicited considerable excitement in Durban, even winning the endorsement of Bill Gates. Under the scheme, Western pharmaceutical companies, like Merck and Glaxo Wellcome, would set different prices for drugs in rich and poor countries. The same pill—say, AZT—could be sold for $4 in New York but only 40 cents in Johannesburg. With tiered pricing, Africans could finally afford the anti-HIV medicines they desperately need, and drug companies could still turn a reasonable profit.

A great idea? Actually, no—at least not by itself. What the enthusiasts seem not to realize is that without adequate health care networks to monitor their distribution, potent new medicines are worse than useless; they're dangerous. Consider Russia's recent experience with anti-tuberculosis drugs. In the 1990s, physicians in the former Soviet Union unleashed a torrent of anti-tuberculosis drugs on the population. The drugs were great, but the patients taking them weren't adequately supervised; in many hospitals, as many as 50 percent of patients strayed from the prescribed regimen. Soon, upwards of five percent of patients in some Russian clinics began to exhibit a strain of tuberculosis completely resistant to all drugs. Subsequently, millions of dollars had to be spent to contain the deadly strain. As Dr. C. Robert Horsburgh, a public health expert from Boston University, recently warned in the *Journal of American Medicine*, "The genie of multi-drug-resistant TB [was] irreversibly out of the bottle."

The HIV genie is even more ominous. HIV's secret—one reason the wispy virus is now a continent-hopping Goliath—is that it mutates rapidly, quickly becoming resistant to drugs. If anti-HIV drugs are not taken properly—a missed capsule here, a forgotten pill there—a low level of viral reproduction continues within the body. And the viruses brewed while the anti-viral medicines are still present in a patient's system can be especially lethal, as they are selected to carry mutations that render them resistant to the original drug. Even in the United States, where an excellent health care network monitors most drug regimens, about ten percent of patients already harbor HIV strains resistant to AZT, the most common anti-HIV drug. And if such potent drugs are dumped unsupervised on Africa—where health care networks cannot afford to be as vigilant—then a virulent, drug-resistant strain of HIV may emerge very quickly and could even boomerang back to the West.

⋅⦿⋅

Fortunately, there is an alternative to the solution hyped . . . in Durban. Since the safety of anti-HIV drugs depends on a country's health care infrastructure, pharmaceutical companies could pay to develop in Africa some of the infrastructure necessary to make sure their anti-HIV drugs are taken properly.

Why would drug companies do something so altruistic? Because it's not altruistic at all. After all, drug companies can only squeeze profits out of Africa by selling Africans their anti-HIV drugs over a long period of time. Right now, with about 22 million Africans infected with HIV, the demand for anti-viral drugs seems inexhaustible. But, if a viral strain immune to a company's drug emerged, the drugmakers would no longer have medicine Africans wanted to buy. Even worse, the resistant virus might spread into more profitable Western markets. Only by making an investment in health care infrastructure—and thus preventing drug-resistant strains of HIV from coming to life—can a pharmaceutical company ensure that its cash-cow drug isn't rapidly made worthless by new mutations.

Glaxo Wellcome, at least, seems to understand this. In May, the company announced it would enter an unusual collaboration with UNAIDS to make sure its discounted anti-HIV drugs would be sold only in selected areas—places that "address[ed] the health care infrastructure and drug distribution aspects" and where there was "access to safe and effective ongoing treatment" for HIV. Glaxo also agreed to foot some of the bill for building these infrastructures through direct training and technical support of AIDS advocacy groups.

No one can be sure the Glaxo-UNAIDS scheme will work, because nothing like it has ever really been tried. There isn't much precedent for such public and private collaborations actually creating safe environments for selling discounted drugs. But, then again, there isn't much precedent for a recalcitrant virus infecting whole swaths of an entire continent. HIV is so deadly because it is enormously resourceful, crafty, and even creative. To defeat it, we will have to be, as well.

POSTSCRIPT

Should International Drug Companies Provide HIV/AIDS Drugs to Africa Free of Charge?

Mukherjee brings up the issue that the health care systems in most African countries are incapable of effectively monitoring the wide-scale use of anti-HIV drugs. While Mukherjee suggests that drug companies have a self-interested stake in supporting these systems, many others have decried the role of structural adjustment programs and debt in facilitating funding cuts to these systems. Structural adjustment programs, largely implemented by the World Bank and the International Monetary Fund, call for the balancing of government budgets through, among other things, reductions in social service spending and contractions in the civil service. Debt (with African countries accounting for 34 of the 41 most indebted countries in the world) also reduces current spending on social services, as a growing proportion of state revenues must be spent on debt service. For more on this topic see Laura Dely, "Aiding and Abetting an Epidemic," in *Sojourners* (November 1999).

Much of the HIV/AIDS work in Africa has been focused on prevention as opposed to treatment. As Mukherjee describes, Uganda is an excellent example of a country that has aggressively pursued prevention strategies and has consequently seen an amazing reduction in the prevalency rate of AIDS. As a result of the exorbitant cost of anti-HIV drugs (even with the price concessions of pharmaceutical companies), some may conclude that investing in drugs that merely prolong the lives of infected individuals is less of a priority than prevention efforts. This position may be tempered by a knowledge of Africa's already large and growing problem of AIDS orphans, that is, children for whom both parents have died from AIDS, as well as the issue of mother-to-child transmission of HIV during birth or through breastfeeding. In the case of the former, drugs that prolong the life of a parent also enhance the life of a child. In the case of the latter, drugs taken shortly before the birth of a child by an HIV-infected mother greatly reduce the chances of transmission. For more on the issue of mother-to-child transmission, see a 2001 World Health Organization technical consultation entitled "New Data on the Prevention of Mother-to-Child Transmission of HIV and Their Policy Implications." A good book on the AIDS orphan problem in Uganda, Zambia, and South Africa is Emma Guest's *Children of AIDS: Africa's Orphan Crisis* (Pluto Press, 2001).

One cannot conclude a discussion of this issue without mentioning George Bush's announcement in his 2003 State of the Union Address that the United States would be committing $15 billion over the next five years to combat AIDS in Africa. Of this money, roughly half will be spent on AIDS drugs with the hope of offering antiretroviral drugs to two million people. This is the largest commitment to fighting AIDS in Africa that we have seen to date. No mention, however, was made concerning the role (if any) of drug companies in this initiative.

ISSUE 13

Is "Overpopulation" a Major Cause of Poverty in Africa?

YES: Partha S. Dasgupta, from "Population, Poverty, and the Local Environment," *Scientific American* (February 1995)

NO: Bernard I. Logan, from "Overpopulation and Poverty in Africa: Rethinking the Traditional Relationship," *Tijdschrift voor Economische en Sociale Geografie* (1991)

ISSUE SUMMARY

YES: Partha S. Dasgupta, a professor of economics at the University of Cambridge, contends that many African families have too many children because the benefits of having an additional child outweigh the costs (especially since the expense of childrearing is subsidized by the greater community). According to Dasgupta, there is a vicious cycle in operation in which overpopulation leads to the depletion of community resources, creating greater poverty and leading individual families to have even more children in an attempt to reverse this poverty.

NO: Bernard I. Logan, a professor of geography at the University of Georgia, argues that Africa is not overpopulated when absolute population numbers are compared to the resource base. The problem is that the terms of exchange between Africa and its European trading partners are unfairly set, diminishing the returns that Africans see on their own resources. This leads to a situation where Africans are subsidizing European consumption at the expense of their own livelihoods.

T he notion that "overpopulation" leads to poverty has been around since at least 1798, when Thomas Malthus evoked it in his tract entitled *Essay on the Principle of Population*. Malthus argued that poor households breed uncontrollably, further contributing to their poverty. Malthus saw it as inevitable that the lower classes would multiply until their numbers were checked by mass starvation. The connection between poverty and population growth was contested by Friedrich Engels, Karl Marx's collaborator, in an 1844 note entitled "Outlines of a Critique of Political Economy." Engels asserted that the capitalist system (a system built on extracting wealth from the working class through cheap wages), rather than population growth, was responsible for generating poverty. Engels believed that since individuals generally produce more than they consume, more people would lead to greater surplus food production. This type of critique has been replicated in the present and is sometimes referred to as the political economy or structuralist perspective on population.

Garrett Harden built on the ideas of Malthus in his famous 1968 essay in *Science* entitled, "The Tragedy of the Commons," by seeking to clarify why rural families have too many children. He stated that households tend to have more children because they reap the benefits of offspring yet do not bear the full cost of their upkeep. Using the metaphor of a village commons where individual farmers have a tendency to graze too many sheep, Harden suggested that individual families have too many children because the greater community bears part of the cost of their upkeep. The tragedy, according to Harden, is that in an attempt to reap greater benefits, households keep degrading the commons and further impoverishing themselves.

In this issue, Partha S. Dasgupta argues that many African families have too many children because those who bear the greatest cost of procreating, women, have relatively little decision-making authority. Furthermore, the benefits of having an additional child outweigh the costs. Children are beneficial to rural families because they provide labor and serve as a social-security system in old age. Similarly to Harden, Dasgupta asserts that the cost of raising children in the African context is at least partially borne by the broader community. Overpopulation, according to Dasgupta, leads to the depletion of community resources, increasing poverty, and individual families having even more children in an attempt to reverse this poverty.

In contrast, Bernard I. Logan maintains that the problem is that the terms of trade between Africa and its European trading partners are unfairly set, diminishing the returns that Africans see on their own resources. This leads to a situation where Africans are subsidizing European consumption by providing cheap commodities in exchange for expensive European manufactured goods to the detriment of their own livelihoods.

Comprehending the concept of *terms of trade* is critical to understanding Logan's argument. Terms of trade can be thought of as the average price of a country's exports over the average price of its imports. The problem for many African countries is that they produce primary commodities (e.g., cotton, cocoa, sugar) for which prices are generally low, and they import manufactured goods for which prices are high. As the profit margins on the latter tend to be much higher, many African nations find that they have low and declining terms of trade. This means that countries could be producing more yet still getting less and less in return. This then, according to Logan, is what allows Europe, with a relatively small resource base, to maintain high population densities and high levels of consumption, while Africa, with a comparatively large resource base, struggles to get by.

Partha S. Dasgupta  **YES**

Population, Poverty, and the Local Environment

As with politics, we all have widely differing opinions about population. Some would point to population growth as the cause of poverty and environmental degradation. Others would permute the elements of this causal chain, arguing, for example, that poverty is the cause rather than the consequence of increasing numbers. Yet even when studying the semiarid regions of sub-Saharan Africa and the Indian subcontinent, economists have typically not regarded poverty, population growth and the local environment as interconnected. Inquiry into each factor has in large measure gone along its own narrow route, with discussion of their interactions dominated by popular writings—which, although often illuminating, are in the main descriptive and not analytical.

Over the past several years, though, a few investigators have studied the relations between these ingredients more closely. Our approach fuses theoretical modeling with empirical findings drawn from a number of disciplines, such as anthropology, demography, ecology, economics, nutrition and political science. Focusing on the vast numbers of small, rural communities in the poorest regions of the world, the work has identified circumstances in which population growth, poverty and degradation of local resources often fuel one another. The collected research has shown that none of the three elements directly causes the other two; rather each influences, and is in turn influenced by, the others. This new perspective has significant implications for policies aimed at improving life for some of the world's most impoverished inhabitants.

In contrast with this new perspective, with its focus on local experience, popular tracts on the environment and population growth have usually taken a global view. They have emphasized the deleterious effects that a large population would have on our planet in the distant future Although that slant has its uses, it has drawn attention away from the economic misery endemic today. Disaster is not something the poorest have to wait for: it is occurring even now. Besides, in developing countries, decisions on whether to have a child and on how to share education, food, work, health care and local resources are in large measure made within small entities such as households. So it makes sense to

From Partha S. Dasgupta, "Population, Poverty, and the Local Environment," *Scientific American* (February 1995). Copyright © 1995 by Scientific American, Inc. Reprinted by permission.

study the link between poverty, population growth and the environment from a myriad of local, even individual, viewpoints.

The household assumes various guises in different parts of the world. Some years ago Gary S. Becker of the University of Chicago was the first investigator to grapple with this difficulty. He used an idealized version of the concept to explore how choices made within a household would respond to changes in the outside world, such as employment opportunities and availability of credit, insurance, health care and education.

One problem with his method, as I saw it when I began my own work some five years ago, was that it studied households in isolation; it did not investigate the dynamics between interacting units. In addition to understanding the forces that encouraged couples to favor large families, I wanted to understand the ways in which a reasoned decision to have children, made by each household, could end up being detrimental to all households.

In studying how such choices are made, I found a second problem with the early approach: by assuming that decision making was shared equally by adults, investigators had taken an altogether too benign view of the process. Control over a family's choices is, after all, often held unequally. If I wanted to understand how decisions were made, I would have to know who was doing the deciding.

Power and Gender

Those who enjoy the greatest power within a family can often be identified by the way the household's resources are divided. Judith Bruce of the Population Council, Mayra Buvinic of the International Center for Research on Women, Lincoln C. Chen and Amartya Sen of Harvard University and others have observed that the sharing of resources within a household is often unequal even when differences in needs are taken into account. In poor households in the Indian subcontinent, for example, men and boys usually get more sustenance than do women and girls, and the elderly get less than the young.

Such inequities prevail over fertility choices as well. Here also men wield more influence, even though women typically bear the greater cost. To grasp how great the burden can be, consider the number of live babies a woman would normally have if she managed to survive through her childbearing years. This number, called the total fertility rate, is between six and eight in sub-Saharan Africa. Each successful birth there involves at least a year and a half of pregnancy and breast-feeding. So in a society where female life expectancy at birth is 50 years and the fertility rate is, say, seven, nearly half of a woman's adult life is spent either carrying a child in her womb or breast-feeding it. And this calculation does not allow for unsuccessful pregnancies.

Another indicator of the price that women pay is maternal mortality. In most poor countries, complications related to pregnancy constitute the largest single cause of death of women in their reproductive years. In some parts of sub-Saharan Africa as many as one woman dies for every 50 live births. (The rate in Scandinavia today is one per 20,000.) At a total fertility rate of seven or more,

Figure 1

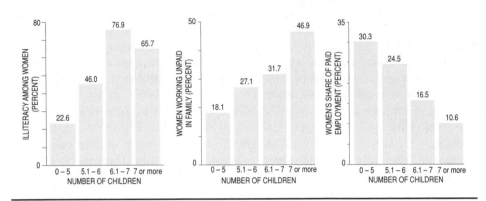

Note: Total fertility rate around the world (the average number of children a woman produces) generally increases with the percentage of women in a country who are illiterate (*left*) or work unpaid in the family (*middle*). Fertility decreases when a larger share of the paid employment belongs to women (*right*). Bringing in a cash income may empower a woman in making decisions within her family, allowing her to resist pressure to bear more children.

the chance that a woman entering her reproductive years will not live through them is about one in six. Producing children therefore involves playing a kind of Russian roulette.

Given such a high cost of procreation, one expects that women, given a choice, would opt for fewer children. But are birth rates in fact highest in societies where women have the least power within the family? Data on the status of women from 79 so-called Third World countries display an unmistakable pattern: high fertility, high rates of illiteracy, low share of paid employment and a high percentage working at home for no pay—they all hang together. From the statistics alone it is difficult to discern which of these factors are causing, and which are merely correlated with, high fertility. But the findings are consistent with the possibility that lack of paid employment and education limits a woman's ability to make decisions and therefore promotes population growth.

There is also good reason to think that lack of income-generating employment reduces women's power more directly than does lack of education. Such an insight has implications for policy. It is all well and good, for example, to urge governments in poor countries to invest in literacy programs. But the results could be disappointing. Many factors militate against poor households' taking advantage of subsidized education. If children are needed to work inside and outside the home, then keeping them in school (even a cheap one) is costly. In patrilineal societies, educated girls can also be perceived as less pliable and harder to marry off. Indeed, the benefits of subsidies to even primary education are reaped disproportionately by families that are better off.

In contrast, policies aimed at increasing women's productivity at home and improving their earnings in the marketplace would directly empower them, especially within the family. Greater earning power for women would also raise for men the implicit costs of procreation (which keeps women from bringing in cash income). This is not to deny the value of public investment in primary and secondary education in developing countries. It is only to say we should be wary of claims that such investment is a panacea for the population problem.

The importance of gender inequality to overpopulation in poor nations is fortunately gaining international recognition. Indeed, the United Nations Conference on Population and Development held in Cairo in September 1994 emphasized women's reproductive rights and the means by which they could be protected and promoted. But there is more to the population problem than gender inequalities. Even when both parents participate in the decision to have a child, there are several pathways through which the choice becomes harmful to the community. These routes have been uncovered by inquiring into the various motives for procreation.

Little Hands Help . . .

One motive, common to humankind, relates to children as ends in themselves. It ranges from the desire to have children because they are playful and enjoyable, to the desire to obey the dictates of tradition and religion. One such injunction emanates from the cult of the ancestor, which, taking religion to be the act of reproducing the lineage, requires women to bear many children [see "High Fertility in Sub-Saharan Africa," by John C. Caldwell and Pat Caldwell; SCIENTIFIC AMERICAN, May 1990].

Such traditions are often perpetuated by imitative behavior. Procreation in closely knit communities is not only a private matter; it is also a social activity, influenced by the cultural milieu. Often there are norms encouraging high fertility rates that no household desires unilaterally to break. (These norms may well have outlasted any rationale they had in the past.) Consequently, so long as all others aim at large families, no household on its own will wish to deviate. Thus, a society can get stuck at a self-sustaining mode of behavior that is characterized by high fertility and low educational attainment.

This does not mean that society will live with it forever. As always, people differ in the extent to which they adhere to tradition. Inevitably some, for one reason or another, will experiment, take risks and refrain from joining the crowd. They are the nonconformists, and they help to lead the way. An increase in female literacy could well trigger such a process.

Still other motives for procreation involve viewing children as productive assets. In a rural economy where avenues for saving are highly restricted, parents value children as a source of security in their old age. Mead Cain, previously at the Population Council, studied this aspect extensively. Less discussed, at least until recently, is another kind of motivation, explored by John C. Caldwell of the Australian National University, Marc L. Nerlove of the University of Maryland and Anke S. Meyer of the World Bank and by Karl-Göran Mäler of the Beijer International Institute of Ecological Economics in Stockholm and me. It stems from children's being valuable to their parents not only for future income but also as a source of current income.

Third World countries are, for the most part, subsistence economies. The rural folk eke out a living by using products gleaned directly from plants and animals. Much labor is needed even for simple tasks. In addition, poor rural households do not have access to modern sources of domestic energy or tap water. In semiarid and arid regions the water supply may not even be nearby. Nor is fuelwood at hand when the forests recede. In addition to cultivating crops, caring for livestock, cooking food and producing simple marketable products, members of a household may have to spend as much as five to six hours a day fetching water and collecting fodder and wood.

Children, then, are needed as workers even when their parents are in their prime. Small households are simply not viable; each one needs many hands. In parts of India, children between 10 and 15 years have been observed to work as much as one and a half times the number of hours that adult males do. By the age of six, children in rural India tend domestic animals and care for younger siblings, fetch water and collect firewood, dung and fodder. It may well be that the usefulness of each extra hand increases with declining availability of resources, as measured by, say, the distance to sources of fuel and water.

. . . But at a Hidden Cost

The need for many hands can lead to a destructive situation, especially when parents do not have to pay the full price of rearing their children but share those costs with the community. In recent years, mores that once regulated the use of local resources have changed. Since time immemorial, rural assets such as village ponds and water holes, threshing grounds, grazing fields, and local forests have been owned communally. This form of control enabled households in semiarid regions to pool their risks. Elinor Ostrom of Indiana University and others have shown that communities have protected such local commons against overexploitation by invoking norms, imposing fines for deviant behavior and so forth.

But the very process of economic development can erode traditional methods of control. Increased urbanization and mobility can do so as well. Social rules are also endangered by civil strife and by the takeover of resources by landowners or the state. As norms degrade, parents pass some of the costs of children on to the community by overexploiting the commons. If access to shared resources continues, parents produce too many children, which leads to greater crowding and susceptibility to disease as well as to more pressure on environmental resources. But no household, on its own, takes into account the harm it inflicts on others when bringing forth another child.

Parental costs of procreation are also lower when relatives provide a helping hand. Although the price of carrying a child is paid by the mother, the cost of rearing the child is often shared among the kinship. Caroline H. Bledsoe of Northwestern University and others have observed that in much of sub-Saharan Africa fosterage is commonplace, affording a form of insurance protection in semiarid regions. In parts of West Africa about a third of the children have been found to be living with their kin at any given time. Nephews and nieces have the

Figure 2

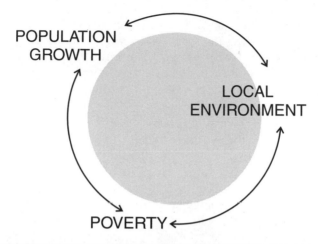

REGION	TOTAL FERTILITY RATE
SUB-SAHARAN AFRICA	6 TO 8
INDIA	4
CHINA	2.3
JAPAN AND WESTERN INDUSTRIALIZED DEMOCRACIES	1.5 TO 1.9

Note: Poverty, population growth and environmental degradation interact in a cyclic pattern (*top*). The chart (*bottom*) shows that fertility is higher in countries that are poorer.

same rights of accommodation and support as do biological offspring. In recent work I have shown that this arrangement encourages couples to have too many offspring if the parents' share of the benefits from having children exceeds their share of the costs.

In addition, where conjugal bonds are weak, as they are in sub-Saharan Africa, fathers often do not bear the costs of siring a child. Historical demographers, such as E. A. Wrigley of the University of Cambridge, have noted a significant difference between western Europe in the 18th century and modern preindustrial societies. In the former, marriage normally meant establishing a new household. This requirement led to late marriages; it also meant that parents bore the cost of rearing their children. Indeed, fertility rates in France dropped before mortality rates registered a decline, before modern family-planning techniques became available and before women became literate.

The perception of both the low costs and high benefits of procreation induces households to produce too many children. In certain circumstances a disastrous process can begin. As the community's resources are depleted, more hands are needed to gather fuel and water for daily use. More children are then produced, further damaging the local environment and in turn providing the household with an incentive to enlarge. When this happens, fertility and environmental degradation reinforce each other in an escalating spiral. By the time some countervailing set of factors—whether public policy or diminished benefits from having additional children—stops the spiral, millions of lives may have suffered through worsening poverty.

Recent findings by the World Bank on sub-Saharan Africa have revealed positive correlations among poverty, fertility and deterioration of the local environment. Such data cannot reveal causal connections, but they do support the idea of a positive-feedback process such as I have described. Over time, the effect of this spiral can be large, as manifested by battles for resources [see "Environmental Change and Violent Conflict," by T. F. Homer-Dixon, J. H. Boutwell and G. W. Rathjens; SCIENTIFIC AMERICAN, February 1993].

The victims hit hardest among those who survive are society's outcasts—the migrants and the dispossessed, some of whom in the course of time become the emaciated beggars seen on the streets of large towns and cities in underdeveloped countries. Historical studies by Robert W. Fogel of the University of Chicago and theoretical explorations by Debraj Ray of Boston University and me, when taken together, show that the spiral I have outlined here is one way in which destitutes are created. Emaciated beggars are not lazy; they have to husband their precarious hold on energy. Having suffered from malnutrition, they cease to be marketable.

Families with greater access to resources are, however, in a position to limit their size and propel themselves into still higher income levels. It is my impression that among the urban middle classes in northern India, the transition to a lower fertility rate has already been achieved. India provides an example of how the vicious cycle I have described can enable extreme poverty to persist amid a growth in well-being in the rest of society. The Matthew effect— "Unto every one that hath shall be given, and he shall have abundance: but from him that hath not shall be taken away even that which he hath"—works relentlessly in impoverished countries.

Breaking Free

This analysis suggests that the way to reduce fertility is to break the destructive spiral. Parental demand for children rather than an unmet need for contraceptives in large measure explains reproductive behavior in developing countries. We should therefore try to identify policies that will change the options available to men and women so that couples choose to limit the number of offspring they produce.

In this regard, civil liberties, as opposed to coercion, play a particular role. Some years ago my colleague Martin R. Weale and I showed through statistical

GREEN NET NATIONAL PRODUCTION

Some economists believe population growth is conducive to economic growth. They cite statistics showing that, except in sub-Saharan Africa, food production and gross income per head have generally grown since the end of World War II. Even in poor regions, infant survival rate, literacy and life expectancy have improved, despite the population's having grown much faster than in the past.

One weakness of this argument is that it is based on statistics that ignore the depletion of the environmental resource base, on which all production ultimately depends. This base includes soil and its cover, freshwater, breathable air, fisheries and forests. No doubt it is tempting to infer from past trends that human ingenuity can be relied on to overcome the stresses that growing populations impose on the natural environment.

Yet that is not likely to be the case. Societies already use an enormous 40 percent of the net energy created by terrestrial photosynthesis. Geoffrey M. Heal of Columbia University, John M. Hartwick of Queens University and Karl-Göran Mäler of the Beijer International Institute of Ecological Economics in Stockholm and I have shown how to include environmental degradation in estimating the net national product, or NNP. NNP is obtained by deducting from gross national product the value of, for example, coal extracted or timber logged.

This "green NNP" captures not only present production but also the possibility of future production brought about by resources we bequeath. Viewed through NNP, the future appears far less rosy. Indeed, I know of no ecologist who thinks a population of 11 billion (projected for the year 2050) can support itself at a material standard of living of, say, today's representative American.

analysis that even in poor countries political and civil liberties go together with improvements in other aspects of life, such as income per person, life expectancy at birth and infant survival rate. Thus, there are now reasons for thinking that such liberties are not only desirable in themselves but also empower people to flourish economically. Recently Adam Przeworski of the University of Chicago demonstrated that fertility, as well, is lower in countries where citizens enjoy more civil and political freedom. (An exception is China, which represents only one country out of many in this analysis.)

The most potent solution in semiarid regions of sub-Saharan Africa and the Indian subcontinent is to deploy a number of policies simultaneously. Family-planning services, especially when allied with health services, and measures that empower women are certainly helpful. As societal norms break down and traditional support systems falter, those women who choose to change their behavior become financially and socially more vulnerable. So a literacy and employment drive for women is essential to smooth the transition to having fewer children.

But improving social coordination and directly increasing the economic security of the poor are also essential. Providing cheap fuel and potable water will reduce the usefulness of extra hands. When a child becomes perceived as expensive, we may finally have a hope of dislodging the rapacious hold of high fertility rates.

Each of the prescriptions suggested by our new perspective on the links between population, poverty and environmental degradation is desirable by itself, not just when we have those problems in mind. It seems to me that this consonance of means and ends is a most agreeable fact in what is otherwise a depressing field of study.

FURTHER READING

POPULATION, NATURAL RESOURCES, AND DEVELOPMENT. Special issue of *Ambio,* Vol. 21, No. 1; February 1992.

AN INQUIRY INTO WELL-BEING AND DESTITUTION. Partha Dasgupta. Oxford University Press, 1993.

POPULATION: THE COMPLEX REALITY. POPULATION SUMMIT REPORT OF THE WORLD'S SCIENTIFIC ACADEMIES, ROYAL SOCIETY, LONDON. North American Press, 1994.

POPULATION, ECONOMIC DEVELOPMENT, AND THE ENVIRONMENT. Edited by Kerstin Lindahl Kiessling and Hans Landberg. Oxford University Press, 1994.

WORLD DEVELOPMENT REPORT. World Bank, annual publication.

POVERTY, INSTITUTIONS AND THE ENVIRONMENTAL RESOURCE BASE. Partha Dasgupta and Karl-Göran Mäler in *Handbook of Development Economics,* Vol. 3. Edited by T. N. Srinivasan et al. North Holland Publishing, Amsterdam (in press).

Bernard I. Logan

Overpopulation and Poverty in Africa: Rethinking the Traditional Relationship

Overpopulation and Underdevelopment in Sub-Saharan Africa

In general, policy-makers, planners and development analysts agree that over-population is a major contributor to, if not the primary cause of, malnutrition, poverty and underdevelopment in Africa and other Third World regions. The proponents of this line of argument support their claims by drawing relation-ships between population growth rates, fertility rates and other demographic indices, on the one hand, and the carrying capacity of the land, on the other. . . . The focus on population growth as an important explanation of poverty rests on the idea that population pressures have reduced the carrying capacity of the land in many areas of Africa. In turn, the population burden severely constrains the ability of African economies (especially the traditional sectors) to obtain high standards of living (World Bank 1984). The association between popula-tion growth and poverty hinges upon an underlying preconception that high rates of population growth are responsible for the overutilization of the natural resource base in Africa.

The problem I address in this article is couched in terms of the argument that emphases on population growth as a critical explanation of poverty and un-derdevelopment are based on premises which may have questionable validity when they are applied to the historical and present contexts of African develop-ment. Principal amongst these premises is the idea that available resources per capita are determined primarily by population size and population growth rates. As such, a decrease in population growth is an initial and necessary condition for enhancing standards of living. The World Bank (1981, pp. 112–113) observes that:

> [t]he consequences of rapid population growth for economic development and welfare are very negative. . . . Rapid population growth slows progress towards universal education and health care. . . . The question of food self-sufficiency become[s] more pressing.

From Bernard I. Logan, "Overpopulation and Poverty in Africa: Rethinking the Traditional Rela-tionship," *Tijdschrift voor Economische en Sociale Geografie,* vol. 82, no. 1 (1991). Copyright © 1991 by The Royal Dutch Geographical Society. References omitted.

Despite statements like these, it has not been documented unambiguously that a drop in population growth rates is a necessary condition, either for reducing the strains on the natural environment, or for improving standards of living in Africa. On the other side of the argument, there is also no empirical basis upon which to build an unequivocal case that an increase in population growth, by itself, is sufficient to reduce standards of living. In fact, it is difficult to discern any significant relationship in African economies between population size, population density and population growth rates versus standards of living or levels of economic development. Considered in isolation, therefore, demographic indicators do not provide adequate insights on present or projected standards of living in the region.

There is no doubt that high rates of population growth exacerbate some of the socioeconomic difficulties that prevail in Africa. Catastrophes like the drought and associated famine in the Sahel [in West Africa] are telling examples of these effects. Yet, it can be argued that the resource crises that have manifested themselves in the recent past in some areas of the continent are symptoms of more fundamental problems, linked to national and global networks of resource control. The population-resource relationship in Africa, then, must be understood within a framework that ventures beyond analyses of population growth, to examine evidences of resource availability and the mechanisms of resource control, resource utilization and resource transfers. An important responsibility of development analysis, therefore, is to place the population issue in a context that recognizes the significance of factors such as: (i) the organization and role of the state in resource management; (ii) the flow of resources within the existing International Economic Order (IEO); and (iii) the role of the international division of labor.

This study touches tangentially upon some of these issues, but it focusses especially upon the second. The following question is central to the discussion: should population growth be held responsible for poverty in Africa, or does population growth merely intensify the negative impacts of resource maldistribution? This question is treated largely and primarily from a conceptual rather than from an empirical perspective. . . .

Poverty and Population Growth in Africa

Conceptual framework and methods.   A thorough assessment of the relationships between population and poverty in Africa would be an extensive empirical undertaking. If it were possible to collect all of the relevant data, then techniques like multiple or stepwise regression, for example, would make it possible to identify the most significant predictors of standard of living, where the latter is measured by surrogates like GNP [gross national product] or GDP [gross domestic product] per capita (although even these measures may have questionable relevance for undeveloped economies). Unfortunately, reliable data with sufficient depth of coverage are unavailable for much of Africa. These data concerns (constraints) preclude a useful empirical investigation of the problem. As a result, much of the discussion in this section is conducted as a conceptual overview.

A limited amount of empirical analysis (simple correlation statistics) is utilized merely to complement the general discussions, and not to determine cause-effect linkages. The conceptual approach has been adopted here primarily to guard against overstretching the inferences from statistical analysis based on generalized aggregate data.

Eight world regions/economic groups are used in the statistical analysis:

(i)   East Bloc (the Soviet Union and East European countries)
(ii) USA and Canada
(iii) Western Europe
(iv) EEC
(v)   EFTA
(vi) Africa (excluding the Republic of South Africa)
(vii) Latin America (South and Central America, including Mexico)
(viii) Asia (excluding Japan and Israel but including the Near and Middle East)

Data are compiled for each region for twelve variables covering the period 1975–1985. There is some overlap both in the regional units and in the data groupings. However, these largely methodological questions do not bear unduly on the discussion.

The following two propositions are used to establish the conceptual framework for the analysis:

- *Proposition 1:* Although population growth and standards of living are directly associated for the developed economies, especially those in Europe, this association does not necessarily hold for Africa.
- *Proposition 2:* Standards of living in Africa are likely to be more closely associated with the mechanisms of domestic and international resource exchange than with demographic parameters such as population growth rates.

Before proceeding with the analysis proper, several caveats must be mentioned. ... In particular, no attempt is made to explain why the results for Asia are almost consistently different from those for Africa and Latin America. Further, generalizations about Africa or Europe in this section are not meant to suggest that conditions on these continents are uniform and monolithic. In the case of Africa, especially, there exists a very wide diversity of environmental and resource conditions, with an equally large array of population characteristics and standards of living. Since it is impossible to accommodate this degree of heterogeneity in a general study like this, localized attributes are distinguished from general patterns only in order to facilitate and clarify the discussion. Finally, some very broad generalizations are made concerning international resource exchange. In examining the aggregated data used in this study, it is understood that the developed economies engage in more trade amongst themselves than with the LDCs [less-developed countries]. The results of the statistical analysis, therefore, are used with caution, and only in a very general way.

Population growth and poverty in Africa.   The argument of the first proposition is that the European experience, which is characterized by a strong association between standards of living and population growth, is not very useful for understanding the poverty dilemma in Africa. Although population and living standards have a historical association for Europe, it is not clear that a similar association should be expected for Africa and other poor regions, or that these expectations provide a realistic or reasonable basis for population control and development planning.

The idea that demographic variables may not reveal a great deal about the general condition of poverty in Africa is not rejected by the correlation results. GDP per capita and population are not associated for Africa (nor for Latin America although they are for Asia). By contrast, the levels of association for the developed regions are generally consistent with orthodox expectations. Population measures (population size, population growth and population density) associate strongly with per capita GDP for all the developed regions.

It is acknowledged (Kasun 1988) that living standards in Europe have often been associated with population growth rates as enunciated by [Thomas] Malthus (1793). Because of this historical background, the fear of population-induced poverty in Europe and America at the turn of the twentieth century was strong enough to have provided the background logic and rationale for Social Darwinism and its concomitant emphasis on population control (Drogin 1979; Kasun 1988). The association between population growth and standards of living is reflected also in the fact that the economies of Western Europe expanded rapidly between the seventeenth and nineteenth centuries during a period of population export (large-scale emigration to the Americas, Africa, Asia, Australia and New Zealand) and raw material import (Hobsbawm 1978; Mabogunje 1981). Colonialism offered Europe access to cheap raw materials, and also a destination for 'excess' population. At present, the European countries must strike a balance between the demands caused by high densities, high life expectancies and high consumption rates by having low rates of population growth and high levels of global resource control. The dynamics between these factors, described below, indicate why population growth rate and standards of living are connected for the West.

European countries are generally densely populated and maintain very high rates of resource consumption. Since they are usually not richly endowed by nature, these countries must depend on continued access to resources from other parts of the world to sustain their economic expansion. Under these conditions, rapid population increase would impede economic growth unless greater quantities of overseas resources could be obtained. European economies, therefore, must either devote larger expenditures to imported raw materials, or they must establish tighter control over the mechanisms of international resource exchange. The structure of international trade in the past two or three decades suggests that the latter option has been chosen. Thus, although the dominant economies of Europe have relinquished direct control over Third World resources, the international trade system remains firmly in their favor. Calls for a new international economic order at the General Agreement on Tariffs and Trade (GATT) and the United Nations Conference on Trade and Development

(UNCTAD) have met with only minor changes on behalf of the competitiveness of LDCs (see Harris & Harris 1981; Greene 1981; Lone 1989).

Although population growth rates in Europe are among the lowest in the world (Population Reference Bureau 1988), the high rates of resource consumption referred to above are also far in excess of those of most other world regions. Given Europe's generally high population densities even with present low rates of natural increase, its state of 'High Mass Consumption', to borrow [Walt W.] Rostow's terminology, can only be maintained with extremely favorable terms of trade. Thus, population growth rates must be at par with: (i) the rate at which new sources of cheap overseas resources are identified and controlled; and/or (ii) the rate of price increase for the raw materials from traditional Third World sources. By and large, these two conditions are met through the ability of the West to ensure that the status quo is destabilized as infrequently as possible. The fact that cheap raw materials are available reduces the need for conservation in the developed economies. These economies, therefore, are extremely sensitive and vulnerable to changes in their favorable terms of trade, as was shown by the impacts of the oil price shocks of the mid-seventies and early eighties.

The strong association between population growth and standards of living which is true for much of Europe may not necessarily hold for other developed regions where population densities are generally low and natural resources are plentiful. The U.S., Canada and Australia, for example, are developed economies which are also net 'importers' of population. Economic expansion is complemented by a process of selective migration which contributes to increased affluence and improved living standards in these countries. Even though they also benefit from the organization of international trade, the wealth of their resource base offers them a wider array of options than is available to the European economies. High rates of consumption, then, are supplemented with domestic resources, many of which are absent in Europe.

The expectation of an inverse association between population growth and living standards is even more questionable for Africa and other less developed regions, primarily because these economies are operating under a completely different set of conditions from those which existed and still prevail for Europe. First of all, the structures of the international political economy (for example, tariff and non-tariff barriers) make economic expansion difficult, if not insurmountable for newcomers. The international trade network is also controlled by global monopolies and oligopolies; and the commodities market in which many African products are traded is a quasi-monopsony. Africa's condition may also be distinguished from Europe's by its generally lower population densities and its general possession of a wider range of natural resources (see, for example, data on regional production of raw materials in the UNCTAD: Handbook of Commodity Statistics, various years).

There is no doubt that poverty and malnutrition are rife in many areas of Africa. However, if it is true that natural resources are generally abundant, and that population densities are generally low (and there is little available evidence to suggest otherwise), then it is difficult to understand how poverty can be traced primarily to population pressures. Two arguments, both of which bear examination, can be made in support of the population pressure thesis: (i) that low densities in some areas of Africa merely reflect low carrying capacity, and that such areas are actually overpopulated; (ii) that there are several areas where very high densities have resulted in overpopulation.

With regard to the first argument, low densities in certain regions (for example, much of the West African Sahel, and Namibia and Botswana in the south-west) can be attributed to the hostility of the environment; this and infertile soils, diseases and pests, inhibitive temperature regimes, and scarce water resources. Since the carrying capacity is low in these areas, population densities must also be low or overpopulation would manifest itself in catastrophes like the recent famines in the Sahel. This line of environmental determinism, however, leaves unanswered some important questions of resource allocation and productivity in Africa. For instance, why do many African countries, including those that are beset by the worst environmental problems, remain net exporters of food? Data from the UNCTAD and FAO [Food and Agriculture Organization] Commodity Yearbooks suggest that aggregate agricultural exports from Africa, including food exports, have either held steady or increased throughout the eighties. It has been reported also, that even at the height of environmental adversity, a number of African countries remained net food exporters (*Dollars and Sense* 1978, p. 5).

> The Sahel [was] a net exporter of barley, beans, peanuts, fresh vegetables and beef, despite protein malnutrition among its children that is about the worst in the world, even in normal years.

The FAO has classified African countries into three groups of food self-sufficiency on the basis of aggregate agricultural output (FAO 1986). Only eight countries (Cape Verde, Mauritania, Burkina Faso, Chad, Angola, Rwanda and Burundi) are identified as not being food self-sufficient. The rest either produce enough food or, in the case of the oil-exporters, can import their required shortfalls. If this classification is based on accurate data, it is clear that Africa should not have a food problem. Yet, it is equally clear that there are many regions on the continent that suffer from acute and chronic cases of famine and malnutrition (Hansen & McMillan 1986; Lofchie 1986). These kinds of evidence tend to question the efficacy of environmental determinism as an explanation of low agricultural productivity and the attendant low standards of living in Africa. Environmental factors, acting alone, cannot possibly have different impacts on export as opposed to domestic food crops. They also cannot explain why food exports are stable or increasing in the midst of famine and poverty.

If poverty is caused primarily by environmental constraints, might it be reasonable to expect that levels of welfare and living standards should be higher in the areas with the least constraining environmental conditions? The Atlas of African Agriculture (FAO 1986) identifies certain regions in Africa which have the most suitable soil and climatic conditions for agricultural development. These include coastal West Africa and much of Central Africa. Although these physiographic zones do not coincide with political units, it is easy to determine that they also do not coincide in any meaningful way with patterns of population distribution, levels of agricultural productivity or standards of living on the continent. It becomes difficult, therefore, to link poverty in Africa directly or primarily to environmental constraints.

The second argument concerns cases of overpopulation in Africa. It points to several 'population islands' as evidence that the carrying capacity has been exceeded in some regions. Gleave & White (1972), Hance (1975), Mabogunje (1981) and, more recently, the FAO (1986) identify several such high-density clusters, among them: the region around Lake Victoria; Uganda, Rwanda and Burundi; the Niger delta; and southern Ghana. The FAO report states that:

> ... Burundi, Botswana and Kenya, are already facing serious shortages [of land] in some areas, with a growing number of landless. Projected population increases by the year 2010 will further reduce the amount of available land per caput. (FAO 1986, p. 52).

The FAO report states also that as a result of rapid population growth, the following sub-Saharan countries now have less than fifty percent of their cultivable land still available for agricultural use: Nigeria, Guinea, Sierra Leone, Gambia, Senegal, Mauritania, Togo, Ethiopia, Somalia, Djibouti, Kenya, Botswana, Lesotho, Rwanda, Burundi, Burkina Faso and Niger. Although these statements suggest that many areas are approaching or may have exceeded their carrying capacity, there still is no incontestable evidence that the 'population islands' or the countries in which they are located are overpopulated. There are no distinguishable patterns to link high densities to low levels of development. The countries on the list include some which are making admirable strides towards development (Kenya and Djibouti), and many which are in grave states of destitution.

Except for the Horn (Ethiopia and Somalia) and some other parts of the Sahel where production problems are exacerbated by drought and warfare (for example, Mozambique and Chad), the difficulties faced by African economies surmount environmental circumstances or levels of population density. To understand the causes of poverty, one must venture further afield. Since land is so central to economic production in Africa, the means of land allocation is one area of fundamental concern. An analysis of the relationship between available land and population size, by itself, does not provide sufficient insight into the problems of productivity, food availability and standards of living. The postcolonial experiences of many LDCs, including those in Africa (Tanzania, Kenya, Ethiopia and Guinea Bissau, to name a few), have shown that land distribution and land use provide a relevant and essential context for understanding standards of living and political stability. By creating a class of landless peasants, the process of land expropriation, both in the colonial and post-independence eras (Helen 1972); the slow pace and sometimes misguided implementation of land reform measures (cf. Low 1989); and the intensive government promotion of export-crop production (Sullivan & Martin 1988) all feature prominently and bear directly on Africa's low standards of living. Some of these concerns are explored in the next section. Meanwhile, to reduce the poverty problem merely to one of population analysis and control is to confuse a symptom, albeit a conspicuous one, with the cause.

Resource exchange and poverty in Africa. The contention of Proposition Two is that low standards of living in Africa can be understood better by exploring the relationships between resource availability—resource control and exchange— rather than by tracing the causes of poverty to a direct relationship between resource availability and population growth. The second approach, which typifies orthodox analyses, implies that resource availability is a state, determined only by the natural occurrence of materials. The idea that resource availability is tied closely to the structures of resource exchange must be integrated fully into the population-poverty discourse.

The terms of trade are used in this section to elaborate on the influences of resource exchange on standards of living in Africa. One part of the argument is that standards of living in poor regions like Africa are not likely to be improved through export earnings, which are usually too small to make a significant difference to the well-being and livelihood of the vast majority of the population. The converse is likely to be true for developed economies. Since the terms of trade are in their favor, high standards of living in developed countries will be based on income from the export of high-priced manufactures and services. The impacts of the costs of imports will have opposite effects on each type of economy. While export earnings for LDCs are likely to be too small to improve standards of living, the costs of imports are likely to cause living standards to deteriorate. The high 'value-added' to manufactured goods and services, coupled with the low price of primary commodities, would reduce and possibly trivialize the benefits of international trade for African economies. The discrepancy between export earnings and the price of imports will worsen as the terms of trade become progressively more biased.

The correlation results for imports and the terms of trade are mixed. However, the correlations for Africa and Asia for the terms of trade do not refute the general contention of Proposition Two. Further, the export measures show a pattern that distinguishes the developed economies from the LDCs. The correlations between GDP per capita and purchasing power of exports (variable 5) are generally weak for the less developed regions (except Asia) and generally strong for the developed regions. This is true also for the value of total exports (variable 9) and growth rates of total exports (variable 10). The Integrated Programme for Commodities (IPC) covers eighteen primary products grouped into three categories: (i) food (bananas, bovine meat, cocoa beans, coffee, sugar, tea and vegetable oils); (ii) agricultural raw materials (cotton, hard fibers, jute, natural rubber and timber); (iii) ores and metals (bauxite, copper, iron ore, manganese ore, phosphate rock and tin). Except for crude oil, these products cover the principal basket of export commodities of most LDCs. The weak correlation between GDP per capita and value of IPC exports (variable 7) for Africa lends some credibility to the conceptual thrust of Proposition Two.

It is not unreasonable to expect that low living standards in Africa should be linked to poor terms of resource exchange because international trade is so central to the living standards of most countries. When the export earnings of an economy are consistently small, it should be expected that conditions will be created which will perpetuate poverty. With terms of trade biased against them, African economies are forced to spend large amounts of national resources to obtain imported requisites for development programs. The high price of African imports relative to exports limits the ability of governments to obtain inputs for critical sectors like agriculture, health and education. Austerity programs and capital constraints impose economic strains that are borne directly by the large population of rural and urban poor. Worsening terms of trade force governments to adopt policies that place increasingly greater stress on these segments of the population.

The processes by which biased terms of trade undermine standards of living through their impacts on export earnings may be pursued by evaluating three interlinked factors: (i) unfavorable primary commodity prices; (ii) biased domestic terms of trade; and (iii) inequitable allocation of resources within the domestic economy.

. . . [U]nfavorable terms of trade reduce the value of national resources below that of available resources for poor regions, because the international price mechanism does not perform the vital function of equating social value effectively between LDCs and DCs [developed countries]. As a result, the social value of African exports in aggregate terms is far in excess of the social value of its imports. As explained above, international trade acts as a means by which the value of manufactures and services are made 'artificially' higher than their actual social value to both trading partners. By contrast, primary products are exchanged at a rate far below their actual social value to both trading partners. If African countries could manufacture many of the goods they have traditionally imported, it is likely that they would need to exploit far smaller amounts of their national resources (materials, time and labor). Unfortunately, import-substitution industrialization has not succeeded in Africa for a variety of reasons that have been identified in the orthodox literature (cf. Mabogunje 1981; Todaro 1989). Primary amongst these are capital scarcity, improper maintenance of equipment, government mismanagement and the absence of a cadre of skilled workers. However, explanations for Africa's industrial failure can be found in other arenas of the political economy. Most of Africa's infant industries have depended on foreign capital; since many of them have been assembly plants, they have also relied on imported parts; and since it is difficult to break the barriers of international trade, many African industries have been forced to depend on their domestic markets which normally have a low purchasing power. The absence of modern industrial development in Africa, therefore, may be tied, at least in part, to the structures of international trade and the international economic order (Wilber 1988).

The difficulties surrounding the establishment of a solid industrial base in the present global system have forced many African economies to remain producers of primary commodities whose value severely limits economic progress. In essence, the price of these exports is so low that it does not generate sufficient national incomes to induce meaningful changes in the domestic economies.

These problems of international exchange have become aggravated in the recent past by the economic adjustment policies sponsored by the World Bank and the International Monetary Fund. The negative impacts of these policies on socio-economic development in Africa have been examined at length elsewhere (Cornia *et al.* 1987; Cox & Owen 1988; Lone 1989). One of the central elements of adjustment is devaluation, and it merits some brief attention here because of its direct impacts on the export earnings of African economies. Devaluation is based on the theoretical assumption that it would increase the attractiveness of African exports in foreign markets. Contrary to World Bank assessments (World Bank/UNDP 1989), the experiences of many African countries in the 1980s (for example, Tanzania, Sudan and Senegal) have shown that, far from initiating or accelerating economic growth, devaluation has actually dampened economic performance (Lone 1989). The negative impacts of devaluation on national earnings ensue from several factors, but two are especially cogent. (i) Many African countries produce similar commodities. When several countries de-value simultaneously, the commodities market becomes a buyers' market, with a concomitant drop in export earnings. These effects are reflected in the fact that the prices of Africa's major export commodities are presently at their lowest in fifty years (Smith 1988). (ii) The problem is complicated by the fact that many of Africa's exports, especially the agricultural products, have a low demand elas-ticity. A drop in real prices, therefore, does not result in a significant increase in demand. Consequently, devaluation has caused Africa's poor terms of trade to deteriorate even further, thereby, aggravating the poverty dilemma.

Africa's low export earnings capacity can be analyzed also within the con-text of the continent's traditional economic production. The traditional rural population, which forms the bulk of Africa's population (Lele 1984; Population Reference Bureau 1988), is engaged predominantly in extensive agricultural production. As shown [elsewhere], orthodox analysts consider the methods of production within this sector to be integral to the poverty dilemma. Undoubt-edly, the technology and modes of production within the traditional sector need serious attention. However, it can be argued that the system can support the continent's population quite adequately with the right kinds of resource support. After all, much of Africa was food self-sufficient until the 1960s (Lofchie 1986), and there is no clear evidence that population growth in the past three decades has absolutely exceeded the food-producing capacity of the continent. As shown earlier, several countries still export food while their popu-lations go hungry, although it is almost certain that the bulk of food exports are produced outside the traditional sector. One is faced, therefore, with a problem of resource allocation in which the agricultural exports subsector is promoted at the expense of the domestic food producing traditional subsector. These con-ditions force one to the conclusion that the difficulties of traditional produc-tion and the accompanying low standards of living in Africa do not arise necessarily from environmental or technological constraints; nor necessarily from the fact that the demands of an increasing population renders the tradi-tional system obsolete and ineffective. It is also not easily attributable to high rates of population growth. An underlying critical problem is that the tradi-tional system has been dismantled by the encroaching needs of the modern sec-tor (Helen 1972; Mazrui 1987), and an effective and productive alternative has yet to be established in its place.

The demands of the modern sector have caused increasingly larger amounts of land to be taken out of rural production. Due partly to unfavorable terms of international trade and partly to internal misallocation of funds, the use of rural lands for mining and plantation agriculture has not spurred economic growth and positive externalities to benefit the rural population. The predicament of Africa's poor, then, comes from two directions. In the first place, they are dispossessed of their productive resources (land and labor), which are used for export production. This immediately causes reduced standards of living through its impacts on food production. In the second place, the earnings from the sale of exports are not sufficient to compensate the poor for the loss of their livelihood. The problem of food production and standards of living within the traditional sector may be summarized, therefore, in a series of interrelated scenarios that are quite different from the views encapsulated earlier in the vicious cycle of population-induced poverty.

1. Land expropriation forces the traditional, food-producing sector to cultivate increasingly smaller amounts of land more intensively (not with more capital inputs, but over longer periods). As more and more land is taken away from this sector, producers are forced to reduce fallow periods from forest fallow through bush fallow to annual cropping. Since levels of inputs are low, productivity declines, and the rural sector loses the ability to feed itself.

2. Land expropriation weakens the rural economy further by creating a class of landless peasants. A portion of this disenfranchised work force provides cheap labor for the export-based enterprises that have been established on their lands. However, many others remain unemployed and become part of a stream of rural-urban migrants who form what has come to be commonly known as the urban underclass. Like the rural poor, this class of society lives at the periphery of the modern economy, at or below subsistence levels (cf. Stren & White 1989). An important reason why this large segment of the population cannot be absorbed into the mainstream economy is that the rate of expansion of African economies is very slow, a condition which can be linked, at least partially, to low net incomes from international trade.

3. Besides unfavorable terms of international trade, the portion of the rural sector that engages in some export crop production is impoverished by unfair domestic pricing policies. A primary mechanism by which the modern sector can obtain a reasonable level of income from export trade is to underpay peasant producers.

These problems are exacerbated by other processes in the internal political economy of Africa. One important factor is the politicization of the resource allocation process. Contrary to orthodox analysis, this process is not inherently bad for development, nor is it necessarily a form of 'government mismanagement'. It tends to work well under certain centralized, authoritarian, patriarchal systems as are found in Zaire and Côte d'Ivoire, where it is used to maintain political stability by compensating client ethnic groups and regions. However, when political considerations override all other planning criteria (such as maximizing social welfare returns), plan implementation can lead to an unrealistic distribution of scarce resources which, in turn, can inhibit economic development. Logan (1985) has shown, for example, that the allocation of funds for rural development in Sierra Leone is often based on the colonial urban administrative hierarchy or on ethnic affinity, both of which may be inappropriate for meeting the needs of the bulk of the rural population. Factors like these seriously reduce the social welfare benefits that could be derived from scarce and limited foreign earnings.

In terms of its impacts on overall standards of living, another important, yet generally ignored element of Africa's internal economic structure is the nature of its markets. Due to the subsistence existence and low purchasing power of Africa's large traditional sector, its needs are hardly reflected in the market. The absence of domestic purchasing power, coupled with the need for foreign earnings caused by the low price of exports, feeds a process of export promotion by which a large proportion of exploited national resources become targeted for overseas markets. The undeveloped nature of African markets limits effective demand far below real demand for critical national resources, including food. Low living conditions, then, are aggravated by the inability of domestic consumers, including even some with purchasing power, to compete with foreign demand for most commodities. This situation goes a long way to explaining why African countries remain net exporters of food even in the midst of widespread domestic famine.

Conclusions

No single factor can be isolated as being totally responsible for Africa's present poverty. However, as developmentalists trace the problem to its root causes, it is incumbent that the process is elevated above ideology and epistemological differences. If development analysis has a *raison d'être,* it is to establish a rational and objective framework within which to attack poverty in LDCs. Obviously, a critical part of that process involves not only the definition of poverty, but also identifying critical factors to which it is related. Only in this way will it become possible to establish the cause of the problem. How these tasks are accomplished has immense implications for the way in which development strategies are formulated and implemented. Even more importantly, it may hold the key to uplifting the standards of living of a vast majority of the world's population.

General wisdom within the dominant orthodox perspective ascribes Africa's poverty to population pressures. This factor cannot be overlooked because rapid rates of population increase can only aggravate existing conditions—but exacerbating poverty is not the same as creating poverty. Standards of living cannot be divorced from resource availability, as has been shown in radical analysis. However, it is also important to link overpopulation to the mechanisms through which resources are controlled and exchanged. Defining overpopulation and poverty only in terms of population growth is unrealistic since it assumes either that resource exchange is an unimportant consideration, or that resource exchange is equitable to the point that it has no immediate importance for standards of living. On the other hand, the radical position that poverty is caused by resource exploitation does not emphasize the direct relationship between resource exploitation and overpopulation.

POSTSCRIPT

Is "Overpopulation" a Major Cause of Poverty in Africa?

One important aspect of the two arguments is that the scale of analysis is different. Dasgupta is largely discussing the problem at the household and community levels, whereas Logan, at least in his empirical analysis, is addressing the issue at the national and international levels.

Another important aspect to note are the connections between different scales that the two authors emphasize. Scale, in this instance, refers to a level of analysis, decision making, or economic activity. Dasgupta is concerned with household-level decision making regarding fertility and the influence that interactions with other households, or the community, may have on these decisions. In contrast, Logan focuses on the influence that economic trading patterns at the international level may have on an African nation's ability to feed itself and prosper.

Another interesting (and related) scale comparision to note is Logan's concern with poor terms of exchange for African countries and Engels's concern (discussed in the introduction to this issue) with low wages for workers. This is the same problem in many ways, only expressed at a different scale.

One factor that Dasgupta explicitly refers to that reduces the cost of child-rearing in some African countries is the practice of child fostering. More recent research on this practice suggests that it may lighten the cost of childrearing in some instances, but not necessarily in others. This is because children are sometimes fostered out not as a way of reducing expenses, but as a way to help other families who need labor. Child fostering in this context casts the exchange as a net loss of labor for the parents rather than a net reduction in expenses. As such, viewing child fostering as a practice that contributes to higher fertility levels is debatable. For more on child fostering, see an article by Sarah Castle in the summer 1996 issue of *Human Organization: The Journal of Applied Anthropology* entitled "The Current and Inter-Generational Impact of Child Fostering on Children's Nutritional Status in Rural Mali."

For more information on this issue from the Malthusian perspective, see a 1993 World Bank Discussion Paper (#189) by S. D. Mink entitled "Poverty, Population, and Environment." A slightly more challenging piece, and one from the political economy perspective, is a classic article by David Harvey in *Economic Geography* (July 1974) entitled "Population, Resources and the Ideology of Science," in which he questions why the scholarly and public policy communities have given considerable attention to the population variable to the exclusion of other potential variables (consumption, technology) in the population-resource-poverty relationship.

ISSUE 14

Is Sexual Promiscuity a Major Reason for the HIV/AIDS Epidemic in Africa?

YES: William A. Rushing, from *The AIDS Epidemic: Social Dimensions of an Infectious Disease* (Westview Press, 1995)

NO: Joseph R. Oppong and Ezekiel Kalipeni, from "A Cross-Cultural Perspective on AIDS in Africa: A Response to Rushing," *African Rural and Urban Studies* (1996)

ISSUE SUMMARY

YES: William A. Rushing, late professor of sociology at Vanderbilt University, explains the high prevalence of HIV/AIDS in Africa in terms of how Africans express and give social meaning to sex. He argues that the confluence of a set of sex-related behavioral patterns and gender stratification explains the HIV/AIDS infection rate. According to Rushing, these behavioral patterns include polygamous marriage practices, weak conjugal bonds, the transactional nature of sexual relations, the centrality of sexual conquest to male identity, and sex-positive cultures.

NO: Joseph R. Oppong, associate professor of geography at the University of North Texas, and Ezekiel Kalipeni, associate professor of geography at the University of Illinois at Urbana-Champaign, take issue with Rushing's conclusions. They contend that his analysis is Americentric, suffers from overgeneralizations, and problematically depicts Africans as sex-positive (and by implication, promiscuous and immoral). They assert that Rushing's cultural stereotypes are far too general to provide any meaningful insight into the AIDS crisis in Africa. An understanding of historical and contemporary migration patterns, as well as associated phenomena, better explain the spread of the virus.

Halting the spread of HIV/AIDS is one of the greatest challenges facing contemporary Africa. Unlike some other diseases, AIDS is particularly problematic because it strikes the working-age population and thus has serious economic and social consequences. In 1999 AIDS became the leading cause of death in Africa, overtaking malaria. As of 2002 the United Nations AIDS program (UNAIDS) reported that 29.4 million adults were infected with the HIV virus in sub-Saharan Africa, which accounts for about 70 percent of the infected population worldwide. Through 2001 approximately 21.5 million Africans had lost their lives to AIDS. The overall infection rate among adults in sub-Saharan Africa is estimated at 8.8 percent (compared with 1.2 percent worldwide), although this rate varies considerably throughout the region. Twelve countries, mostly in eastern and southern Africa, have infection rates above 10 percent. Botswana leads the continent with an infection rate of 38.8 percent, while Zimbabwe, Swaziland, and Lesotho also have infection rates above 30 percent. South Africa, Zambia, and Namibia have rates between 20 and 25 percent. West Africa has been hit less hard by the disease. Notable exceptions are Ivory Coast with an infection rate of 9.7 percent and Nigeria with a rate of 5.8 percent. The case of Nigeria is particularly worrisome because it is the most populous country in Africa.

The spatial pattern of the disease has changed over time. Initially, it was distributed along major highways and in the urban centers of eastern and central Africa. More recently, the epicenter of the virus has moved to southern Africa. In Africa, the disease is largely transmitted through unprotected heterosexual sex and unsafe medical practices. Truck drivers, prostitutes, and military personnel all have above-average infection rates and are believed to play a significant role in the spread of the virus. Women comprise 58 percent of those infected in Africa.

In the following selections, William A. Rushing argues that it is the confluence of a set of sex-related behavioral patterns and gender stratification that explains the HIV/AIDS infection rate in Africa. Joseph R. Oppong and Ezekiel Kalipeni maintain that historical and contemporary migration patterns, as well as associated phenomena, are a major reason for the HIV/AIDS epidemic in Africa.

A nice aspect of the readings selected for this issue is that Oppong and Kalipeni wrote their selection directly in response to the chapter in Rushing's book dealing with Africa. An additional benefit is that the writings capture the two main lines of argumentation used to interpret the spread of HIV/AIDS in Africa, that is, those based on cultural factors and those based on political/economic factors.

William A. Rushing

 **YES**

The AIDS Epidemic: Social Dimensions of an Infectious Disease

The Cross-Cultural Perspective

AIDS in Africa

Statistics show that since 1985, AIDS has been increasing faster in Africa than in any other region of the world (Mann et al., 1992:893–901). Surveys indicate that as many as 5 percent of the populations of Uganda, Rwanda, and Ivory Coast are infected. More than 20 percent of pregnant women in some urban areas are infected (Mann et al., 1992:41–47, 65), and extrapolation to the general population in one city (Kigali) gives an estimated *annual incidence* of 3–5 percent (Bucyendore et al., 1993). The International AIDS Center of the Harvard School of Public Health estimated that in 1992 more than 8 million Africans were infected, which would be about 65 percent of all cases worldwide (Mann et al., 1992:89–90). HIV-AIDS is about evenly distributed between males and females (Mann et al., 1992:76).

Several factors account for the pattern of high rates and balanced sex distribution. Poor nutrition and many infectious diseases compromise immune systems and possibly make people susceptible to HIV. Medical technology to screen donor blood for HIV and disposable hypodermics are quite limited. HIV-contaminated needles may be reused without being properly sterilized (Mann et al., 1992:433–434; Root-Bernstein, 1993:301–309).

In addition, STDs [sexually transmitted diseases] are exceedingly prevalent in Africa. For example, in a review of STDs in developing countries, Robert Brunham and Alan Ronald (1991:61) concluded that in some African countries STDs are one of the top five diseases for persons seeking health services. STDs are especially high for persons (men and women) infected with HIV (Allen et al., 1991:1660; Plummer et al., 1991:236; Plourde et al., 1992:89–90; Dallabetta et al., 1993:40). An epidemic of STDs appeared in the years preceding the AIDS epidemic in Africa (Arya and Bennet, 1976; Osoba, 1981; Mann et al., 1992:167), just as it did among American gays. One study in the early 1980s revealed that as many as 20 percent of Zimbabwe's urban population had an

STD (Ungar, 1989:331). Another study showed that 10 percent of Ugandans had gonococcal infections in 1981; in comparison, only 0.4 percent of American and Europeans had these infections, and differences of similar magnitude for syphilis and other STDs were reported (Root-Bernstein, 1993:165, 301–303). Given the presence of HIV, the high rate of STDs assures that HIV will be widely transmitted in heterosexual intercourse, which may account for 90 percent of HIV infections (Williams, 1992:46).

But high rates of STDs do not act alone. Polygamous behavior spreads STDs and hence HIV, whereas monogamous behavior limits the spread. Most experts agree that polygamous behavior is a major factor in African HIV-AIDS, and from a sociological point of view, it is the most relevant factor. It is also the focus of this [selection].

That STDs are so widespread is indirect evidence that polygamous behavior is also widespread. Ethnographic studies leave no doubt that having multiple sexual partners is a common cultural practice in many groups in Africa (Caldwell et al., 1989:205–216; see also Southall, 1961:52; Gregersen, 1983:190; Hrdy, 1987; Larson, 1989; Bledsoe, 1990). For this reason many Westerners claim Africans are "promiscuous." This view is ethnocentric. Sexual behavior varies widely across societies (Davenport, 1977; Gregersen, 1983; Becker, 1984), and there is simply no universal cultural standard against which sexual practice can be judged as promiscuous and excessive (or repressive); what is appropriate in one society may be deviant in others. Sociologically, if the customs of a particular society approve polygamous behavior, such behavior is normal and appropriate by the standards of that society. To call it promiscuous simply reflects the view of persons who belong to sex-negative societies in which monogamous relations are the normative ideal and having multiple partners is viewed as deviant and immoral. Such moralizing does nothing to enhance our understanding of the practice. Analysis based on the cross-cultural perspective does.

The Cross-Cultural Perspective and Sex

The central idea in the cross-cultural perspective is that behavior patterns that exist in one society but not in others or in varying degrees in different societies are the result of differences in the way societies structure and give meaning to behavior. This is true for sexual behavior no less than other types. Eroticism and, for heterosexual sex, reproduction are universal. But although the biological dynamics of sex are universal because the biology of sex does not vary with society, the social dynamics—how society structures the way sex is expressed and gives it social meaning—are not universal. Therefore, to understand why polygamous behavior is so common in Africa (and a major reason for the pattern of HIV-AIDS in Africa), we must understand why sexual expression is structured this way in African societies, or "tribes," and the social meaning sex has for members of most of these societies.

. . . Commonalities in the social dynamics of sex derive from (1) the traditional marriage institution and kinship ties, (2) cultural norms and beliefs about sexual expression (sexual culture), and (3) gender stratification. The descriptions that follow are for rural tribal societies, even though HIV is more serious in urban areas, because sexual practices in towns and cities are patterned on rural customs. In addition, the rates of HIV-AIDS in rural Africa are low only when compared to rates in urban Africa. In comparison to most places in the world, the prevalence of HIV-AIDS in many rural areas in Africa is very high.

Marriage and Kinship

Polygyny. Traditionally, in most African societies polygamy has been the preferred form of marriage. Polygyny is still widespread, though polyandry is (and traditionally has been) rare (Gregersen, 1983:190; see also Southall, 1961:52; Molnos, 1968:50–51; Caldwell et al., 1989:201; Bledsoe, 1990:117). In certain regions in Europe and Asia, polygamous (mostly polygynous) marriage is also normatively approved, but only about 3–4 percent of marriages are in this form, whereas according to the World Fertility Survey, up to 30–50 percent of all marriages in Africa are polygamous (Caldwell et al., 1989:201). Consequently, the proportion of all women involved in a polygynous marriage at some stage during their lives is very high. Since multiple wives increase the spread of STDs in a population, polygyny obviously contributes to the high HIV-AIDS rate as well as the balanced rate between males and females.

Patrilineage. Typically, after marriage the couple lives embedded in a compound or adjacent huts usually belonging to the husband's extended family, subclan, or clan. This makes for cohesive kinship ties, which are valued more than marital ties. Children are usually descended patrilineally and belong to the husband's kinship unit. They are valued as economic assets (to perform field labor and take care of parents and other relatives in old age) (Molnos, 1968:50; Lamb, 1987:33–34; Ungar 1989:184). Children also increase the numerical strength of clan and tribe. And a man's personal status within the clan-tribe rises as his progeny increases; traditionally, "the key measure of a man's wealth [has been] the number of dependents in his household" (Henn, 1984:5). Patrilineage thus gives men strong social incentives to acquire many wives (Southall, 1961:52; Caldwell et al., 1989:202).

It also promotes polygamous behavior outside marriage. Wives are valued largely for their potential as "baby machines" (Lamb, 1987:39). The respect they receive in the clan-tribe depend on the number of children they bear. Children are social and economic assets to women no less than to men. Consequently, the physical and emotional aspects of sex in marriage are subordinate to childbearing (Molnos, 1968:58, 79). This makes for weak conjugal bonds (Caldwell et al., 1989:199–189, 200; see also Radcliffe-Brown, 1950:51–54; Larson, 1989:722; Bledsoe, 1990:117; O'Connor, 1991:51), so that extramarital sex is common, normal, and even expected, though more so for men than women (Caldwell et al., 1989:212). And the fact that children born from such unions belong to the husband and his kinship group is an incentive for men to engage in extramarital affairs. Thus, since patrilineage encourages men to acquire wives and to engage in polygamous nonmarital behavior, it is a major etiological factor in the African pattern of HIV-AIDS.

Sexual Culture

Sexual culture refers to the cultural beliefs, attitudes, and norms regarding sex. In contrast to Americans, who usually view sex morally and think that people who have multiple partners (even if unmarried) are immoral and unfaithful, most Africans do not judge sexual behavior in such terms at all. They experience little guilt about sex, and they enter into sex more casually and have more sexual partners than Westerners do. The cultural beliefs and norms that do bear most directly on sex are the transactional element in sexual relations, a masculine sexual ideology, and sex-positive beliefs.

The transactional element in sexual relations. In general, Africans view sex as an ordinary activity, much like work (Caldwell et al., 1989:194, 209, 218). Traditional sexual ethics are similar to those that regulate other services, namely, the ethic of exchange (Caldwell et al., 1989:203). Sexual relations are characterized by a "transactional element" (Caldwell et al., 1989:202–205), which is especially explicit in the traditional marriage.

Marriage is primarily an arrangement between kinship groups rather than individuals (Radcliffe-Brown, 1950:41–54; Little, 1971:17–20, 1974:4–8; La Fontaine, 1974:112). In return for loss of the daughter's labor as well as for her sexual favors and the children she will produce for the husband and his clan, the wife's family receives a bride-price (traditionally in the form of cattle). The transaction is negotiated by family elders who have little, if any, concern for the wishes and passion of the couple (Lamb, 1987:37). (Indeed, in some instances marriages are arranged early in the couple's life and sometimes before birth). This approach to marriage contributes to weak conjugal bonds, so that divorce, separation, and desertion are common (Caldwell et al., 1989:201; Larson, 1989:720); see also Pankhurst and Jacobs, 1988).

In extramarital and premarital affairs, men are expected to give women money and gifts as an expression of affection, respect, and gratitude (Larson, 1989:723; Caldwell, et al., 1989:203; see also Shoumatoff, 1988:155). And for women, such affairs are important sources of income. Extramarital sex is less common for women than premarital sex (Larson, 1989:721), though a wife sometimes has several lovers, serially or concurrently (Obbo, 1980:151). This (or the threat thereof) may give her leverage over the husband and thus access to his economic resources (Caldwell et al., 1989:204). (This tactic may well backfire, however, especially if the husband has several wives.)

In general, then, in most sexual relations—whether casual liaisons, premarital relations, extramarital affairs, or marriage—"there is an economic core," with women "exchanging sexual favours and often also reproductive potential for economic benefits" (Barnett and Blaikie, 199:77). This exchange is explicitly acknowledged and normatively accepted. In some groups mothers may actually encourage their daughters to trade sexual favors for money and gifts as a way to provide for themselves (Caldwell et al., 1989:203–204).

Although many Westerners have difficulty seeing how this arrangement differs from prostitution, Western anthropologists who study African societies disagree on how prostitution should even be defined in the African context

(Molnos, 1968:79; Little, 1973:84; Caldwell et al., 1989:218–219; Dirasse, 1991:10). The usual definition of prostitution holds that a prostitute is a woman whose income is derived more or less exclusively from payment for brief impersonal encounters with all comers, most of whom are strangers, in which no services or activities beyond the sex act are involved. (Many Westerners use "prostitute" more loosely to refer to all women who use sex to elicit money and other material favors from men.) By this definition, prostitution in Africa appears to have been most limited; in fact, prior to colonization in most groups prostitutes as a category of women were not even recognized (Gregersen, 1983:15; Hrdy, 1987:1112; Caldwell et al., 1989:220–221). Even so, the transactional element in sex increases the spread of HIV, as the following example reveals.

In a study of lakeside trading villages in the Rakai district of Uganda, Tony Barnett and Piers Blaikie (1992:78) observed that women commonly support themselves through commodity trading. However, they also "set up independent households," with "one or more regular lovers who help them financially." These woman are obviously involved in sexual transactions (financial help for sex) but they are not prostitutes. Although lovers may give them money, non-sexual activity is central to the relationship; sexual activity is simply integrated in a round of other activities that the partners share. In addition, such relationships are not the sole (or even primary) source of women's livelihood. And these relationships involve more than a single sexual encounter, the partners are not strangers, and a degree of affection may be assumed. Little (1973:81) observed the same pattern in African towns. Similar arrangements do exist in the West, or course, though they are less common and less open. More significantly, the arrangement in the West in almost always limited to one partner extending over a period of time. In contrast, in Africa "the rate of partner change can be assumed to be fairly rapid" (Barnett and Blaikie, 1992:78). Although many Americans would consider this immoral, even as prostitution, Africans do not; this arrangement is best viewed as part of the transactional cultural norm in sex rather than as either promiscuity or prostitution. The implications for the transmission of HIV are clear. Barnett and Blaikie (1992: 32, 69) reported that in the Rakai district, around 40 percent of men and women between twenty and thirty are seropositive for HIV.

Masculine sexual ideology and the status of men.   In most African societies a plurality of sexual partners is a male right (Southall, 1961:52; Molnos, 1968:66; Caldwell et al., 1989:202; Larson, 1989:721; Barnett and Blaikie, 1992:77–78). This right is related to polygyny (if only as a rationalization for it) (Davenport, 1977:125). But it goes beyond polygyny since the right does not end with marriage. It is socially acceptable for a married man to have mistresses and "outside wives" (concubines) (Little, 1974b:17–18; Obbo, 1980:89; Larson, 1989:720). Men's extramarital relations are "taken for granted" and simply "expected of the normal man" (Caldwell et al., 1989:212). Indeed, sexual conquest and fatherhood are central to male identity, and children enhance a man's social status. This reinforces women's use of sex for material gain (Barnett and Blaikie, 1992:44, 77–78).

Masculine sexual ideology combines with the transactional element such that the more wealth a man has, the more sexual partners he can get. For example, a bride-price must be paid for a wife. Thus, traditionally "the key measure of a man's wealth [has been] the number of [wives and children] in his household" (Henn, 1984:5). Some chiefs are known to have hundreds of wives (Molnos, 1968:50). The association of number of female partners with male wealth suggests that HIV infection rates are higher among men with higher status than among men with lower status.

Sexual relations and a sex-positive culture. Most groups in Africa have sex-positive cultures. Sex is viewed as a part of courtship and a form of recreation, and relations between lovers are viewed as affairs between friends (Larson, 1989:723, 727). Despite the emphasis on sex for reproduction, *New Yorker* reporter Alex Shoumatoff (1988:154–155), who is married to a Rwandan, stated that most African societies "are unquestionably sex-positive." This is especially so for men, who tend to be "womanizers." Women "put up with it or participate in it depending on how much freedom they are allowed by their culture." (See also Barnett and Blaikie, 1992:77–78). And by all accounts, as the existence of the transactional element in sex would indicate, in many groups women are allowed considerable freedom indeed (Molnos, 1968:58). Premarital sex for females is accepted (Molnos, 1968:58–59; Caldwell et al., 1989:195, 197, 203–205; Larson, 1989:727), as is female adultery (Caldwell et al., 1989:197, 199, 212; Larson, 1989:723). According to Laketch Dirasse (1991:57), in at least one tribe (the Borana), men allow their wives "to have as many lovers as they want" (see also Bledsoe, 1990:123). Wife sharing is also reported for a number of societies, which permit or even require a wife to have sex with persons besides her husband, most frequently distant relatives of the husband's clan, sometimes as a form of hospitality to guests (Gregersen, 1983:190; Caldwell et al., 1989:213). In some groups a widow is inherited by one of her husband's brothers (levirate), and a woman may have sex with each brother to see which one would please her most in case her husband dies (Schuster, 1979:14).

In sum, for females as well as males, "fairly permissive . . . sexual attitudes are found generally across sub-Saharan Africa" (Caldwell et al., 1989:222). Many scholars of African society have observed that a "wide range of all types of unstable and occasional [sexual] unions" are widespread in Africa (Molnos, 1968:79). Beyond polygamous relations, sex is simply "regarded . . . positively [and as] normal and good for the health, and which, if not experienced, might well result in ill health" (Caldwell et al., 1989:209). In short, sex is viewed in sex-positive terms.

At the same time, sex is socially regulated. That the traditional African marriage is an economic arrangement between families limits the choices individuals have in mate selection. Family decisions are constrained by the custom of exogamy, and lovers usually must also be selected from outside the clan, sub-clan, or tribe (Davenport, 1977:125; Lamb, 1987:11). In other instances groups stipulate that the wife may have sex only with her husband's relatives or members of his age group (Gregersen, 1983:190; Caldwell et al., 1989:213). The nature of village life puts women under the surveillance of the husband's relatives and clan, and any deviation from the duties of wife and mother are apt to be quickly detected, as are deviations from restrictions on the tribal affiliation of sex partners. Even so, for most of Africa social norms permit and even encourage sex with multiple partners. Polygamous sexual relations are thus widespread for the married no less than the unmarried (Gregersen, 1983:186). This is the hallmark of a sex-positive culture. It also facilitates the spread of HIV.

**Joseph R. Oppong and
Ezekiel Kalipeni**

A Cross-Cultural Perspective on AIDS in Africa: A Response to Rushing

Introduction

Data compiled by the World Health Organization and from various surveys seem to indicate that HIV/AIDS in Africa has reached epidemic proportions, particularly in the so-called AIDS epicenter in central African countries. It is usually noted that anywhere from 5 to 10 percent of the urban population may be infected. Rates of infection are assumed to be even higher for certain sectors of society such as commercial sex workers, pregnant women in urban areas, and truck drivers. The share of the disease between the sexes is equal and hetero-sexual contact is considered the main mode of transmission. Over the past decade research has shown that HIV/AIDS infections have increased at a faster rate in Africa than anywhere else. It is estimated that in excess of eight million Africans are believed to be infected with the deadly virus and more than 1.5 million have full-blown AIDS (Tastemain and Coles 1993). This is about 65 percent of all cases worldwide. Although Africa's share of HIV-infected persons is ex-pected to decrease from the current two-thirds to one-half of the world's total by the year 2000, Good (1995) points out that this relative decrease cannot be celebrated, because the "long wave" character of the AIDS epidemic will con-tinue to spiral out of control in many African countries. It is undeniable that the socioeconomic and demographic consequences of the epidemic are serious, since it strikes at the heart of the work force, mostly young men and women in their prime productive and reproductive years. Yet most Western approaches to the study of the factors that facilitate the transmission of the disease tend to be largely overgeneralizations and too simplistic. Indeed, as [Angus] Nicoll and [Phyllida] Brown (1994) point out, the images of Africa conjured in Western minds, perpetuated by the biased media, have been those of an oversimplified exotic place variously depicted as a game park or an apocalyptic vision of famine and civil war. Recent ominous accounts by notable journalists such as [R. D.] Kaplan (1994) have tended to perpetuate such stereotypes.

In the medical and epidemiological arena, the different pattern of AIDS infection exhibited by African countries has resulted in the development of a plethora of research on AIDS in Africa which, as [Randall M.] Packard and [Paul] Epstein (1991) note, resembles earlier narrow-minded colonial efforts to understand the epidemiological patterns of TB and syphilis. Indeed, current research on the AIDS epidemic in Africa has tended to focus on why Africa exhibits a different epidemiological pattern than that found in the West and elsewhere. Explanations of the different pattern invariably lay blame on the peculiarities of African customs, traditions, and behaviors that relate to issues of sexuality and reproduction at the expense of a range of other equally significant factors such as the colonial historical context, poverty, dependency, and underdevelopment (Packard and Epstein 1991).

In this [selection] we take issue with [William A.] Rushing's recent work titled "The Cross-Cultural Perspective of AIDS in Africa," which appears as chapter 3 in his book *The AIDS Epidemic: Social Dimensions of an Infectious Disease,* published in 1995 by Westview Press. The discussion contained in Rushing's paper follows the overgeneralization syndrome, as overly Americentric, and depicts Africans as being sex-positive and, by implication, promiscuous and immoral while Americans are sex-negative and morally upright and hence less akin to HIV infection. In our response to Rushing's stereotypical assertions, we argue that Rushing's cultural stereotypes are far too general to be of any meaningful use to the understanding of the AIDS epidemic in Africa, and that such careless propositions tend only to encourage a premature narrowing of research questions as happened during the colonial era in the cases of tuberculosis (TB) and syphilis (see for example, Fendell 1963; Packard 1987; Packard and Epstein 1991).

The Overgeneralization Syndrome

[Benjamin] Ofori-Amoah (1995) has correctly observed that overgeneralization frequently characterizes research on Africa. First, studies based on specific national or people groups assume an African or sub-Saharan Africa title when it comes to publication of results. Thus, a study on AIDS in Uganda with a special focus on Rakai District or Buganda region is called *AIDS in Africa* (Barnett and Blaikie 1992). Second, despite the rich cultural mosaic and differences in geographical, economic, and historical experiences, Africa is portrayed as culturally homogenous. Rushing's (1995) work is replete with several excellent examples of overgeneralization as he portrays a sex-positive African culture, which sees sex as a recreation activity, and thus, "guides women into prostitution more easily than Western culture does" (73–74) as a major causal factor of AIDS in Africa. While generalizing results is a critical step for theory development in the search for universalism in social science (Nachimas and Nachimas 1981), loose generalizations can be inappropriate.

Rushing creates the impression that Africans are homogeneously promiscuous, that perhaps sexual promiscuity is the one attribute common to Africa's numerous cultures. Africa is discussed as though it were one country, not some 50 nations with hundreds of different ethnic groups, each with a complex set of traditions in as far as issues of sexuality, reproduction, and inheritance are concerned. No serious consideration is given to the many cultural influences

African peoples have been subjected to during the colonial era. His discussion ignores the reality that on top of the many traditions, one finds Western and in some cases Islamic cultural influences, all operating within the same locality, a fact which [Ali P.] Mazrui (1986) calls Africa's *triple heritage*. . . .

The Cross-Cultural Perspective

Rushing begins his paper by advancing four main factors that contribute to the proliferation of the AIDS epidemic in Africa. . . . The fourth factor, which Rushing considers to be the most important factor and hence the preoccupation of his paper, is the omnipresent polygamous behavior among all African societies. . . . Drawing upon outdated ethnographic studies such as those by [Alfred Reginald] Radcliffe-Brown (1950), [Aidan] Southall (1961), and a few recent ones by prominent Western Africanist scholars such as [John C.] Caldwell et al. (1989), and [Caroline] Bledsoe (1990), Rushing concludes that such studies leave no doubt that having multiple sexual partners is a universal cultural practice in many African societies and that such behavior is culturally determined and considered normal and appropriate by Africans, and hence the proliferation of HIV/AIDS. In so doing Rushing invokes the behavioral paradigm to account for the widespread nature of the HIV/AIDS epidemic in Africa. It is this kind of rush to find a cause, i.e., "the sexual life of the natives" and to prescribe an immediate solution, i.e., "modification of sexual behavior," that obscures the real risk factors, namely, the historical, social, political, and economic contexts within which such risk behaviors are played out.

Once the central factor in the proliferation of HIV/AIDS in Africa is identified as promiscuous behavior, Rushing tries to rationalize it through the cross-cultural framework. The central idea in this framework is that "behavior patterns that exist in one society but not in others or in varying degrees in different societies are the result of differences in the way societies structure and give meaning to behavior." In the context of African societies, exotic or almost primitive marriage, and sexual and kinship arrangements are highlighted as the culprits for the proliferation of AIDS in both rural and urban settings. . . .

Marriage and Kinship

. . . In terms of polygyny, it is noted that in most traditional societies the preferred form of marriage is polygamy whose prevalence is given as being 30 to 50 percent of all marriages, according to the World Fertility Surveys of the 1970s, compared to 3–5 percent elsewhere in the world. Based on this fact it is concluded that since multiple wives increase the spread of STDs [sexually transmitted diseases] in a population, polygyny contributes to the high HIV/AIDS rate as well as the balanced rate between males and females. If indeed polygamy was such a potent means of proliferation, then the Islamic societies of northern Africa should exhibit larger than usual HIV rates since polygamy is the normative form of marriage in these societies. Yet evidence suggests otherwise. In North Africa and the Middle East HIV is spreading, but more slowly than elsewhere in the world (Tastemain and Coles 1993). In short a culture can be polygamous, but that does not automatically result in the rapid transmission of sexually transmitted diseases; it all depends on marital norms. Furthermore, the 30–50 percent prevalence rate of polygamy is not broken down by age. Recent data from censuses and the Demographic and Health Surveys (DHS) of the mid-1980s and early 1990s indicate much lower rates of polygamous unions in most African countries. For example, in most southern and east African countries, marriage for women is universal but mostly monogamous. In Malawi census data show that only 20 percent of men over 40 years of age have more than one wife (World Bank 1992). In the DHS data only 9 percent of men reported having more than one wife and 20 percent of women reported being in polygamous unions (Malawi National Statistical Office 1994). In a number of African countries such as Ghana, Malawi, Zambia, Namibia and Botswana, the proportion of husbands in a polygamous union is around 10 to 15 percent among those less than age 30 and about 40 percent for those over age 50 (Ghana Statistical Service 1989; Malawi National Statistical Office 1994; Gaisie 1993; Lesetedi et al. 1989; Katjiuanjo et al. 1993). Younger men and women are more likely to be in monogamous unions than polygamous ones. Paradoxically, HIV/AIDS rates are much lower among the most polygamous age group, those over 50 years, in comparison to the less polygamous group, those below age 40.

In typical traditional society, women in polygamous unions were expected to be faithful to their husbands and so were husbands expected to be faithful to their wives, since not to do so would result in tragedy of one sort or another such as the husband or child dying. For example, among the Chewa found in central Malawi, Zambia, and parts of Mozambique, there is a disease called *Tsempho* which has been discussed at great length in ethnographic studies (Hodgson 1913, p. 129–31; Rangeley 1948, pp. 34–44; Williamson 1956; Marwick 1965, pp. 66–68). *Tsempho* is considered to be a particular kind of wasting disease which may have fatal consequences, and it is related to promiscuous sexual relationships, or to the indulging of sexual intercourse by spouses during prohibited periods. The disease affects both men and women, as well as young children, and is caused by another's social transgression. Thus wives and husbands were forewarned of the dangers of promiscuity. There are certain variations of the *tsempho* sexual taboo among other ethnic groups of central, southern and eastern Africa such as the Bemba, Luapula, Tonga, Ndembu, etc. (Richards 1956; Cunnison 1959; Colson 1958, 1960; Colson and Gluckman 1961; Maxwell 1983; Gibbs 1988; Ouma 1996). In other words, traditional societies were not predominantly promiscuous or sex-positive as Rushing and others would like us to believe.

Underlying polygamy and promiscuity, the patrilineal kinship system is blamed as one of the major etiological factors in the African pattern of HIV/AIDS. All of Africa is considered to be patrilineal in descent, and matrilineal societies are relegated to a tiny footnote as an insignificant group. It is argued that patriarchy gives men the incentive to acquire as many wives as possible because of the value placed in children as economic assets and, as such, promotes polygamous behavior outside of marriage. But when one reads ethnographic studies such as those relied upon by Rushing, it becomes clear that many groups in central Africa and parts of West Africa such as Ghana, practice the matrilineal system of descent. Often men have to move to the wife's village, sometimes in a subservient position. In these societies women have been known to rise to positions of power in society, and generally enjoy some autonomy in comparison to women in patrilineal societies (Chilivumbo 1975). In both patrilineal and matrilineal societies premarital sex is taken very seriously. Although rules of chastity differ radically, in patrilineal societies (such as the Tumbuka, Ngonde, Sukwa, and Ngoni of central Africa) stress is laid on a girl's chastity before marriage. In the orthodox form of chastity rules, the girl's virginity on the eve of a wedding determines the value of bridewealth (Southall 1961; Chilivumbo 1975). Among the Bemba and Chewa (matrilineal groups) girls go through an initiation ceremony called *Chisungu* or *Chinamwali,* a puberty rite that initiates girls through the symbols of life and death in order to give social form and meaning to their sexuality (Richards 1956; Yoshida 1993; Helitzer-Allen 1994; 1997). One of the worst things that could happen to a girl in Bemba traditional society was to bear a child before she had been initiated, a sign that premarital sex was not condoned. Such a child was considered a creature of ill-omen and both the father and mother could be banished.

In short, there are several social and religious sanctions on sexual behavior in both patrilineal and matrilineal societies. Sex is not merely a transactional process or an ordinary activity in which men are expected to give women money and gifts as an expression of affection, respect, and gratitude for sexual favors. Sex is never considered as a form of recreation as Rushing claims; it is much more than this, a sacred undertaking in most traditional societies. Premarital sex, pregnancy out of wedlock, homosexuality, and other forms of sexual deviance are considered to be abhorrent behaviors and are never encouraged. Thus to argue, as Rushing does, that Americans usually view sex morally and think that having multiple partners constitutes immorality and unfaithfulness, while most Africans do not judge sexual behavior in such terms and have a positive-sex culture, is indeed to go against reality. If anything, it could be argued that Americans are more sex-positive than Africans. In the United States and Europe, commercials on TV and billboards often utilize seminude if not completely nude models. In the United States and other Western nations, business in pornographic movies, sexually explicit TV programs, and pornographic magazines is lucrative and vigorous. If indeed most Americans were morally upright in matters of sex are concerned, such activities would have been banned long ago. The French scientist Luc Montagnier, who was the first to isolate HIV, supports the theory that a mycoplasma, a bacterium-like organism, is the trigger that turns a slow-growing population of AIDS viruses into killers. Montagnier believes that the explosion of *sexual activity* (our emphasis) in the United States during the 1970s fostered the spread of a hardy, drug resistant strain of the mycoplasma and that the AIDS epidemic began when the mycoplasma got together with HIV, which had been dormant in Africa (Ungeheuer 1993). . . .

Toward a Reformulation

. . . We suggest that more plausible explanations of the HIV/AIDS epidemic in Africa may be found in the migrant labor thesis (Hunt 1996; Bassett and Mhloyi 1991; Jochelson et al. 1991) which provides a social, historical, and economic context for urban sexuality and unsafe medical practices. The migrant labor thesis asserts that the establishment of wage labor on the continent, particularly in eastern, central and southern Africa, to support mining, agricultural plantations, and other economic activities of the colonialists, created a situation where migrant labor, mostly young men were contracted for long-term work (1 to 3 years). Because families were not encouraged to accompany the laborer, farming and traditional activities such as raising, feeding, and educating children became the responsibility of women (without husbands) in rural areas. The women's limited ability to increase productivity of the land led to declining fertility of the land and, ultimately, absolute shortage of food and malnutrition. When agricultural production becomes untenable as a means of survival, such women often migrate to the city where some engage in one of the very few options open to them—prostitution—"an activity that with a morbid irony serves as their only lifeline to economic survival in the short term at the same time that it poses an enormous risk of curtailing their survival in the long term" (Craddock 1996).

Male migrant workers, away from their families for long periods of time, use alcoholism and frequent visits to prostitutes to deal with their loneliness and boredom. For men and women in eastern, central, and southern Africa, these long separations lead to breakdown in the stability of the family, divorce, and a definitive increase in the numbers of sexual partners for both men and women. Thus, the relatively high number of sexual partners is not the result of a long-standing cultural attribute of Africans, or an innate craving of Africans for more sex, or promiscuity, but is directly a consequence of the economic and labor markets. Definitely, women do not turn to prostitution because African culture "encourages prostitution more easily than Western culture does" as Rushing argues. As [Susan] Craddock (1996) shows, women living in households where there is chronic undernutrition, who are most nutritionally at risk, are the ones most likely to seek other options including prostitution, for survival. . . .

Conclusion

The issue is not that HIV does not spread through polypartner sexual activity, whether as polygamy or simply multiple-partner sexual activity within or outside a marriage relationship. Wherever polypartner sexual activity is practiced, whether in Africa, Australia, or the United States, participants are undisputedly at an increased risk of getting the HIV virus. What is at stake is whether Africans are polysexual by nature, and whether African culture (if we can think about one homogenous African culture) promotes and rewards polysexual behavior. Is being promiscuous and polysexual an African culture trait or is polypartner sexual activity a survival response dictated and enhanced by a vicious political economic system? Our approach to solving the problem hinges critically on our answer to this question. Rushing's work incorrectly assumes and attempts to prove the former is true.

We have argued that this view is erroneous, based on ethnocentrism and overgeneralization. It fails to explain the geographical variations in the incidence of HIV/AIDS across the continent. The simplistic solutions that result from such explanations that advise on behavior modification and condom distribution that may provide little help to rural women, who out of necessity are unable to turn away paying customers who refuse to use condoms. They do not reach rural residents who get HIV while getting injections from an IDV or bush doctor, the only source of health care, to cure malaria. They do not protect the many people privileged to receive services in urban health facilities who face unsterile procedures including injections, blood tests, and transfusions daily. Moreover, this approach fails to stop the spread of HIV.

Why do such ridiculous explanations as Rushing's persist? Obviously, there is an urgent need to find quick answers that can be easily translated into programs. After all, it is a lot easier to distribute condoms in urban Africa than to change the sociopolitical and economic contexts that condition the spread of HIV. Nevertheless, any interventions that fail to address the broader issues of African social and economic life, not merely the sexual promiscuity of Africans, are bound to fail.

POSTSCRIPT

Is Sexual Promiscuity a Major Reason for the HIV/AIDS Epidemic in Africa?

The two sides presented in this issue clearly have little to agree on. However, it is interesting to note that while Oppong and Kalipeni attack Rushing's cultural characterizations as overly generalized and ethnocentric, they do not raise issue with his concern that gender stratification (namely limited empowerment of women in many instances) may be contributing to the problem. Furthermore, they do not categorically dismiss cultural factors as unimportant in the spread of the disease (they just disagree with Rushing's analysis and characterization of these factors). The reality is that a thorough understanding of the spread of HIV/AIDS in Africa probably should include an analysis of cultural and political/economic factors. An interesting point, however, and one brought up by Oppong and Kalipeni at the end of their selection, is that much of the analysis and associated HIV/AIDS prevention programming has focused on cultural and behavioral issues as opposed to political/economic issues. This may be because donors believe that there is a better chance of changing behavioral patterns than political/economic conditions. It might also be related to the ideological orientation of donors that causes them to focus on one set of factors over another.

There are other factors contributing to the spread of HIV/AIDS in Africa that neither of the selections spend much time focusing on. For example, part of the reason that HIV/AIDS may be spreading less rapidly in some West African countries (e.g., Senegal and Guinea-Bissau) is that a different strain of the virus predominates there (HIV-2), as opposed to HIV-1, which is the major strain in other parts of Africa. Unlike HIV-1, HIV-2 does not appear to be spread as easily, is less likely to convert to full-blown AIDS, and tends to attack an older segment of the population. Also frequently mentioned as contributing to the AIDS problem in Africa are high levels of poverty, food insecurity, and armed conflict in some areas. In 2000 South African President Thabo Mbeki drew a firestorm of criticism when he appointed a committee of scientists to re-examine the AIDS epidemic in Africa. Included on the committee were several dissident specialists who argued that HIV is not the cause of AIDS, but factors such as malnutrition and poor hygiene are. It is the former assertion (that HIV does not cause AIDS) that was roundly condemned by the scientific community. For more on this issue, see several articles in the *New York Times*, including Rachel Swarns, "South Africa in a Furor Over Advice About AIDS" (March 19, 2000) and "In Debate on AIDS, South Africa's Leaders Defend Mavericks" (April 21, 2000).

It may be easy for the student to become depressed or overwhelmed when studying the AIDS crisis in Africa. As such, it is worth pointing out that there have been successful attempts at combating the disease in Africa, most notably in Uganda. While many African leaders were denying that HIV/AIDS was a problem in the 1980s and 1990s, President Yoweri Museveni (in power since 1986) jumped on this issue early. He constantly spoke about the issue in public addresses, putting considerable stress on prevention and health education. The result is that Uganda's infection rate dropped from around 30 percent to 6 percent. This decline, however, is not only related to a decreasing rate of new infections. There is also a decrease in the death of those who had the disease. Furthermore, it should be noted that Uganda has also experienced an economic recovery during this period, suggesting that improving political/economic conditions may also have contributed to the improved situation.

For more general information on the AIDS situation, see "AIDS Epidemic Update," *UNAIDS/WHO* (December 2002), which has a significant number of pages devoted to Africa.

ISSUE 15

Is the Use of European Languages as the Medium of Instruction in African Educational Institutions More Negative Than Positive?

YES: Grace Bunyi, from "Rethinking the Place of African Indigenous Languages in African Education," *International Journal of Educational Development* (1999)

NO: Véronique Wakerley, from "The Status of European Languages in Sub-Saharan Africa," *Journal of European Studies* (June 1994)

ISSUE SUMMARY

YES: Grace Bunyi, a researcher at Kenyatta University in Nairobi, Kenya, argues that educating children in their own language is pedagogically more effective, and that mass education in indigenous African languages is the surest way to increase literacy rates and to further develop the human capital necessary for economic development.

NO: Véronique Wakerley, a professor of modern languages at the University of Zimbabwe, sees a role for local language instruction at the primary level, especially if the student is unlikely to go further in his or her studies. However, she asserts that instruction in a language such as English is more effective at higher levels because it allows access to the international community.

The role that policymakers see for education in African development has changed over time, and this has contributed to the rising clamor in recent years for local language education in Africa. At independence, a number of African nations were left with inadequately staffed and underqualified civil services. As such, it was often a top priority in the early independence period to train a cadre of civil servants. Given this objective, it is not surprising that African governments invested heavily in secondary and tertiary education, the levels of schooling that were needed for most civil service positions. As the civil services became fully staffed (or overstaffed in some instances), the notion of a guaranteed job at the end of schooling became a fading reality. Without such a guarantee, peasant families became less and less likely to make the sacrifices necessary to put a child through school. This rural disenchantment with education, combined with a growing body of scholarly research indicating that basic literacy and numeracy skills contributed to general development, led donor and African governments to rethink their approach to education. This change in African educational policy resulted in a renewed emphasis on primary education that included a concern to make this level of schooling more useful to those not bound for further studies and to use local languages as the medium of instruction.

The way in which development itself is conceptualized has also led some social change theorists to call for a new approach to education. In contrast to a focus on development as economic growth, some have argued that development should be seen as a process of empowering people to meet their potential. They suggest that the education system inherited from the colonial authorities was anything but empowering, as it entailed instruction in the language of the colonial power, the use of texts that directly or indirectly asserted the superiority of European culture, and an emphasis on rote learning. This system tended to produce individuals who looked down on anything traditional or indigenous, were concerned with style over substance, and were elitist in their attitudes. Researchers such as Paulo Friere (see *Pedagogy of the Oppressed*, Continuum, 1970) have emphasized the importance of empowerment in human development. Empowerment entails a sense of self-worth and pride in one's own culture, as well as the critical thinking skills necessary to solve problems.

Others would argue that many of the issues expressed here could be addressed by a revamped educational approach that still uses a foreign language as the medium of instruction. The concerns associated with using local languages in African schooling number at least four. First, a primary concern has always been that in African nations with several ethnic groups and associated languages (the vast majority of cases), one language must be emphasized to unify the nation, and the use of a foreign language would not favor any one group over another. Second, communication with the outside world and access to international scholarship require knowledge of a European language. Third, students set on advancing to higher levels of education will be academically less well prepared if they receive schooling at the primary level in a local language. Fourth, it would be prohibitively expensive to conduct class and develop teaching material for the myriad of local languages that exist in Africa.

In the following selections, Grace Bunyi states that mass education in indigenous African languages will increase literacy rates and will also promote economic development. In contrast, Véronique Wakerley asserts that foreign language instruction allows access to the international community and promotes national unity by steering people away from linguo-tribal affiliations.

Grace Bunyi

 **YES**

Rethinking the Place of African Indigenous Languages in African Education

Introduction

The problems that confront the African continent today are enormous and permeate all areas of human life. [George] Ayittey (1991) has summarised the African situation in the following terms:

> "Once a region with rich natural resources as well as bountiful stores of optimism, and hope, the African continent now teeters perilously on the brink of economic disintegration, political chaos, institutional and social decay" (p. xiii).

While some might see Ayittey's assessment as being too strong an indictment, few would disagree with his general observation that Africa has been experiencing regression rather than progress in the social, economic and political spheres. It is within this regression that education in Africa is being called upon to contribute to the social transformation of African societies. On the other hand, education in Africa is part of the social systems that require transformation in order to be of value to other social systems and hence the need for a rethinking of African education.

Western education was brought to Africa by the European colonial powers and as such eurocentricism was part of its baggage. . . . In order for education to liberate itself from the eurocentric colonial legacy, calls have been made for the grounding of African education in African indigenous cultures as primary vehicles for social transformation (Dei, 1994). Since language is both part of the culture and the medium through which culture is transmitted, serious deliberations about the policy as regards the place of indigenous languages in education must constitute part of the rethinking of education in Africa. This is particularly so because the policies we adopt will have an impact on curriculum, pedagogy and access. . . .

Language, Education and Development

I believe that development in any country has to do with the improvement of the social, cultural, economic and political lives of the people of that country. However, I will confine the following remarks to language, education and economic development. The relationship between language, education and economic development lies in the link between language and education on the one hand and education and development on the other. No one would deny that language plays a very central role in education; it is mainly through the interactions between the teacher and the learners and among the learners themselves that knowledge is produced and acquired in the teaching–learning process. Indeed, language learning itself is seen by many as an important part of what becoming educated is all about. Consequently, much of a child's early years in school are spent on developing his/her linguistic skills. Such years are said to be spent on literacy development.

At its most basic level, literacy is defined as the ability to read and write. Going by this definition, there is a logical relationship between language and literacy. Literacy presupposes language since one becomes literate in a language or languages. However, mere ability to decode and encode words is seen to be an inadequate criterion for literacy. Critical educators such as [Paulo] Freire, for example, argue that only critical literacy has the power to transform individuals and make them desirous of transforming their societies. . . .

According to Freire (1970), critical literacy is attained through the use of language in dialogue between the teacher and the students. In multilingual post-colonial Africa, we need to examine what type of literacy is supported by our policies as regards indigenous languages in education.

The relationship between literacy and economic development has already been established and UNESCO [United Nations Educational, Scientific and Cultural Organization] has concluded that "illiteracy has a close correlation with poverty" (cited in Bamgbose, 1991, p. 38). According to the UNESCO figures for 1972, the developed countries of Europe and North America have the lowest adult illiteracy rates, while the less developed and poor Africa had the highest adult illiteracy rates. In 1995, the adult illiteracy rate in Africa stood at 43.8% (UNESCO, 1997). It is suggested that in order to attain economic development, the literacy rates in Africa must be raised. I believe that our policies as regards indigenous languages in education will have an impact on our success (or lack of) in this area. In considering our options, we may do well to note that:

> No developed or affluent nations, though many of these have minority languages, utilise a language for education and other national purposes which is of external origin and the mother tongue of none, or at most few of its people (Spencer, 1985, p. 390).

African Indigenous Languages and African Education

A Sociolinguistic Overview

Africa is a highly multilingual continent. [John] Spencer (1985) has observed that "Africa, particularly Sub-Saharan Africa, is probably the most linguistically complex area of the world, if population is measured against languages" (p. 387). At the same time, individual African countries are rich in languages. For example, Nigeria has 400, Zaire 206, Ethiopia 92, Tanzania 113, Sudan 21 (Bamgbose, 1991), and individual multilingualism is also common in Africa.

African societies were multilingual long before European colonization. With colonization, powerful colonial languages such as English, French and Portuguese were introduced. At the same time, disparate groups of people were brought together under different European powers at conference tables in Berlin in 1885. This resulted in the need for common languages to act as lingua francas [UNESCO defines this term as a language which is used habitually by people whose mother tongues are different in order to facilitate communication between them] in the newly created colonial states. The end result of all this is that most African nations are linguistically very complex. In most countries, there are the various indigenous languages and the lingua franca(s) which may be a widely spoken indigenous language such as Kiswahili in Kenya and Tanzania or the language of the former colonial power such as English in Uganda. Where a widely spoken indigenous language is the lingua franca, there is also the former colonial power's language. Consequently, the question of the educational role of the various languages (African indigenous languages, the lingua franca and the colonial languages) has attracted the attention of policy makers and educators alike.

However, historically, whether through practice—as languages of serious study and as media of instruction—or by attitudes, as the languages that the Africans have craved, it is the colonial languages that have enjoyed pride of place in formal education in Africa. This is despite the unequivocal views that many hold with regard to education in indigenous languages. These views were expressed very strongly by the 1951 UNESCO meeting of specialists. The meeting stated:

> It is axiomatic that the best medium for teaching a child is his mother tongue. Psychologically, it is the system of meaningful signs that in his mind works automatically for expression and understanding. Sociologically, it is a means of identification among members of the community to which he belongs. Educationally, he learns more quickly through it than through an unfamiliar linguistic medium (UNESCO, 1953, p. 11).

Ironically, it is African indigenous languages that have been and continue to be neglected in the formal education of African children. . . .

Academic Achievement

The academic consequences of an educational innovation are often considered an important indicator of the value of the innovation. This is only to be expected since students, parents and educators do want academic success. In North America for example, the lack of academic success among some groups of non-native speakers of English is the driving force behind bilingual education programmes. Such programmes have reported academic gains (Cummins, 1996).

Where indigenous languages have been used under experimental conditions in Africa, the results as regards academic achievement have been encouraging. The best known example is the Six Year Primary Project in Nigeria, where an experimental group was taught in Yoruba (except for English which was taught by a specialist teacher of English) throughout the six years of primary school while a control group was taught in Yoruba for three years and later in English. Similar materials were used with both groups. [Ayo] Bamgbose (1991) reports that systematic evaluation results showed that the experimental group outperformed the control group in all subjects including English. Bamgbose concludes that "the superior performance of the experimental groups . . . could only have been due to the use of Yoruba as the medium of instruction" (Bamgbose, 1991, p. 86).

For those worried about the "falling" standards of English in Kenya, it may be useful to emphasize here that learning through an indigenous language, Yoruba in this case, had beneficial effects on the learning of English, the reasons for which [Jim] Cummins (1996) elaborates upon.

From the foregoing discussion, it is clear that the teaching of and in indigenous languages can only lead to many of the educational goals that we hold dear. . . .

Costs

It is often argued that because of the multiplicity of indigenous languages, the cost of providing teaching–learning materials in all these languages is prohibitive. However, [Anna] Obura (1986); Bamgbose (1991) have argued that the cost argument has not been proven and that it might be impossible to prove, especially if such factors as poor performance, school drop-out rates, time invested and use of culturally inappropriate materials are taken into account.

At the same time there is evidence that producing teaching–learning materials even for minority languages (languages with very small speech communities) need not be too expensive. The Rivers Readers Project which began in 1970 in Nigeria with the aim of producing literacy materials in 20 minority languages showed that by making use of uniform formats and illustrations and by using cheaper materials, it was possible to reduce the production costs of teaching–learning materials considerably (Williamson, 1976).

Another perspective from which to assess the cost of an education system is to examine the extent to which it achieves societal goals. Education in Kenya and elsewhere in Africa has been called upon to contribute to national development. . . . [T]he use of English as the medium of instruction in all primary schools in Kenya makes it very difficult for non-elite background children to do well in school. However, for Kenya to develop, we need the participation of the masses. As Bamgbose (1991) has noted, education for development should concern itself with "the liberation of human potential for the welfare of the community" (p. 75). The role of indigenous languages in such an education is right in the centre. This is because the majority of Kenyan children do not remain in the educational system long enough to learn and learn in English well enough. At the same time, the majority of Kenyans continue to participate interactionally in the development process through their indigenous languages.

Indigenous Languages Are Underdeveloped

Those who argue against the teaching of and in Kenyan indigenous languages point out that many of these languages have yet to acquire orthographies and that indigenous languages lack adequate terminology to teach and learn school subjects in. While this may be true as of now, the potential for development of orthographies and the linguistic development of these languages does exist. As regards the issue of orthographies, nineteenth century European missionaries with less training and other resources than are to be found in Kenya today, were able to develop orthographies in some Kenyan languages. I believe that as we enter the twenty-first century, African Kenyans can do the same for the other languages. Indeed there are reports that when the need for written materials in these languages arises, development of orthographies can be achieved. In Nigeria, in the Rivers Readers Project referred to earlier, orthographies for the 20 minority languages in which the materials were produced were developed through the collaboration of linguists and influential native speakers working in Language Committees (Williamson, 1976). In Kenya, Obura (1986) cites the example of how the initiative of a single researcher led to community collaboration in the production of Kiembu literacy materials. Clearly the development of orthographies and teaching–learning materials can be done fairly cheaply and in a way in which local communities are transformed into knowledge producers. This leads to a breaking down of the barriers that separate the researchers and educators from local communities, and contributes to the empowerment of communities.

As regards indigenous languages' inadequacy in handling school subjects, linguists tell us that all languages have the capacity to develop to meet all the communicative needs of their users. One of the reasons why indigenous languages lack appropriate vocabulary for school knowledge is because they have not been used in those situations. The development of Kiswahili in Tanzania has allowed it to function as the instructional medium for the entire primary school level. As [Robert] Phillipson and [Tove] Skutnabb-Kangas (1994) have noted about indigenous languages in former colonies, it is because of the presence and favouring of English in education that indigenous Kenyan languages have not developed to meet school subjects' linguistic needs.

Tied with the question of the adequacy of indigenous languages and school subjects is the question of indigenous languages and science and technology. The argument is that Kenya is dependent on the West for science and technology which is in English. It is suggested that to benefit from this science and technology, Kenyans must be educated in English. Whereas no one would deny that English controls much of the science and technology in the world today or that Kenya is dependent on the West for science and technology, there is still the question whether the majority of Kenyans who live and work in their own indigenous languages in the rural areas can benefit from Western science and technology in English. Research from Kenya by Eisemon, Eisemon and Patel, and Eisemon, Patel and Ole Sena cited in Cleghorn et al. (1989) suggests that learners are not able to use scientific knowledge acquired (or at the very least) taught in English in the solving of practical problems in their everyday life.

Indigenous Languages and National Unity

It has been argued that current boundaries in Africa are "concocted" rather than "natural" (Laitin, 1992), having been arrived at in Berlin. Therefore, like other countries in Africa, Kenyan society is made up of peoples of varied ethnolinguistic backgrounds. Consequently, an emphasis on indigenous languages in education might lead to divisiveness. My argument here is that whereas indigenous languages may divide people along ethnic lines, English divides them along class lines. At the same time, I do not think linguistic differences of themselves lead to social and political differences. [D. P.] Pattanayak (1998)(p. 380) is right when he states "'languages' do not quarrel. When representatives of languages do, the reasons are mostly extra-linguistic". More often than not, the reasons have to do with the sharing of power and resources among the various linguistic groups in the country. In addition, I believe we should adopt a positive attitude towards linguistic and cultural diversity in Kenya and in Africa. Different languages represent different cultural knowledges and skills. Consequently, linguistic and cultural multiplicity is enriching.

There are compelling reasons why indigenous African languages should form an important part of the education of African children. However, the sociolinguistic situation is very complex and therefore does not allow for easy solutions. For example, in Kenya, we have three competing languages: the indigenous languages, Kiswahili and English. As the national language, Kiswahili should also receive emphasis in education. As far as policy goes, I would recommend that we extend the period of instruction through the indigenous languages to five years so that the switch to English or even Kiswahili takes place in Standard 6. I would also recommend that we introduce a gradual switch so that some subjects, such as social studies and literature in the form of oral literature, can be taught in the indigenous languages for the whole of the primary school level. I would also recommend that the indigenous languages be examined in the national examination. This would ensure that indigenous languages are not neglected.

Implementation of the policy would depend on the availability of teaching–learning materials and appropriately trained teachers. Consequently, I would recommend that we start the implementation under experimental conditions in a few schools. Due to the devaluation that indigenous languages have suffered for a long time, their strengthening could meet with some resistance.

I believe the way to deal with such resistance is by demonstrating the value of these languages in education through a few well resourced and supported classes.

However, it is important to acknowledge that a uniform policy will not be workable. For example, while some languages such as Gikuyu have large speech communities and have had orthographies for some time and therefore can be used as media of instruction for much of the primary school level, those that have very small speech communities and no orthographies can only be used for initial literacy development as soon as the materials are made available. At the same time, while it is possible to use indigenous languages as media of instruction in linguistically homogenous rural areas, this is not possible in the urban and in the formerly White settled areas. However, I would recommend that arrangements be made so that even children in these areas can learn their indigenous languages. This should not be impossible to organise; those children who share a language can get together for the indigenous language lessons.

In this [selection], I have argued that indigenous Kenyan languages should be given a more central role in the education of Kenyan children. I have backed my argument with a discussion of . . . academic achievement justifications, and critically analyzed the arguments that are often advanced against indigenous languages in education. I have, however, conceded that strengthening indigenous languages will not be easy because of Kenya's sociolinguistic complexity, and have offered some suggestions on how a start could be made.

Véronique Wakerley

The Status of European Languages in Sub-Saharan Africa

It would be naïve in the extreme to try to divorce current linguistic policy in Africa from the history of the continent. While the world in some ways becomes more and more homogenized, the question of language is one which strikes at the very roots of cultural identity. . . . 'Language is not only the mind of a culture, it is also its most exclusive vehicle.' Historically speaking, the recognition of the importance of language as the means of expression of a society's distinctive self has sometimes led to the attempted suppression of the use of such languages in the colonizing process. Equally, some colonizers recognized the importance of language as a barrier between ethnic groups, and, by their linguistic policies, whether intentionally or not, kept those barriers in place.

The Africa which has emerged in the last thirty years, as each country gained its independence, has had to face huge problems of identity, and one of the key factors in solving these problems has been language. The question is not only a political one, though that has carried considerable weight in the decision-making process, but also a cultural one, latterly much more divorced from political considerations than thirty years ago, and, most importantly, an educational one. The decision in each country regarding the language(s) of instruction has greatly exercised the experts, because of the many variables which influence the decision. Broadly speaking, however, most nations have opted to retain the language of the departing colonial power at some educational level. . . .

Post-Colonial Africa: Choices and Solutions

The difficulties faced by post-colonial African governments in regard to language are well known. The decision as to what language(s) should be accorded national and/or official status cannot be merely a question of pragmatics. There are political and ideological considerations too, which must affect the status of the languages of the former colonial powers. The multilingual situation which pertains not merely within the continent of Africa, but within individual nations, has, in

From Vèronique Wakerley, "The Status of European Languages in Sub-Saharan Africa," *Journal of European Studies,* vol. 24 (1994). Copyright © 1994 by Richard Sadler, Ltd. Reprinted by permission of Sage Publications, Ltd. References omitted.

some cases, complicated the decision-making process still further. No less important is the educational question of what language should be the medium of instruction in schools. Since on the quality of education received depends the future of a nation, this is an extremely important question. I should like to take as an example of this dilemma the situation of Namibia, Africa's most recently emerged state, and a nation which had had ample opportunity to observe the effects of language policy in other independent African states.

It was decided by SWAPO as early as 1981 to introduce English as the official language of the about-to-become-independent state of Namibia, although independence was still several years away, at that time. This was, on the face of it, a curious decision, because English was an extra-ethnic language, and not one which had previously had any importance in the country. The problem was similar to that already faced by other emergent nations; there were seven identifiable main local languages, plus three imported languages, Afrikaans, German and English. Of these three, only Afrikaans is widely-spoken, but for political reasons, this language was no longer acceptable to the potential new administration. In fact, it is the lingua franca of Namibia, since it is used by speakers from different language groups in order to bridge the gap between them. English, in 1980, was the mother-tongue of only about 16% of the population; but it was increasingly being used by expatriates and visitors, and opened the country to a world where English, with 350 million speakers, holds undisputed sway as universal lingua franca. Also, one of the aims of introducing English as the official language was to steer the people away from linguo-tribal affiliations and differences and to create a climate conducive to national unity without resorting to what was seen as the language of the oppressor. Similar discussions have recently been taking place in South Africa, which has opted to accord official status to no fewer than eleven languages, including Afrikaans and English, the two previous official languages, and nine other indigenous languages. The rationale of the A.N.C., which has always conducted its meetings in three languages, English, Afrikaans and Zulu, is that every language group should have the right to insist on being addressed and understood in his or her own language. There is no doubt that this is an extremely sensitive and important issue, relating to cultural and political identity. Practically speaking, however, it is unlikely that more than two or three of these languages will be used on any significant scale.

The formulation of a language policy in these two countries serves to highlight and reinforce what has already occurred in the rest of Africa. Consideration of the question revolves firstly around whether to choose an indigenous language; however, the choice of one language over another might be construed as being founded on tribal preference or bias. Also, as regards the uniting of all administrative functions, the choice of one indigenous language is unsatisfactory. Thus, as Fishman observes:

> Instead of trying to cope with hundreds of local languages as instruments of government, education, industrialization, etc, most African states have decided to assign all of them equally to their respective home, family and neighbourhood domains and utilise a single major, European language [usually

English or French] for all formal, statusful and specialized domains. This approach tends to minimize internal divisiveness since it does not place any indigenous language at an undue advantage as the language of nationhood.

With very few exceptions, this is, in fact, what has occurred all over sub-Saharan Africa. A country such as Nigeria, with 410 languages, saw the adoption of English as its official language as the logical way to unify the country—any other choice would have proved divisive. But there is a case for a common indigenous language acting as a unifying force, which is the situation in Tanzania, where Swahili has been adopted as the official language. The rationale for this is that it may be able to cut across tribal and ethnic boundaries and mobilize the people in the common goal of nationhood as no exogenous language could. Also, it can be used to interpret the Black experience in a way that it is not possible for a second language to do. For many black Africans, this is a question of cultural identity and authenticity. However, with much of a country's resources already channeled into the promotion of English or French, or some other language of wider communication (LWC), there is often not much left over for the promotion of indigenous languages. Nevertheless, governments are becoming concerned to improve the status of local languages and are beginning to ensure that they have some examinable status on the school curriculum.

The Namibian experience is particularly interesting because it incorporates so many of the elements that are troubling black Africans in the post-colonial era. The question that is being asked is what place, if any, is to be given to foreign languages in Africa. Such languages are perceived as potentially threatening to African cultural identity and this includes the language of broadcasting. The head of the African National Congress's Department of Arts and Culture, Wally Serote, observed at the Union of National Radio and Television Organizations of Africa's (Urtna) 30th anniversary celebrations held in Nairobi in June (1993) that the use by African media organizations of English or French 'means that we have played a role in the marginalization of our people'. In the subsequent discussion, former President Kaunda of Zambia said that, in his country, English is preferred as the broadcasting medium because of its ability to unite the nation. 'This view was confirmed in interviews with delegates from Ghana, Benin, Egypt, Ivory Coast and Senegal, who said the use of African languages was not an issue in their respective countries, and that former colonial languages were preferred by local media.'

The key issue, however, is the language of education; it is impossible to over-estimate the importance of such a decision. Language policy must be closely associated with educational objectives. In *Language Policy, Literacy and Culture* (op. cit., (Note 2)) Francis R. Whyte, in his introduction, observes that there exist dual and concurrent tendencies to meet increasing demands for linguistic competence in one or more of the international languages as well as strengthening minority or local languages for literacy and school purposes, and as a means of preserving local and regional identity and the languages themselves. The desire to become proficient in a foreign language frequently conflicts with mother-tongue literacy, a major element in building the individual's identity and self-esteem.

A UNESCO conference on the *Use in Education of African Languages in Relation to English* concluded that 'ideally the medium of instruction for a child living in its own language environment should be the mother tongue and that a child should be educated in the mother tongue for as long as possible.' This view has been taken to heart in the Anglophone former colonies, where in fact the general practice in colonial days had been to conduct education through the medium of the vernacular; but in the former French and Portuguese colonies, primary education continues to be administered through the medium of the respective foreign languages.

The question must be posed as to what is the function of primary schools, because it is against the background of that declared purpose that the language of instruction can be best seen in perspective. If a primary school education is to be seen as complete in itself and universal, as opposed to preparing pupils for further education and therefore not necessarily for the majority, then the preferred medium will almost certainly be the vernacular, so that teaching may be centred on the acquisition of skills useful for becoming productive citizens. It is generally agreed that children learn new concepts far more readily through a language with which they are already familiar, rather than attempting to acquire two skills simultaneously—that of the subject under study and that of the language in which they are being instructed.

A number of countries, in particular, Kenya, Nigeria and Zambia, have conducted experimental programmes to monitor the effects on cognitive and conceptual development of learning through the medium of a foreign language and through a local language; the results would seem to support the argument for early concept formation being learnt through the mother tongue.

However, the debate rages on, since many are clamouring to be educated through the medium of a foreign language, or at least a Language of Wider Communication, which, in effect, means English or French. They would argue that to be introduced to the LWC only at secondary level severely handicaps the learner in his attempts to acquire a profound knowledge of the language being studied, and that this in turn limits the options available to the learner in his future career. We must take into consideration the concerns of many African governments regarding the possible adverse effects of substituting the LWC as medium of instruction on cultural authenticity, social structure and, ultimately, national identity. Indeed, one of the most pressing reasons for introducing mother-tongue instruction could well be the tradition of orality to which many indigenous languages belong. If they are not transcribed and used in written form, there is a danger that they may disappear. As Wardhaugh observes:

> Promoting vernaculars is a long and sometimes perilous undertaking. Many are first used by missionaries in their work and in this way gain alphabets and a basic literature. Then they become used increasingly in the media, particularly on the radio and for information-giving purposes. Informal uses for them tend to develop in education, religion and politics. As demand increases, the languages find their way into the schools on a more regular basis [. . .] but only when the language has achieved a full range of functions and no stigma is attached to its use has it 'arrived'.

The Current Situation in Education

French and English are widely recognized, albeit in some cases reluctantly, as official languages in Africa. Although some states have opted to vernacularize their education system to a much greater degree than others, these two languages play some role in the education process of almost every country. Only Tanzania did not include English or French among its official languages at independence, choosing instead Swahili, the lingua franca of West Africa. Most former colonies continue to be administered in the language of the colonizer.

Wardhaugh states that at independence, almost no use was made of the vernaculars; Senegal, Mali, Niger, Upper Volta (now Burkina Faso), Ivory Coast, Togo and Benin all used French, whereas Sierra Leone, Liberia, Gambia, Ghana, which lays more stress on English as the medium of instruction than the British did in the colonial era, and northern Nigeria took English for their official language. Indeed, the disputed use of Ibo and the perceived favouritism towards those who spoke it was the cause of a very bloody civil war in Nigeria. Since that time English has been used as the language of national unity, though Hausa, Yoruba and Ibo are promoted regionally as the languages of early education.

Whereas English arrears to have been adopted for largely pragmatic reasons, unity, accessibility of outside world, French, on the other hand, has a perceived cultural superiority, which some francophone states have identified with to a certain extent. Under the leadership of President Léopold Senghor, Senegal pursued a course of active use of French, even though a large proportion of the population can use Wolof. Senghor was one of the instigators of the policy of Francophonie, which propagates French not merely as a means of communication, but as the vehicle of an entire civilization. Thus there have tended to be very different attitudes towards the assistance in teaching English and French on the part of anglophone and francophone countries outside Africa.

Ivory Coast, Mali, Niger, and the Central African Republic all use French as the language of instruction, and it is widely spoken, albeit often at the centre of the continuing controversy regarding linguistic purity. Guinea, which tried to promote local languages initially, has reverted to the use of French in education. The Malagasy Republic uses French at secondary and tertiary level, but Malagasy in primary schools. The former Belgian colonies of Zaïre and Burundi both have French as an official language, but have promoted their respective indigenous languages, with varying degrees of success. Cameroon, which has given French and English equal status, has attempted to foster bilingualism in these two languages, though in practice 80% of the population inhabits the francophone provinces. Nevertheless, the unofficial main language seems to be a kind of pidgin English rather than French or any indigenous language, of which there are over 200 in Cameroon.

The following table shows what languages are used in which countries as the medium of instruction at lower primary(1), upper primary(2) and post primary(3) levels.

Table 1

English:	Gambia, Kenya, Sierra Leone, Liberia, Cameroon (1, 2, 3) Malawi, Uganda, Ghana, Zambia, Lesotho, Swaziland Botswana, Nigeria (2,3) Ethiopia, Tanzania, Zimbabwe (3)
French:	Mali, Burkina-Faso, Niger, Chad, Burundi, Central African Republic, Benin, Rwanda, Guinea, Mauritania, Côte d'Ivoire, Cameroon, Congo, Gabon (1,2,3) Zaïre, Togo, Madagascar, Senegal (2,3)
Portuguese:	Guinea-Bissau, Angola, Mozambique (1,2,3)
Arabic and Amharic:	Sudan (1,2,3), Chad (1,2), Ethiopia (1)
Local languages:	Somalia, Tanzania, (1,2,3) Zimbabwe (1, 2) Chad, Zaïre, Malawi, Togo, Madagascar, Rwanda, Kenya, Ghana, Senegal, Zambia, Lesotho, Swaziland, Botswana, Nigeria (1)

In every country listed in the table . . . , English, French and Portuguese are shown to be spoken by a very low proportion of the population as a mother tongue, by still fewer as second language, in the respective countries where they are used as the medium of instruction. Nevertheless: 'Eleven of 15 former French colonies and all three former Portuguese colonies officially begin instruction in the national language from the first day of primary school. In contrast, 13 of 15 former British colonies begin instruction in one or more African languages and teach English at first as a subject; only later is English introduced as the medium of instruction'. In an article entitled 'National Languages in Teaching', Joseph Poth continues:

> The latest statistics [1989] show that, for example, 37 out of 47 African countries are committed to using one or more of their national languages in the curriculum and four others have embarked upon a study phase prior to doing so. So 41 states have given, or are about to give, full teaching status to national languages in their educational systems [. . .] Furthermore, the use of these languages is not to the detriment of national or international tongues. The school systems of the Third World are moving towards institutional bilingualism.

Even as a second language, they do not score very highly, with French in Côte d'Ivoire and Portuguese in Angola coming in top with 35%. In fact, however, the table shows that such data was not available in many cases. . . .

Conclusion

The state of foreign (European) language teaching in sub-Saharan Africa cannot be defined in a word, or even a sentence. There are an enormous number of variables to be taken into consideration, and the changing political and economic status of many countries in Africa makes it impossible to predict what will happen even in the immediate future. There are, however, certain factors which, while not remaining constant, must always be considered.

In the past thirty years, the situation of almost every country in Africa has changed from being under the dominion of a foreign colonial power to being independent; independence has brought with it the responsibility of educating millions of people who had not been offered the opportunity before. The question of 'what language' is of more than academic interest; there has, in some countries, been a resistance to what is seen as a form of neo-colonialism, particularly in the form of francophonie. Though English is not perceived as carrying with it the same cultural baggage, which may seek to undermine or at least compete with African cultural authenticity, African governments have tended to be wary of offers of help from European powers to help financially in the promotion of their languages for it is feared that such assistance might lead to cultural imperialism, and be part of a larger plan to gain some form of ascendancy over independent states.

Certainly, the educational question of mother tongue or foreign medium instruction has been taken into careful consideration in most newly independent countries, but it is always combined with the vexed question of national and cultural identity, which, as most experts agree, is intimately linked with mother-tongue use. Thus, even in the francophone former colonies, the trend is now towards early education in the mother tongue, so that concepts can be formed with reference to the learner's national and ethnic identity.

The place for foreign languages then, is tending to be a less privileged one than in colonial days. Languages join other subjects on the curriculum as a choice to be made according to students' interests and ambitions. In addition, in every African country, English, French or Portuguese appear at some level, generally upper primary, as the medium of instruction. It seems likely, in fact, that the situation as regards foreign language instruction is healthier than it was in colonial days, for now, not only are pupils at some point in their school career given instruction in a foreign (albeit often official) language, but very often they have the opportunity to learn one or more other languages as well, though it is true that a relatively small proportion of pupils ever reach an educational level where this is possible. In any case, while many pupils are anxious to obtain a functional knowledge of a Language of Wider Communication, very few are interested in pursuing language studies for their own sake, taking on the whole the pragmatic view that school studies, afforded to a minuscule proportion of the youth of Africa, must lead to a well-paid and secure profession.

The situation is not entirely bleak, however. The importance of intercontinental communication is now becoming recognized in Africa, and as African states become more secure in their national identity, it is likely that foreign languages will lose the stigma of colonial domination attached to them. Practically speaking, the shortage of trained local teachers, money for equipment and textbooks, and access to mother-tongue speakers, are serious handicaps. But it seems that African and European authorities are working now to overcome some of these difficulties together. Obviously, the economic factors of increasing trade and aid links play a large part, but there is a growing tendency to exchange cultural information with a view to promoting greater understanding between the former colonizers and colonized. Perhaps, in the phrase of R. Amonoo, it will lead not to domination of one culture by another, but to 'a healthy symbiosis of cultures which can only lead to a deeper appreciation of mankind'.

POSTSCRIPT

Is the Use of European Languages as the Medium of Instruction in African Educational Institutions More Negative Than Positive?

It is important to note that to date, the debate about local language education in Africa has largely focused on the primary level. However, Tanzania and Somalia use local language instruction at all levels (primary through tertiary). The call for more applied and appropriate education is often referred to as the *basic education movement.*

Another relevant historical factor to consider is that the use of local languages at the primary level was generally accepted, and often promoted, by the British during the colonial era. The approach taken by the French colonizers was very different. They saw the teaching of French to the local population as an integral part of the colonizing process. Considerable support for French language education in Africa continues today in states who are members of "La Francophonie," an association of former French colonies. Despite these policy differences, the irony is that the proportion of the population in former British colonies that is literate in English is much higher than the fraction that speaks French in the former French colonies. It is likely that the global prominence of English has much to do with this. Others would suggest that the heavy-handed cultural imperialism of the French has produced an adverse reaction in some instances.

It is finally worth pointing out that prior to the colonial period, certain local languages had evolved to become *lingua franca,* or languages used to communicate between people where several languages are spoken. Trade between groups often played a significant role in the spread and development of these languages. Swaheli, a hybrid of Arabic and indigenous African tongues, is probably the most well-known example of a lingua franca that was and is widespread in East Africa.

For those who are interested in further study of this issue, there are a number of nongovernmental organizations that have sponsored local language education projects in Africa. For example, Save the Children USA has a Village School Program that offers a practical education to rural children in their local language. See their Web site at http://www.savethechildren.org/edu/sb_primary_edu.shtml. UNESCO also has a very informative Web site on education initiatives in Africa at http://www.dakar.unesco.org/education_en/index.shtml.

ISSUE 16

Are Women in a Position to Challenge Male Power Structures in Africa?

YES: Richard A. Schroeder, from *Shady Practices: Agroforestry and Gender Politics in The Gambia* (University of California Press, 1999)

NO: Human Rights Watch, from "Double Standards: Women's Property Rights Violations in Kenya," A Report of Human Rights Watch (March 2003)

ISSUE SUMMARY

YES: Richard A. Schroeder, an associate professor of geography at Rutgers University, presents a case study of a group of female gardeners in The Gambia who, because of their growing economic clout, began to challenge male power structures. Women, who were the traditional gardeners in the community studied, came to have greater income-earning capacity than men as the urban market for garden produce grew. Furthermore, women could meet their needs and wants without recourse to their husbands because of this newly found economic power.

NO: Human Rights Watch, a nonprofit organization, describes how women in Kenya have property rights unequal to those of men, and how even these limited rights are frequently violated. It is further explained how women have little awareness of their rights, that those "who try to fight back are often beaten, raped, or ostracized," and how the Kenyan government has done little to address the situation.

As is the case in other parts of the world, African women suffer from discrimination and inequality. According to the World Bank, the female adult illiteracy rate in Africa in 2001 was 46 percent as compared to 38 percent for the general population. Girls also continue to attend primary school in lower numbers than boys in many African countries (although this ranges from near equality in nations such as South Africa, Zimbabwe, and Namibia to great disparities in countries like Benin, Chad, and Guinea). Despite these disadvantages, women are the backbone of the rural economy in many African settings where it is estimated that they produce, on average, 70 percent of the food supply.

The inequities faced by many women in the African context led to the rise of the women in development (WID) movement in aid circles in the late 1970s and 1980s. These WID programs were also instigated because of a general recognition that many aid programs had not addressed the needs of women or had excluded them entirely. Many agricultural development programs catered almost exclusively to men. In many instances, such programs exacerbated economic disparities between men and women. WID programs were specifically designed to counteract these problems, including a number of initiatives related to gardening, income generation, health care, and education.

In promoting these initiatives, development agencies occasionally exploited the image of African women as a downtrodden class of people who undertake a disproportionate share of the work, yet are severely disadvantaged in terms of access to education, health care, land, and legal protection. While there may be some truth to this generalization, it is problematic because it denies African women "agency." In other words, it could negate or understate the ability of African women to change their situation. This is not to say that African women should not form alliances with outside groups to work for transformation, but those peddling the assistance need to be careful that they are not trafficking images and stereotypes that may be disempowering.

In this issue, Richard A. Schroeder presents a case study of a group of female gardeners in The Gambia. These women, who were the traditional gardeners in the community, benefited from outside funding for fencing and wells during the heyday of WID programming in the 1980s. They eventually came to have greater income-earning capacity than men as the urban market for garden produce grew, and they adeptly intensified production. As a result, men were often forced to turn to their wives for loans. This allowed women to "purchase" freedom of movement and social interaction. Furthermore, women with growing economic clout could challenge male power structures because they were capable of meeting their needs and wants without recourse to their husbands. They also were less susceptible to the threat of divorce (which historically implied the nearly impossible obligation of repaying one's bride-price) because women were now capable of repaying their bride-price with their gardening income.

In contrast, the selection from Human Rights Watch contains the assertion that in Kenya "discriminatory property laws and practices impoverish women and their dependents, put their lives at risk by increasing their vulnerability to HIV/AIDS and other diseases, drive them into abhorrent living conditions, subject them to violence, and relegate them to dependence on men and social inequality." The Kenyan government is castigated for having done little to address this situation.

Richard A. Schroeder  **YES**

Shady Practices: Agroforestry and Gender Politics in The Gambia

Introduction

Some sixty kilometers upriver along the North Bank of The River Gambia lies the Mandinka-speaking community of Kerewan (ke´-re-wan). The dusty headquarters of The Gambia's North Bank Division is located on a low rise overlooking rice and mangrove swamps and a ferry transport depot that facilitates motor vehicle transport across Jowara Creek (Jowara Bolong), one of The River Gambia's principal tributaries. Since the Kerewan area was dominated by opposition political parties throughout the nearly thirty-year reign of The Gambia's first president, Al-Haji Sir Dawda Jawara (1965–1994), it became something of a developmental backwater. Before 1990, Kerewan town had no electricity or running water beyond a few public standpipes. For a community of 2,500 residents, there were no restaurants and only a poorly stocked market that lacked fresh meat. Indeed, from the standpoint of the civil servants assigned to the North Bank Division, Kerewan was considered a hardship post. Mandinka speakers sarcastically referred to the divisional seat as "Kaira-wan," a place where "peace" (Mandinka: *kaira*) reigned to the point of overbearing stagnation. Neighboring Wolof speakers, meanwhile, disparaged the community by dubbing it "Kerr Waaru"—"the place of frustration."

Kerewan's reputation was only partially deserved, however, for the community was actually the center of a great deal of productive economic activity. Over two decades beginning in the mid-1970s, the town's women transformed the surrounding lowlands into one of the key sites of a lucrative, female-controlled, cash-crop market garden sector. A visitor to Kerewan as recently as 1980, when I made my first trip to The Gambia, would have found that vegetable production on the swamp fringes ringing Kerewan on three sides was decidedly small-scale. Most gardeners, virtually all of whom were women, worked single plots that were individually fenced with local thorn bushes or woven mats. Outside assistance in obtaining tools, fences, and wells was minimal. Seed suppliers were not yet operating on a significant scale, and petty commodity production was largely confined to tomatoes, chili peppers, and onions. The market season, accordingly, stretched only a few weeks, and sales outlets were all but nonexistent. Most Kerewan produce was sold directly to end users in the nearby Jokadu District by women who transported their fresh vegetables by horse or donkey cart and then toted them door to door on their heads (a marketing strategy known as *kankulaaroo*).

By 1991, when I completed the principal phase of research for [my] book, large gardens on the outskirts of Kerewan had come to dominate the landscape. Each morning and evening during the October–June dry season, caravans of women plied the footpaths connecting a dozen different fenced perimeters to the village proper. Over the course of nearly twenty years, the number of women engaged in commercial production rose precipitously from the 30 selected to take part in a pilot onion project in the early 1970s to over 400 registered during an expansion project in 1984, and some 540 recorded in my own 1991 census. The arrival of the first consignments of tools and construction materials donated by developers for fencing and wells in 1978 initiated an expansion period which saw the area under cultivation more than triple in size, growing from 5.0 ha to 16.2 ha in ten years. Between 1987 and 1995, a second wave of enclosures nearly doubled that area again. At least a dozen separate projects were funded by international NGOs, voluntary agencies, and private donors. These donations were used toward the construction of thousands of meters of fence line and roughly twenty concrete-lined irrigation wells. In addition, there were some 1,370 hand-dug wells and nearly 4,000 fruit trees incorporated within Kerewan's garden perimeters. Growers purchased seed, fertilizer, and other inputs directly from an FAO [Food and Agriculture Organization]-sponsored dealership in the community and sent truckloads of fresh produce to market outlets located up and down the Gambia-Senegal border, which thrived on the vegetable trade. In sum, the Kerewan area developed over two decades into one of the most intensive vegetable-producing enclaves in the country. . . .

Theories Connecting Gender, Development, and the Environment

The image most widely used to capture the "plight" of Third World women is that of an African peasant woman toting an improbably large and unwieldy bundle of firewood on her head. She may or may not have a young child tied to her back, but the image is always meant to convey that she has traveled a great distance to gather her load. As a metaphor, this feminine icon suggests the incredible burdens women shoulder, and the great lengths they go to, to satisfy the multiple and competing demands society and their families place on them. The implication is that women suffer these conditions universally, as a class, and that pure, selfless motives drive them to undertake routinely dull, repetitive, and ultimately thankless tasks. At the same time, the graphic portrayal of firewood collectors is meant to underscore the idea that close connections exist between women and the natural environment. It suggests that women forced to gather wood from the countryside lead a hand-to-mouth existence, where knowledge of the landscape is bred of necessity and deep personal experience, and where the vagaries of climate and ecology have profound and immediate implications for human well-being. Thus, by virtue of their collective lot in a singular division of labor, women mediate the relationship between nature and society, and they feel the brunt of natural forces as a consequence.

Such images convey a stark reality: life for peasant women is often filled with considerable toil and drudgery. Yet if these women suffer a common plight, it resides not in any particular niche in some all-encompassing division of labor but in the countless ways the range and variety of their lived experiences are distorted in the words and images conveyed by outsiders. The wood-gathering icon represents Third World women as Africans, African women as peasants, and peasant women as a single type. There is no geographical detail at either localized or macropolitical scales that might serve as an explanation for the plight thus portrayed. Moreover, to render such women as beasts of burden, dumb, stolid, unwavering in their support of their families, unstinting in their service of same, is to acquiesce in the notion that they are perpetual victims, steeped in need, and incapable or disinclined to contest their lot creatively.

This tension between images of women as victims and women as autonomous actors traces back to the earliest efforts of developers to promote Women in Development (WID) programs in the Third World. The United Nations–sponsored convocation in Mexico City in 1975 proclaimed an International Decade for Women and initiated efforts within the major development agencies to address a broad agenda of issues deemed especially pertinent to Third World women. . . .

Gone to Their Second Husbands: Domestic Politics and the Garden Boom

One of the offshoots of the surge in female incomes and the intense demands on female labor produced by the garden boom was an escalation of gender politics centered on the reworking of what [Anne] Whitehead once called the "conjugal contract." In Kerewan, the political engagement between gardeners and their husbands can be divided into two phases. The first phase, comprising the early years of the garden boom, was characterized by a sometimes bitter war of words. In the context of these discursive politics, men whose wives seemed preoccupied with gardening claimed that gardens dominated women's lives to such a degree that the plots themselves had become the women's "second husbands." Returning the charge, their wives replied, in effect, that they may as well be married to their gardens: the financial crisis of the early 1980s had so undermined male cash-crop production and, by extension, husbands' contributions to household finances, that gardens were often women's only means of financial support during this period.

As the boom intensified, so, too, did intra-household politics. The focus of conflict in the second phase—which extended into the mid-1990s—was the role of garden income in meeting household budgetary obligations. Several studies have examined "non-pooling" households in Africa, that is, households in which men and women tend to engage in distinctly different economic activities and control their own incomes from these enterprises. The garden boom offers a case study in which women, by virtue of their new incomes, entered into intra-household negotiations over labor allocation and income disposition with certain economic advantages. The upshot of these negotiations was not, however, quite so simple. In terms of budgetary obligations, women in the garden districts assumed a broad range of new responsibilities from their husbands. Moreover, they frequently gave their husbands part of their earnings in the form of cash gifts. This outcome appears in some respects as a capitulation on the part of gardeners. I argue, . . . however, that it can also be read as symbolic deference designed to purchase the freedom of movement and social interaction that garden production and marketing entailed. In effect, gardeners used the strategic deployment of garden incomes to win for themselves significant autonomy and new measures of power and prestige, albeit not always at a price of their own choosing.

. . . Before the garden boom, men in Mandinka society had powerful economic levers at their disposal which they could, and did, use to "discipline" their wives. They controlled what little cash flowed through the rural economy due to their dominant position in groundnut production and were able to fulfill or deny a range of their wives' expressed needs at will. These included such basic requirements as clothing, ceremonial expenses (naming ceremonies, circumcisions, and marriages for each individual woman's children), housing amenities, and furnishings. The power vested in control over cash income was only enhanced by polygamous marital practices and the opportunities they afforded to play wives off against one another. A second advantage was derived from the husband's rights in divorce proceedings. In the event of a divorce, Mandinka customary law requires that the bride's family refund bridewealth payments. Consequently, when marital relations reach an impasse, divorce is not automatic; the financial arrangement between the two families must first be undone. Typically, the woman flees or is sent back to her family so that they can ascertain to their own satisfaction whether she has made a good faith effort to make her marriage work. The onus is on the woman to prove her case, however, and she is not infrequently admonished by her own family to improve her behavior before being returned to her husband.

The advent of a female cash-crop system reduced the significance of both these sources of leverage, not least because women's incomes had outstripped their husbands' in many cases. A rough comparison of the garden incomes of women in Kerewan and Niumi Lameng and the earnings their husbands reported from groundnut sales showed that 81 percent and 47 percent of women in the Niumi Lameng and Kerewan samples, respectively, earned more cash than their husbands from sales of these crops. This reversal of fortunes changed fundamentally the way male residents of the garden districts dealt with their wives:

> *Before gardening started here, if you saw that your wife had ten dalasis you would ask her where she got it. At that time, there was no other source of income for women except their husbands. . . . But nowadays a woman can save more than two thousand dalasis while the husband does not even have ten dalasis to his name. So now men cannot ask their wives where they get their money, because of their garden produce.*

Gardener's husband

Indeed, the garden boom reduced male authority ("If she realizes she is getting more money than her husband, she may not respect him"), and the extent of gardeners' economic influence expanded proportionately. The simple fact that women could largely provide for themselves ("If we join [our husbands] at home and forget [our gardens in] the bush, we would all suffer. . . . Even if he doesn't give you [what you want], as long as you are doing your garden work, you can survive") constituted a serious challenge to the material and symbolic bases of male power. In the first phase of conflict brought on by the boom, men openly expressed their resentment in pointed references to female shirking and selfishness. Their feelings were also made plain in actions taken by a small minority who forbade their wives to garden, or agitated at the village level to have gardening banned altogether. In the second phase, men dropped their oppositional rhetoric, became more generally cooperative, and began exploring ways to benefit personally from the garden boom. Sensing the shift in tenor of conjugal relations, women, accordingly, began a prolonged attempt to secure the goodwill necessary to sustain production on a more secure basis.

. . . Survey data show that both senior members of garden work units and women working on their own took on many economic responsibilities that were traditionally ascribed to men. Fifty-six percent of the women in the Kerewan sample, for example, claimed to have purchased at least one bag of rice in 1991 for their families. The great majority bought all of their own (95%), and their children's (84%), clothing and most of the furnishings for their own houses. Large numbers took over responsibility for ceremonial costs from their husbands, such as the purchase of feast day clothing (80%), or the provision of animals for religious sacrifice. Many paid their children's school expenses. In a handful of cases, gardeners undertook major or unusual expenditures such as roofing their family's living quarters, providing loans to their husbands for purchasing draught animals and farming equipment, or paying the house tax to government officials. There are, once again, unfortunately no baseline data that could be used to gain historical perspective on this information. Nonetheless, several male informants stated unequivocally that, were it not for garden incomes, many of the marriages in the village would simply fail on the grounds of "non-support." . . .

Gone to Their Second Husbands

It is fair to say that domestic budgetary battles did not originate with the garden boom in Mandinka society; nor are they wholly unique to either The Gambia or Africa. Nonetheless, the Gambian garden boom clearly produced dramatic changes in the normative expectations and practices of marital partners in the country's garden districts. . . .

The price of autonomy notwithstanding, women in The Gambia's garden districts succeeded in producing a striking new social landscape—by embracing the challenges of the garden boom, they placed themselves in a position to carefully extricate themselves from some of the more onerous demands of marital obligations. Indeed, in a very real sense, they won for themselves "second hus-bands" by rewriting the rules governing the conjugal contract. Thus the prod-uct of lengthy intra-household negotiations brought on by the garden boom was not the simple reproduction of patriarchal privilege and prestige; it was in-stead a new, carefully crafted autonomy that carried with it obligations and con-siderable social freedoms.

Double Standards: Women's Property Rights Violations in Kenya

Summary

Shortly after Emily Owino's husband died, her in-laws took all her posses-sions—including farm equipment, livestock, household goods, and cloth-ing. The in-laws insisted that she be "cleansed" by having sex with a social outcast, a custom in her region, as a condition of staying in her home. They paid a herdsman to have sex with Owino, against her will and with-out a condom. They later took over her farmland. She sought help from the local elder and chief, who did nothing. Her in-laws forced her out of her home, and she and her children were homeless until someone offered her a small, leaky shack. No longer able to afford school fees, her children dropped out of school.

—Interview with Emily Owino, Siaya, November 2, 2002

When Susan Wagitangu's parents died, her brothers inherited the family land. "My sister and I didn't inherit," said Wagitangu, a fifty-three-year-old Kikuyu woman. "Traditionally, in my culture, once a woman gets married, she does not inherit from her father. The assumption is that once a woman gets married she will be given land where she got married." This was not the case for Wagitangu: when her husband died, her brothers-in-law forced her off that homestead and took her cows. Wagitangu now lives in a Nairobi slum. "Nairobi has advantages," she said. "If I don't have food, I can scavenge in the garbage dump."

—Interview with Susan Wagitangu, Nairobi, October 29, 2002

Women's rights to property are unequal to those of men in Kenya. Their rights to own, inherit, manage, and dispose of property are under constant at-tack from customs, laws, and individuals—including government officials—

who believe that women cannot be trusted with or do not deserve property. The devastating effects of property rights violations—including poverty, disease, violence, and homelessness—harm women, their children, and Kenya's overall development. For decades, the government has ignored this problem. Kenya's new government, which took office in January 2003, must immediately act to eliminate this insidious form of discrimination, or it will see its fight against HIV/AIDS (human immuno-deficiency virus/acquired immune deficiency syndrome), its economic and social reforms, and its development agenda stagger and fail.

This report recounts the experiences of women from various regions, ethnic groups, religions, and social classes in Kenya who have one thing in common: because they are women, their property rights have been trampled. Many women are excluded from inheriting, evicted from their lands and homes by in-laws, stripped of their possessions, and forced to engage in risky sexual practices in order to keep their property. When they divorce or separate from their husbands, they are often expelled from their homes with only their clothing. Married women can seldom stop their husbands from selling family property. A woman's access to property usually hinges on her relationship to a man. When the relationship ends, the woman stands a good chance of losing her home, land, livestock, household goods, money, vehicles, and other property. These violations have the intent and effect of perpetuating women's dependence on men and undercutting their social and economic status.

Women's property rights violations are not only discriminatory, they may prove fatal. The deadly HIV/AIDS epidemic magnifies the devastation of women's property violations in Kenya, where approximately 15 percent of the population between the ages of fifteen and forty-nine is infected with HIV. Widows who are coerced into the customary practices of "wife inheritance" or ritual "cleansing" (which usually involve unprotected sex) run a clear risk of contracting and spreading HIV. The region where these practices are most common has Kenya's highest AIDS prevalence; the HIV infection rate in girls and young women there is six times higher than that of their male counterparts. AIDS deaths expected in the coming years will result in millions more women becoming widows at younger ages than would otherwise be the case. These women and their children (who may end up AIDS orphans) are likely to face not only social stigma against people affected by HIV/AIDS but also deprivations caused by property rights violations.

A complex mix of cultural, legal, and social factors underlies women's property rights violations. Kenya's customary laws—largely unwritten but influential local norms that coexist with formal laws—are based on patriarchal traditions in which men inherited and largely controlled land and other property, and women were "protected" but had lesser property rights. Past practices permeate contemporary customs that deprive women of property rights and silence them when those rights are infringed. Kenya's constitution prohibits discrimination on the basis of sex, but undermines this protection by condoning discrimination under personal and customary laws. The few statutes that could advance women's property rights defer to religious and customary property laws that privilege men over women. Sexist attitudes are infused in Kenyan society: men that Human Rights Watch interviewed said that women are untrustworthy, incapable of handling property, and in need of male protection. The guise of male "protection" does not obscure the fact that stripping women of their property is a way of asserting control over women's autonomy, bodies, and labor—and enriches their "protectors."

Currently, women find it almost hopeless to pursue remedies for property rights violations. Traditional leaders and governmental authorities often ignore women's property claims and sometimes make the problems worse. Courts overlook and misinterpret family property and succession laws. Women often have little awareness of their rights and seldom have means to enforce them. Women who try to fight back are often beaten, raped, or ostracized. In response to all of this, the Kenyan government has done almost nothing: bills that could improve women's property rights have languished in parliament and government ministries have no programs to promote equal property rights. At every level, government officials shrug off this injustice, saying they do not want to interfere with culture.

As important as cultural diversity and respecting customs may be, if customs are a source of discrimination against women, they—like any other norm—must evolve. This is crucial not only for the sake of women's equality, but because there are real social consequences to depriving half the population of their property rights. International organizations have identified women's insecure property rights as contributing to low agricultural production, food shortages, underemployment, and rural poverty. In Kenya, more than half of the population lives in poverty, the economy is a disaster, and HIV/AIDS rates are high. The agricultural sector, which contributes a quarter of Kenya's gross domestic product and depends on women's labor, is stagnant. If Kenya is to meet its development aims, it must address the property inequalities that hold women back.

Unequal property rights and harmful customary practices violate international law. Kenya has ratified international treaties requiring it to eliminate all forms of discrimination against women (including discrimination in marriage and family relations), guarantee equality before the law and the equal protection of the law, and ensure that women have effective remedies if their rights are violated. International law also obliges states to modify discriminatory social and cultural patterns of conduct. Kenya is violating those obligations.

With a new government in office and a new draft constitution containing provisions that would enhance women's property rights set for debate, this is a pivotal time for Kenya to confront the deep property inequalities in its society. It must develop a program of legal and institutional reforms and educational outreach initiatives that systematically eliminates obstacles to the fulfillment of women's property rights.

Conclusion

Women's property issues touch deeply the ways people live, think, and organize their social and economic lives. It's not just a matter of getting a few women in parliament. People feel threatened.

—Professor Yash Pal Ghai, chairman, Constitution of Kenya Review
Commission, Nairobi, October 23, 2002

Property rights abuses inflicted on women in Kenya should be recognized for what they are: gross violations of women's human rights. Discriminatory property laws and practices impoverish women and their dependents, put their lives at risk by increasing their vulnerability to HIV/AIDS and other diseases, drive them into abhorrent living conditions, subject them to violence, and relegate them to dependence on men and social inequality.

Despite the slow recognition that property rights violations harm not just women and their dependents but Kenya's development as a whole, little has been done to prevent and redress these violations. Averting these abuses in a country where dispossessing women is considered normal will be difficult. A concerted effort is needed not just to improve legal protections, but to modify customary laws and practices and ultimately to change people's minds. With extreme poverty, a moribund economy, rampant violence, and catastrophic HIV/AIDS rates, Kenya can no longer afford to ignore women's property rights violations. Eliminating discrimination against women with respect to property rights is not only a human rights obligation; for many women, it is a matter of life and death.

POSTSCRIPT

Are Women in a Position to Challenge Male Power Structures in Africa?

In many ways, the viewpoints presented in this issue get at a deeper debate about social change and the best way to improve the situation of women in Africa. The selection by Human Rights Watch presents the local situation for women in Kenya as deplorable and intractable, suggesting that a top-down, legislative solution is the best course of action. Critics of this approach might argue that, while this is all well and good, it is largely ineffectual as the reach of government is fairly limited in many African contexts. The case study presented by Schroeder about women in The Gambia provides ammunition for those who suggest that a bottom-up approach that is focused on economic empowerment is the best avenue to greater gender equality in Africa. Imagine, for example, what type of social change might occur in the United States if women earned more on average than men (the situation in Kerawan). This can be compared with U.S. Bureau of Labor Statistics survey results showing that American women earned 77 percent of their male counterparts' salaries in 1999. However, it should be noted that a reading of Schroeder's entire volume (of which a small portion was excerpted for this issue) reveals that the situation was later constrained because women's access to land for gardening was somewhat tenuous. Many men who had temporarily loaned land to women for gardening began reasserting their rights to these plots in the 1990s. As such, it may be that both top-down (i.e., legislative) and bottom-up approaches are needed in order to improve the situation of women in Africa.

The case presented by Schroeder is not an isolated incident of economically empowered women in Africa. Another classic example concerns the "Nanas-Benz" of Togo who are wealthy cloth merchants. They are emblematic of how successful women can be in the West African marketplace. *Nanas* means "established woman" or "woman of means." *Benz* refers to the type of auto preferred by these market women. The most successful of these merchants can turn over about $600,000 in cloth per month. They act as agents between importers and wide-ranging clientele in West Africa. Successful Nanas-Benz make sure that their children attend university. The girls study economics, management and administration while the boys become architects, teachers, and bankers. The business is often passed down to a woman's female children.

While WID programs still exist today, there has been an effort to move beyond stand-alone programs focused on women to attention and awareness of the situation and needs of women in all types of programs and policy initiatives.

This broader approach is often simply referred to as "gender" or "gender and development." For examples of the WID and gender approaches, see relevant sections of the Web sites of the United States Agency for International Development http://www.usaid.gov/wid/links.htm and the World Bank http://www.worldbank.org/gender/.

On the Internet . . .

AllAfrica Global Media

AllAfrica Global Media is the largest provider of African news online with offices in Johannesburg, Dakar, Abuja, and Washington, D.C. Over 700 stories are posted daily in French and English, in addition to multimedia content.

http://allafrica.com

African Governments on the WWW

African Governments on the WWW catalogues Internet links to many government organizations in Africa.

http://www.gksoft.com/govt/en/africa.html

Africa Action

Africa Action provides information on political movements related to a wide variety of African issues. This site hosts its own information and contains links to other Internet sites.

http://www.africaaction.org/index.php

Centre for Democracy and Development

The Centre for Democracy and Development is a United Kingdom–based non-governmental organization (NGO) promoting democracy, peace, and human rights in Africa.

http://www.cdd.org.uk

African Union

The Web site of Africa's premier pan-African organization, the African Union (formerly the Organization of African States), contains information on the organization's special programs, news, and events.

http://www.africa-union.org

PART 4

Politics, Governance, and Conflict Resolution

*T*he terrain of politics, governance, and conflict resolution is simultaneously one of the most hopeful and distressing realms in contemporary African studies. While more contested elections have been held in the last 10 years than at any other time in the postcolonial period, the African continent also suffers from more instances of civil strife than other world regions. Scholars and commentators intensely debate the connections between contemporary political developments, historical patterns of governance, global geopolitics, and local traditions of decision making, public discourse, and conflict resolution.

- Are Multi-Party Democratic Traditions Taking Hold in Africa?

- Is Foreign Assistance Useful for Fostering Democracy in Africa?

- Are African Governments Inherently Disposed to Corruption?

- Are International Peacekeeping Missions Critical to Resolving Ethnic Conflicts in African Countries?

ISSUE 17

Are Multi-Party Democratic Traditions Taking Hold in Africa?

YES: Michael Bratton and Robert Mattes, from "Support for Democracy in Africa: Intrinsic or Instrumental?" *British Journal of Political Science* (July 2001)

NO: Joel D. Barkan, from "The Many Faces of Africa: Democracy Across a Varied Continent," *Harvard International Review* (Summer 2002)

ISSUE SUMMARY

YES: Michael Bratton, professor of political science at Michigan State University, and Robert Mattes, associate professor of political studies and director of the Democracy in Africa Research Unit at the University of Cape Town, find as much popular support for democracy in Zambia, South Africa, and Ghana as in other regions of the developing world, despite the fact that the citizens of these countries tend to be less satisfied with the economic performance of their elected governments.

NO: Joel D. Barkan, professor of political science at the University of Iowa and senior consultant on governance at the World Bank, takes a less sanguine view of the situation in Africa. He suggests that one can be cautiously optimistic about the situation in roughly one-third of the states on the African continent, nations he classifies as consolidated democracies and as aspiring democracies. He asserts that one must be realistic about the possibilities for the remainder of African nations, countries he classifies into three groups: stalled democracies, those that are not free, and those that are mired in civil war.

There was a great deal of enthusiasm among Africanists in the early 1960s when more than 40 African nations gained independence and formed popularly elected governments. This enthusiasm was tempered when a large proportion of these countries succumbed to one-party rule or military regimes by the end of the decade. The 1970s and 1980s were largely characterized by the persistence of undemocratic forms of governance. Lacking popular support, undemocratic regimes and guerilla movements often sought Soviet or American patronage within the context of the cold war. The United States, in the name of anti-communism, financially and militarily backed a number of unsavory political leaders and guerilla insurgents during this period. These ranged from Mabuto Sese Seko in former Zaire to UNITA rebels in Angola. Seko, perhaps one of the most corrupt of African dictators, plundered his country for over 20 years during the cold war with the full support of the United States. In Angola, the U.S. and then-white-ruled South Africa sustained a bloody civil war by supporting UNITA rebels in the late 1980s.

The end of the cold war largely led to the end of perverse outside intervention in African affairs. Combined with this change in the external environment was a groundswell of internal support for political reform, which some commentators attribute to the democratic changes occurring in Eastern Europe in the late 1980s that many Africans observed through the international media. The result has often been referred to as Africa's "second wave" of democratization in which, between 1991 and 2000, multiparty elections were held in all but 5 of Africa's 47 states.

In this issue, Michael Bratton and Robert Mattes find as much popular support for democracy in Zambia, South Africa, and Ghana as in other regions of the developing world. However, citizens of these countries tend to be less satisfied with the performance of their elected governments than those in comparable non-African nations. The authors interpret these results to mean that support for democracy in Africa is more intrinsic (an end in itself) than instrumental (a means to an end—such as improving material standards of living). This finding highlights the importance that Africans attach to the basic political rights afforded by democracy. It also contradicts other research indicating that governments in new democracies mainly legitimate themselves through economic performance.

In contrast, Joel D. Barkan takes a more sober view of the situation in Africa. He states that those assessing the political situation in Africa roughly break down into two camps, the optimists and the realists. These two groups tend to draw very different conclusions because they focus their attention on different countries in Africa. Barkan contends that one can be cautiously optimistic about the situation in roughly one-third of the states on the African continent, nations he classifies as consolidated democracies (Botswana, Mauritius, and South Africa) and as aspiring democracies (15 countries). The prospects for the remainder of African nations are much more uncertain. Barkan classifies these states into three groups, the stalled democracies (13 countries), those that are not free (10 countries), and those that are mired in civil war (roughly 6 countries). As a realist, he believes that there is a tendency to overcelebrate progress in the first two groups and to retreat from the challenges in the third, fourth, and fifth groups.

Michael Bratton and Robert Mattes  **YES**

Support for Democracy in Africa: Intrinsic or Instrumental?

Popular support for a political regime is the essence of its consolidation. By voluntarily endorsing the rules that govern them, citizens endow a regime with an elusive but indispensable quality: political legitimacy. The most widely accepted definition of the consolidation of democracy equates it squarely with legitimation. In a memorable turn of phrase, Linz and Stepan speak of democratic consolidation as a process by which all political actors come to regard democracy as 'the only game in town'. In other words, democracy is consolidated when citizens and leaders alike conclude that no alternative form of regime has any greater subjective validity or stronger objective claim to their allegiance.

This article explores how the general public in new multiparty political regimes in sub-Saharan Africa is oriented towards democracy. What, if anything, do Africans understand by the concept? Do they resemble citizens in new democracies elsewhere in the world in their willingness to support a regime based on human rights, competing parties and open elections? And beyond democracy as a model set of rights and institutions, are citizens in Africa satisfied with the way that elected regimes operate in practice? All of these questions are coloured by the fact that many of Africa's democratic experiments are taking place in countries with agrarian economies, low per capita incomes and minuscule middle classes. Under such unpropitious conditions, observers have every reason to wonder whether elected governments have the capacity to meet citizen expectations and, if they cannot, whether citizens may therefore quickly lose faith in democracy.

We assume that citizens will extend tentative support to neo-democracies, if only because they promise change from failed authoritarian formulae of the past. But what is the nature of any such support? Is it *intrinsic*, based on an appreciation of the political freedoms and equal rights that democracy embodies when valued as an end in itself? Or does support reflect a more *instrumental* calculation in which regime change is a means to other ends, most commonly the alleviation of poverty and the improvement of living standards?

The resolution of this issue has direct implications for regime consolidation. Intrinsic support is a commitment to democracy 'for better or worse'; as such, it has the potential to sustain a fragile political regime even in the face of economic downturn or social upheaval. By contrast, instrumental support is conditional. It is granted, and may be easily withdrawn, according to the temper of the times. If citizens evaluate regimes mainly in terms of their capacity to deliver consumable benefits or to rectify material inequalities, then they may also succumb to the siren song of populist leaders who argue that economic development requires the sacrifice of political liberties.

Let us be clear. We do not dispute that evaluations of democracy in new multiparty regimes are likely to be based in good part on the performance of the government of the day. After all, it is very unlikely that citizens in neo-democracies would possess a reservoir of favourable affective dispositions arising from a lifetime of exposure to democratic norms. If democracy is a novel experience, how could such socialization have taken place? Instead of bestowing 'diffuse support', citizens fall back on performance-based judgements of what democracy actually does for them.

We wish to divide regime performance, however, into distinct baskets of goods: an *economic* basket (that includes economic assets, jobs and an array of basic social services) and a *political* basket (that contains peace, civil liberties, political rights, human dignity and equality before the law). The African cases provide a critical test of the importance of political goods to evaluations of democracy. If the denizens of the world's poorest continent make 'separate and correct' distinctions between 'a basket of economic goods (which may be deteriorating) and a basket of political goods (which may be improving)', then citizens everywhere are likely to do so. And if political goods seem to matter more than economic goods in judging democracy, then we can cast light on the 'intrinsic v. instrumental' debate. If democracy is valued by citizens as an end in itself in Africa, then this generalization probably holds good universally.

In this study we find that citizen orientations to democracy in Africa are most fully explained with reference to both baskets of goods. With one interesting country exception, satisfaction with democracy (the way elected governments actually work) is driven just as much by guarantees of political rights as by the quest for material benefit. Support for democracy (as a preferred form of government) is rooted even more deeply in an appreciation of new-found political freedoms, a finding that runs counter to the conventional view that the continent's deep economic crisis precludes regime consolidation. At least so far, new democratic regimes in Africa have been able to legitimate themselves by delivering political goods.

Scope of the Study

Our substantive focus is intentionally restricted—to attitudes to democracy, among masses rather than elites—because our geographical coverage is broader than most studies in Africa. This article uses standard survey items to compare political attitudes in Ghana, Zambia and South Africa, thus bridging the major regions of the sub-Saharan sub-continent and situating public opinion in Africa in relation to other new democracies in the world.

All three countries underwent an electoral transition to multi-party democracy during the last decade but their political trajectories have since diverged. Both of South Africa's competitive polls (in April 1994 and June 1999) were ruled substantially free and fair by independent observers. By contrast, Zambia's founding elections of October 1991 were far more credible than its dubious second contest of November 1996. For its part, Ghana experienced improved electoral quality, with flawed elections in November 1992 being followed by a December 1996 poll that drew almost universal praise. Thus, with reference to the institution of elections alone, South Africa's democracy has stabilized, Ghana's is gradually consolidating, and Zambia's is slowly dying.

In reality, democracy is a fragile species throughout Africa. It is far from clear that a pervasive political culture exists to promote and defend open elections, let alone any other democratic institution. Regime transitions in Africa commonly resulted from intense struggles between incumbent and opposition elites, whose interest in self-enrichment was sometimes more palpable than their commitment to democracy. Even elected leaders have tampered with constitutional rules in order to prolong a term of office or to sideline rivals. And the armed forces continue to lurk threateningly in the wings: about half a dozen of Africa's new democracies succumbed to military intervention within five years of transition. Only in places like South Africa in 1994 (and possibly Nigeria in 1999), where transitions were lubricated by pacts among powerful insiders, are there signs that a culture of compromise and accommodation has penetrated the ranks of the political elite.

The extent to which a commitment to democracy has radiated through the populace is also open to question. After all, regime transitions in Africa were sparked by popular protests that were rooted in economic and political grievances. While the protesters had clear ideas about what they were *against* (the repressions and predations of big-man rule) they did not articulate an elaborate or coherent vision of what they were *for*. Judging by the issues raised in the streets, people seemed to want accountability of leaders and to eliminate the inequities arising from official corruption. To be sure, these preferences loosely embodied core democratic principles. And multiparty elections quickly became a useful rallying cry for would-be political leaders. But, during the tumult of transition, relatively little attention was paid to the institutional design of the polity. Emerging from life under military and one-party rule, citizens could hardly be expected to have in mind a full set of democratic rules or to evince a deep attachment to them.

This article takes stock of what has been learned from the first generation of research on political attitudes in new African democracies in the 1990s. . . .

The Meaning of Democracy in Africa

In considering the meaning of 'democracy' in Africa, the first possibility is that the term has not entered popular discourse, especially where indigenous languages contain no direct semantic equivalent. Some cultural interpretations emphasize that the word changes its meaning in translation, sometimes even signifying consensual constructs like community or unity. Or, because African languages borrow new terminology from others, a phonetic adaptation from a European language (like 'demokrasi') may have become common currency.

In one form or another, democracy seems to have entered the vocabulary of most African citizens. When the 1997 Ghanaian survey asked respondents 'What is the first thing that comes to mind . . . when you think of living in a democracy?', 61.5 per cent were able to provide a meaningful response, rising to 75 per cent in 1999. Interestingly, even more respondents felt that Ghana was a democracy in 1997, implying that some people who could not specify a meaning for democracy could nevertheless recognize one if they saw one. In both countries, the salience of the concept was a function of education, with democracy having meaning in direct proportion to a respondent's years of schooling.

Contrary to cultural interpretations, we contend that standard liberal ideas of civil and political rights lie at the core of African understandings of democracy. In Zambia in 1993 and 1994, participants in two rounds of focus groups were asked 'What does democracy mean to you?' In the ensuing discussions, democracy was most commonly decoded in terms of the political procedure of competitive elections in which 'people are free to vote if they want to' and 'have a right to choose their own leaders'. Informants described how they resented having been forced to vote for the former ruling United National Independence Party (UNIP) and decried the political intimidation exerted by the party's youth wing. They favourably compared a choice of candidates under a multi-party regime with the system of 'appointed representatives' under a one-party state.

An open-ended question in the 1999 survey in Ghana about 'the first thing that comes to your mind . . . when you hear the word "democracy"' elicited the following responses, in frequency order: civil liberties and personal freedoms (28 per cent of all respondents), 'government by the people' (22 per cent), and voting rights (9.2 per cent). The only other major response was 'Don't know' (24.8 per cent) and very few respondents offered a materialistic interpretation (2.5 percent). These findings seem to suggest that Ghanaians view democracy almost exclusively in political terms, with an emphasis on selected civil liberties (especially free speech), collective decision-making and political representation.

Survey findings point to a much more materialistic world view in South Africa. In 1995, South Africans were asked to choose from a list of diverse meanings (both political and economic) that are sometimes attached to democracy. At the top of the popular rankings, 91.3 per cent of respondents equated democracy with 'equal access to houses, jobs and a decent income' (with 48.3 per cent seeing these goods as 'essential' to democracy). This earthy image of democracy far outstripped all other representations: for example, regular elections (67.7 per cent), at least two strong parties (59.4 per cent), and minority rights (54.5 per cent). To be sure, a majority of South Africans did associate democracy with procedures to guarantee political competition and political participation, but their endorsement of these political goods was far less ringing than the almost unanimous association of democracy with improved material welfare. Tellingly, only small minorities found it 'essential' to democracy to hold regular elections (26.5 per cent) or guarantee minority rights (20.6 per cent).

Because South Africa is a deeply divided society with mutually reinforcing fault lines of race and class, one would expect that various social groups would hold disparate views of democracy. We have noted elsewhere 'massive racial differences in agreement with regime norms'. Whites are much more likely than blacks to agree that regular elections, free speech, party competition and minority rights are essential to democracy. This procedural interpretation of democracy most likely reflects their own minority status and their reliance for protection on constitutional and legal rules. South African blacks, for their part, attach just as much or more importance to narrowing the gap between rich and poor. And while many South Africans of all races say they accept the necessity of redistributing jobs, houses and incomes, blacks seem to focus more on 'equality of results' while whites stress 'equality of opportunity'.

We reach four working conclusions based on recent research on citizen conceptions of democracy in three African countries. First, Africans here are more likely to associate democracy with individual liberties than with communal solidarity, especially if they live in urban areas. Secondly, popular conceptions of democracy have *both* procedural *and* substantive dimensions, though the former conception is more common than the latter. Thirdly, citizens rank procedural and substantive attributes in different order across countries. Zambians place political rules at the top of the list of democratic attributes, whereas South Africans relegate such guarantees behind improvements in material living standards. Finally, rankings differ even within the category of political goods: whereas Zambians (and to a lesser extent South Africans) grant primacy to elections, Ghanaians elevate freedom of speech to the top of their own bill of democratic rights.

These cross-national differences can be interpreted in terms of the life experiences of citizens under each country's old regime. Zambians may regard democracy mainly in terms of competitive multi-party elections because of their disappointing experiences with the ritual of 'elections without choice' under Kenneth Kaunda's one-party state. Ghanaians, for their part, emphasize freedom of speech as a reaction against the tight controls over communication imposed by the previous military regime, whose populist ideology was the only approved form of political discourse. Finally, South Africans place socio-economic considerations at the heart of their notion of democracy because of the integrated structure of oppression experienced under apartheid. Impoverished under the old regime, they see the attainment of political freedom as only the first step in rectifying manifold inequalities in society. In this conception, democracy has an inclusive meaning; it is as much a means to social transformation as a politically desirable end in itself.

Support for Democracy in Africa

The best way to ask questions about popular support for democracy is in concrete terms and in the form of comparisons with plausible alternatives. Since democracy has motley meanings, it is not useful to ask whether people support it in the abstract. It is far better to elicit opinions about a real regime with distinctive institutional attributes, such as a 'system of governing with free elections and many parties'. And if citizens support democracy as the 'least worst' system (the so-called 'Churchill hypothesis'), it is worth testing their levels of commitment against other regime forms that they have recently experienced or could conceive of encountering in the future.

Table 1 reports results of survey questions of this sort from various world regions, with sub-Saharan Africa represented by Ghana, Zambia and South Africa. In so far as these countries are representative of the region, Table 1 shows that the level of public commitment to democracy is much the same in Africa as in other regions of the world that have recently undergone regime change. Excluding Southern Europe, almost two out of three citizens in new democracies extend legitimacy to elected government as their preferred political regime: the relevant mean figures are 65 per cent for East and Central Europe, 63 per cent for South America, and 64 per cent for the three countries of sub-Saharan Africa. Indeed, the average level of support in Africa (64.3 per cent) is virtually identical to the combined mean for Latin America and post-Communist Europe (64.2 per cent).

Moreover, deviation in support for democracy around the regional mean is lower for the three African countries than for other parts of the world. The countries with the lowest and highest levels of support for democracy are separated

Table 1

Public Attitudes to Democracy: Preliminary Cross-National Comparisons

	Support democracy	Satisfied with democracy	Supportive and satisfied	Supportive but not satisfied
European Union	78	53	—	—
Southern Europe	84	57	79	11
Greece	90	52	84	11
Portugal	83	60	77	9
Spain	78	60	75	12
East and Central Europe	65	60	72	6
Czech	77	56	70	8
Poland	76	61	70	4
Romania	61	77	68	4
Bulgaria	66	61	75	2
Slovakia	61	49	62	14
Hungary	50	53	79	4
South America	63	50	45	22
Uruguay	80	54	57	29
Argentina	77	53	55	28
Chile	53	48	38	17
Brazil	41	46	32	16
Sub-Saharan Africa	64	48	41	18
Ghana (1997)	74	53	46	13
Zambia (1996)	63	53	49	14
South Africa (1997)	56	38	29	13
South Africa (blacks)	61	45	35	11
South Africa (whites)	39	7	5	18

Note: Regional means are raw estimates, uncorrected for proportional population size of countries. Further fnn. to Table 1 can be found in the electronic version of the journal available at www.cup.cam.ac.uk

by just 18 per centage points in the African cases, but by 27 points for Eastern Europe and 39 points in South America. We interpret this to mean that authoritarian regimes have been widely discredited across the continent. Although the citizens of Ghana and Zambia may not have committed themselves to democracy as firmly as the citizens in Uruguay and the Czech Republic, they evince less nostalgia for hardline rule than citizens in Hungary and Brazil. Once again, though, South Africa is an exception. And we would need many more confirming cases before we could be sure that legitimating sentiments are evenly spread across all African countries.

Indeed, variations are evident within Africa in the extent to which citizens support new regimes. Of the three cases under review, Ghana displays the highest levels of citizen commitment to democracy. In 1997, fully 73.5 per cent of citizens thought it somewhat or very important for Ghana to 'have at least two political parties competing in an election'. The intensity of this support appears to be strong, as reflected by the 55.9 per cent of respondents who thought these institutions 'very important'. And the quality and depth of this support is underlined by the even higher proportions who granted importance to the right of citizens to form parties representing diverse viewpoints (82.5 per cent), to the openness of the mass media to political debate (89.3 per cent), and to the regular conduct of honest elections (92.7 per cent). While there is some possibility that respondents are acquiescing here to non-controversial 'motherhood' questions, Ghanaians nonetheless appear to consistently favour a full basket of liberal political rights.

Among the countries considered, legitimation of the new regime was lowest in South Africa, where citizens do not yet feel a widespread attitudinal commitment to democracy. A 1997 survey asked respondents to choose between the following statements: '[When] democracy does not work . . . some say you need a strong leader who does not have to worry with elections. Others say democracy is always best'. Since only a bare majority chose the democratic option (56.3 per cent, up from 47 per cent in 1995, but dropping back again below 50 per cent in 1998), support for democracy appeared to be weaker there than in the other African countries. Other responses underscore the shallowness of democratic legitimacy and the appeal of authoritarian alternatives in South Africa. In 1997, about one-third of the population thought that, under democracy, 'the economic system runs badly' (29 per cent), order is poorly maintained (30.2 per cent), and leaders are 'indecisive and have too much squabbling' (35.1 per cent). And more than half of all South Africans (53.8 per cent) stated that they would be 'willing to give up regular elections if a non-elected government or leader could impose law and order and deliver jobs and houses'.

Thus, the potential constituency for forceful rule appears to be larger in South Africa than in South America, where an average of just 15 per cent of citizens considers that 'in some circumstances an authoritarian government can be preferable to a democratic [one]'. Sentiments for a strong man were higher in South Africa (30.8 per cent) than in Chile (19 per cent) and Brazil (21 per cent), where authoritarian nostalgia is usually considered to be high. Question wording may have had a significant effect, with the cue of higher material living standards ('jobs and houses') inducing even some of democracy's supporters to abandon it. But, at minimum, this finding draws attention to the role of instrumental calculations in the assessments of democracy by many South African citizens.

South Africa's deviance is explicable again, however, in terms of its cultural diversity. White South Africans were much less likely to judge that 'democracy is always best' (39 per cent) than the country's African citizens (61 per cent). And, while 'coloureds' situated themselves between blacks and whites when granting such support to democracy (53 per cent), Asian South Africans were the least supportive of all (27 per cent). Thus the cautious, even retrogressive, attitudes of ethnic minorities tended to depress overall levels of commitment to democracy in South Africa. Examined alone, African citizens can be seen to support this form of regime at the highest level of any ethnic group in South Africa (61 per cent), a level not too different from citizens in Zambia (63 per cent) and the sub-Sahara region as a whole (64 per cent).

In Zambia, the question on support for democracy differed slightly, while still focusing on a political system featuring elections and posing a comparison with a realistic alternative regime. Respondents were asked to choose: Is 'the best form of government . . . a government elected by its people' or 'a government that gets things done'? On the assumption that support erodes as regimes mature, especially if citizens' expectations are not fully realized, we thought that support for 'elected government' would decline over time. To date this has not happened in Zambia. Public support for democracy held steady, at 63.4 per cent in 1993 and 62.9 per cent 1996. As in Ghana, other related items bespoke an electorate with a relatively firm syndrome of democratic commitments. In 1996, 73 per cent preferred 'a choice of political parties and candidates' to 'a return to a system of single-party rule'.

Satisfaction With Democracy in Africa

Democracy looks better in theory than in practice. In elected regimes worldwide, more citizens support democracy as their preferred form of government than express satisfaction with the way that it actually works. This generalization holds true not only for Third Wave neo-democracies but, even more so, for the established regimes of Western Europe. . . .

Unlike support for democracy, satisfaction with democracy is not as widespread in the three African countries as it is in South America and Eastern Europe. Satisfaction lags support by a wider margin in the sub-Saharan region (16 per centage points) than in the other two world regions (13 and 5 per centage points respectively). Substantively, fewer than half (48 per cent) of the citizens in these new African democracies report satisfaction with key aspects of the performance of elected regimes. Once more, the African average is pulled down by South Africa, with Ghana and Zambia displaying popular approval of regime performance at levels similar to consolidating democracies like Uruguay and Argentina. Although different racial groups in South Africa again evince distinct levels of satisfaction (45 per cent for blacks versus just 7 per cent for whites), black South Africans in this instance trail their fellow citizens elsewhere on the continent in their contentment with democracy in practice (39 per cent). Instead, they tend to more closely resemble the citizens of Brazil (41 per cent), more of whom are unhappy with democracy than are satisfied with it. . . .

Explaining Satisfaction With Democracy

. . . Satisfaction with democracy among African citizens appears to depend upon their assessment of the performance of government, particularly its performance at delivering *both* economic *and* political goods. Taken together, these factors explain between a quarter and two-fifths of the variance in expressed satisfaction in three African countries. Apart from social background, no set of factors—whether general performance, economic goods or political goods—can be discarded without a significant loss of explanatory power. Any ecumenical explanation of satisfaction with democracy in Africa must make reference to government performance in *both* its political *and* economic dimensions.

But what about the relative weight of economic and political explanations? We note that the delivery of economic goods sometimes has large independent effects on satisfaction with democracy in individual countries. Cross-nationally, however, such effects are rather inconsistent. We therefore conclude that economic effects are subject to the exigencies of time and place, such as gradual economic recovery in Ghana and persistent economic crisis in Zambia. We therefore doubt that a general explanation of satisfaction with democracy can be constructed from economic data alone. At the same time, we note that the effects of political factors, while occasionally weaker than those of economic factors, prevail more consistently across countries. This observation suggests that the delivery of political goods is a more reliable general predictor of satisfaction with democracy and a more promising foundation on which to construct a theory of democratic consolidation. This line of argument is explored further in the next section.

Explaining Support for Democracy

We turn, finally, to explain support for democracy as a preferred regime type. As Table 2 shows, our analysis accounts for 12 to 17 per cent of the variance in popular support (see adjusted R^2 statistics). Our explanation was less complete in this case perhaps because of the impalpability of the issue at hand: citizens may find it more difficult to assess the qualities of abstract constitutional rules than the concrete performance of actual governments. In any event, public opinion in Africa seems to be less fully formed, and more contradictory, when it comes to support for democracy.

Nevertheless, Table 2 does reveal interesting findings. First, it reconfirms that attitudes to democracy cannot be inferred from standard social background characteristics. Again, gender and age were irrelevant to the legitimation of democracy in all countries studied and education had a positive impact only in Zambia. These findings are consistent with the observations that 'the more education a person has, the more likely he or she is to reject undemocratic alternatives' but that, overall, 'social structure [has] little influence . . . on atti-

Table 2

Multiple Regression Estimates of Support for Democracy

	S. Africa			Ghana			Zambia		
	B	(s.e.)	Beta	B	(s.e.)	Beta	B	(s.e.)	Beta
Social background factors									
Gender			0.028			-0.014			0.048
Age			-0.019			0.000			0.042
Education			-0.009			0.001	0.064	(0.010)	0.200***
General performance factors									
Approval of government performance	0.084	(0.014)	0.145***	0.368	(0.064)	0.169***	0.122	(0.030)	0.128***
Satisfaction with democracy	0.082	(0.011)	0.196***	0.393	(0.052)	0.195***			-0.010
Economic factors									
Assessment of current economic conditions			-0.013			0.038			0.001
Assessment of current personal QOL			-0.017			0.023			-0.021
Assessment of future personal QOL	0.035	(0.011)	0.077***			-0.023			0.067
Support for market reforms			0.038			-0.006			0.017
Delivery of economic goods			-0.002	0.066	(0.031)	0.062***			0.059
Political factors									
Interest in politics	0.054	(0.015)	0.077***			0.015	0.115	(0.025)	0.144***
Trust in government institutions			0.032			-0.003			0.018
Perception of government responsiveness			0.031	-0.140	(0.072)	-0.008			0.012
Perception of official corruption			-0.003			-0.045*			-0.020
Delivery of political goods			0.028	0.158	(0.025)	0.153***	0.461	(0.080)	0.181***
N	3,500			2,005			1,182		
R	0.349			0.417			0.382		
R²	0.122			0.174			0.146		
Adjusted R²	0.120			0.171			0.142		

(*)Significant at 0.05

(**)Significant at 0.01

(***)Significant at 0.001

tudes towards the new regime'. If African societies do not contain entrenched pockets of generational or gender-based resistance to democratization, then the prospects for the consolidation of democratic regimes would seem to be slightly brighter than is sometimes thought.

Secondly, regime legitimacy in Africa depends upon popular appraisals of government performance. Consistently, in all three countries, support for democracy was strongest among citizens who felt that elected governments were generally doing a good job. But approval of government performance was closely connected to party identification, with supporters of the ruling party in each country being much more approving. Thus we must investigate further whether citizens are accrediting government performance—and thereby supporting democracy—out of 'knee-jerk' loyalty to a ruling party rather than a rational calculation that democratic governments deserve legitimation because they are more effective.

In Ghana and South Africa, support for democracy also was accompanied by expressions of satisfaction with democracy. We take this as further evidence that regime legitimation in Africa rests squarely upon performance considerations. On the up-side, popular demand for government performance increases the likelihood that citizens will make use of the rules of democratic governance to hold their leaders accountable. On the down-side, it also raises the possibility that citizens may conflate the performance of governments (that is, the achievements of incumbent groups of elected officials) with the performance of regimes (that is, the rules by which governments are constituted). The risk thus arises that, faced with continued mismanagement by ineffective governments, Africans may throw the baby out with the bath-water. By punishing government under-performance, they may inadvertently dismiss democracy.

Thirdly, and notwithstanding what has just been said about performance, we find little systematic evidence from Africa that citizens predicate support for democracy on the delivery of economic goods. Generally speaking, the legitimation of democratic regimes does not depend on citizen assessments of personal or national economic conditions, either now or for the future. Only in South Africa are assessments of future personal conditions linked positively to support for democracy. Strikingly too, when other relevant factors are controlled for, citizen perceptions of economic delivery have no discernible effects on the endorsement of democracy in either Zambia or South Africa. The delivery of economic goods only seems to matter in Ghana, though the influence of this instrumental consideration is far from the strongest in the Ghana model.

Instead, we are led back again to the impact of political factors. For the first time, we find that citizen interest in politics had a positive effect on attitudes to democracy in two out of the three countries (Zambia and South Africa). It stands to reason that democracy will not consolidate where citizens remain disinterested in, and detached from, the political process; before people can actively become democracy's champions, they must orient themselves towards involvement in political life. One wonders why Ghanaians, who display the highest levels of interest in politics among the Africans surveyed, do not automatically support democracy. The answer appears to lie, at least in part, in the popular perception of rampant official corruption in that country. Many persons who are predisposed by their interest in politics to become active citizens are 'turned off' from democracy by what they see as the illicit machinations of civilian politicians. As one would expect, perceptions of official corruption are negatively associated with support for democracy in all three African countries; only in Ghana, however, is this relationship statistically significant.

Finally, and most importantly, the delivery of political goods bears a strong and significant relationship to the popular legitimation of democracy. In judging democracy, the Africans that we surveyed think of government performance first and foremost in political terms. Unlike the delivery of economic goods, a factor that is relevant in only one country, this relationship holds in at least two country cases. . . .

Conclusion

In this [selection,] we have established that levels of popular support for democracy are roughly similar in three neo-democracies of sub-Saharan Africa as in other Third Wave countries. Almost two-thirds of eligible voters in these African countries say that they feel some measure of attachment to democratic rules and values. Under these circumstances, the popular consolidation of democracy in at least some African countries does not seem an entirely far-fetched prospect.

Yet the African cases stand apart from other new democracies in terms of lower levels of mass satisfaction with actual regime performance. The fact that African survey respondents support democracy while being far from content with its concrete achievements suggests a measure of intrinsic support for the democratic regime form that supersedes instrumental considerations. But, although support for democracy may be quite broad, we cannot confirm that it is deep. We do not yet know if citizens will vigorously defend the political regime if economic conditions take a decisive turn for the worse or if rulers begin to backtrack on hard-won freedoms.

Joel D. Barkan

The Many Faces of Africa: Democracy Across a Varied Continent

A decade ago, seasoned observers of African politics including Larry Diamond and Richard Joseph argued that the continent was on the cusp of its "second liberation." Rising popular demand for political reform across Africa, multiparty elections, transitions of power in several countries, and negotiations toward a new political framework in South Africa led these experts to conclude that the prospects for democratization were good. Today, these same observers are not so sure. They describe Africa's current experience with democratization in terms of "electoral democracy," "virtual democracy," or "illiberal democracy," and are far more cautious about predicting what is to come. What is the true state of African democracy? And what is its future?

Governance Before the 1990s

Africa's first liberation was precipitated by the transition from colonial to independent rule that swept much of the continent, except the south, between 1957 and 1964. The West hoped that the transition would be to democratic rule, and more than 40 new states with democratic constitutions emerged following multiparty elections that brought new African-led governments to power. The regimes established by this process, however, soon collapsed or reverted to authoritarian rule—what Samuel Huntington has termed a "reverse wave" of democratization. By the mid-1960s, roughly half of all African countries had seen their elected governments toppled by military coups.

In the other half, elected regimes degenerated into one-party rule. In what was to become a familiar scenario, nationalist political parties formed the first governments. The leaders of these parties then destroyed or marginalized the opposition through a combination of carrot-and-stick policies. The result was a series of clientelist regimes that served as instruments for neo-patrimonial or personal rule by the likes of Mobutu Sese Seko in Zaire or Daniel Arap Moi in Kenya—regimes built around a political boss, rather than founded in a strong party apparatus and the realization of a coherent program or ideology.

This pattern, and its military variant (as with Sani Abacha in Nigeria), became the modal type of African governance from the mid-1960s until the early 1990s. These regimes depended on a continuous and increasing flow of patronage and slush money for survival; there was little else binding them together. Inflationary patronage led to unprecedented levels of corruption, unsustainable macroeconomic policies that caused persistent budget and current account deficits, and state decay, including the decline of the civil service. Most African governments still struggle with this structural and normative legacy, which has obstructed the process of building democracy.

Decade of Democratization?

Africa's second liberation began with the historic 1991 multiparty election in Benin that resulted in the defeat of the incumbent president, an outcome that was replicated in Malawi and Zambia in the same year. The results of these elections raised expectations and created hopes for the restoration of democracy and improved governance across the continent. By the end of 2000, multiparty elections had been held in all but five of Africa's 47 states—Comoros, the Democratic Republic of Congo, Equatorial Guinea, Rwanda, and Somalia.

Along with the new states of the former Soviet Union, Africa was the last region to be swept by the so-called "third wave" of democratization, and as with many of the successor states of the former Soviet Union, the record since has been mixed. In stark contrast to the democratic transitions that occurred in the 1970s and 1980s in Southern and Eastern Europe and Latin America (excluding Mexico), most African transitions have not been marked by a breakthrough election that definitively ended an authoritarian regime by bringing a group of political reformers to power. While this type of transition has occurred in a small number of states, most notably Benin and South Africa, the more typical pattern has been a process of protracted transition: a mix of electoral democracy and political liberalization combined with elements of authoritarian rule and, more fundamentally, the perpetuation of clientelist rule. In this context, politics is a three-cornered struggle between authoritarians, patronage-seekers, and reformers. Authoritarians attempt to retain power by permitting greater liberalization and elections while selectively allocating patronage to those who remain loyal. Meanwhile, patronage-seekers attempt to obtain the spoils of office via electoral means, as reformers pursue the establishment of democratic rule. The boundaries between the first and second of these groups, and sometimes between the second and third, can be blurred because political alignments are very fluid. Liberal democracy is unlikely to be consolidated until reformers ascend to power.

The result is what Thomas Carothers has termed a "gray zone" of polities, describing countries where continued progress toward democracy beyond elections is limited and where the consolidation of democracy, if it does occur, will unfold over a long period, perhaps decades. This characterization does not necessarily mean that the third wave of democratization is over in Africa. Rather, we should expect Africa's democratic transitions to be similar to those of India or Mexico. In the former, the party that led the country to independence did not lose an election for three decades, and periodic alternation of power between parties did not occur until after 40 years. In the latter, the end of one-party rule and its replacement by an opposition committed to democratic principles played out over five elections spanning 13 years rather than a single founding election. Such appears to be the pattern in Africa, where two-thirds of founding and second elections have returned incumbent authoritarians to power, but where each iteration of the electoral process has usually resulted in a significant incremental advance in the development of civil society, electoral fairness, and the overall political process.

That many African polities fall into the gray zone is confirmed by the most recent annual *Freedom in the World* survey conducted by Freedom House. Of the 47 states that comprise sub-Saharan Africa, 23 were classified by the survey as "partly free" based on the extent of their political freedoms and civil liberties. Only eight (Benin, Botswana, Cape Verde, Ghana, Mali, Mauritius, Namibia, and South Africa) were classified as "free" while 16, including eight war-torn societies (Angola, Burundi, the Democratic Republic of Congo, Ethiopia, Eritrea, Liberia, Rwanda, and Sudan) were deemed "not free."

The overall picture revealed by these numbers is sobering. Less than one-fifth of all African countries were classified as free, and of these, only two or three (Botswana, Mauritius, and perhaps South Africa) can be termed consolidated democracies. On the other hand, if one excludes states in the midst of civil war, one-fifth of Africa's countries are free, one-fifth not free, and three-fifths fall in-between. That is to say, four-fifths of those not enmeshed in civil war are partly free or free, a significant advance over the continent's condition a decade ago. Only a handful are consolidated democracies, but few are harsh dictatorships of the type that dominated Africa from the mid-1960s to the beginning of the 1990s. As noted by Ghana's E. Gyimah-Boadi, "Illiberalism has persisted, but is not on the rise. Authoritarianism is alive in Africa today, but is not well."

Optimists and Realists

The current status of democracy in Africa varies greatly from one country to the next, and one should resist generalizations that apply to all 47 of the continent's states; one size does not fit all. Notwithstanding this reality, those who track events in Africa have divided themselves into two distinct camps: optimists and realists. Those in the United States who take an optimistic view—mainly government officials involved in efforts to promote democratization abroad, former members of President Bill Clinton's administration responsible for Africa, members of the Congressional Black Caucus, and the staff of some Africa-oriented nongovernmental organizations—trumpet the fact that multi-party elections have been held in nearly 90 percent of all African states. They note that most African countries have now held competitive elections twice and that some, including Benin, Ghana, and Senegal, have held genuine elections three times, at least one of which has resulted in a change of government. The optimists further note that the quality of these elections has improved in some countries, both in terms of efficiency and of fairness. Electoral commissions seem to have been more independent, even-handed, and professional in recent elections than in the early 1990s. Opposition candidates and parties have greater freedom to campaign and have faced less harassment from incumbent governments. The presence of election observers, both foreign and domestic, is now widely accepted as part of the process. Perhaps most significant, citizen participation in elections has been fairly high, averaging just under two-thirds of all registered voters.

Recognizing that elections are a necessary but insufficient condition for the consolidation of democracy, the optimists also point to advances in several areas, listed below in their approximate order of accomplishment. First, there has been a re-emergence and proliferation of civil society organizations after their systematic suppression during the era of single-party and military rule. Second, an independent and free press has also re-emerged, spurred on by the privatization of broadcast media in several countries. Third, members of the legislature have increasingly asserted themselves in policymaking and overseeing the executive branch. Fourth, the judiciary and the rule of law have been strengthened in countries such as Tanzania, and human rights abuses have also declined. Fifth, there have been new experiments with federalism—the delegation or devolution of authority from the central government to local authorities—to enhance governmental accountability to the public and defuse the potential for ethnic conflict, most notably in Nigeria but also in Ethiopia, Ghana, Tanzania, Uganda, and South Africa. One or more of these trends, especially the first two and perhaps the third, can be found in most African countries that are not trapped in civil war.

Optimists also point to less exclusive membership in the governing elite, which has expanded into the upper-middle sector of society far more than during the era of authoritarian rule. In country after country, repeated multi-party elections have resulted in significant turnover in the national legislatures and local government bodies, sometimes as high as 40 percent per cycle. While the quality of elected officials at the local level remains poor, members of national legislatures are younger, better educated, and more independent in their political approach than the older generation they have displaced. Although further research is needed to confirm any major change in the composition of these bodies, new politicians and legislators also appear more likely than their predecessors to be democrats and to focus on issues of public policy and less likely to be patronage seekers.

Finally, public opinion across Africa appears to prefer democracy over any authoritarian alternatives. Surveys undertaken for the Afrobarometer project in 12 African countries between 1999 and 2001 found that a mean of 69 percent of all respondents regarded democracy as "preferable to any other kind of government," while only 12 percent agreed with the proposition that "in certain situations, a non-democratic government can be preferable." Moreover, 58 percent of all respondents stated that they were "fairly satisfied" or "very satisfied" with the "way democracy works" in their country.

Realists—who criticize what they contend was a moralistic approach to US foreign policy by the Clinton administration and disparage the use of democratization as a foreign policy goal—take a far more cautious view of what is occurring on the continent. Considering the same six developments that optimists cite as examples of democratic advances, realists note that all six are present in fewer than six countries. They also see much less progress than the optimists when nations are considered one by one. First, regular multi-party elections across the continent have resulted in an alternation of government in only one-third of the countries that have held votes. Moreover, only about one-half of these elections have been regarded as free and fair, with results accepted by those who have lost. It is also debatable in most of these countries whether recent elections have been of higher quality than those held in the early 1990s.

Second, although the re-emergence of civil society and the free press is a significant advance from the era of authoritarian rule when both were barely tolerated or systematically suppressed, civil society remains very weak in Africa compared to other regions and is concentrated in urban areas. Political parties are especially weak and rarely differentiate themselves from one another on the basis of policy. Apart from the church, farmers' organizations, or community self-help groups in a smattering of countries (such as Kenya, Côte d'Ivoire, and Nigeria), civil society barely exists in rural areas where most of the population resides. The press, especially the print media, is similarly concentrated in urban areas and thus reaches a relatively small proportion of the entire population. Only the broadcast media penetrates the countryside, but it is largely state-owned. Although private broadcasting has grown in recent years, especially in television and FM radio, stations cater almost exclusively to urban audiences. With a few exceptions, AM and short- and medium-wave radio—the chief sources of information for the rural population—remain state monopolies.

Third, while the legislature holds out the promise of becoming an institution of countervailing power in some countries, it remains weak and has rarely managed to effectively check executive power. Fourth, the judicial system in most countries is ineffectual, either because its members are corrupt or because it has too few magistrates and too poor an infrastructure to keep pace with the number of cases. Human rights abuses also continue, though less frequently and with less intensity than a decade ago. Fifth, Africa's experiments with federalism, though apparently successful, are confined to six states. Finally, the extent to which Africans have internalized democratic values is hard to judge. Although the Afrobarometer surveys indicate broad support for democracy, the results also suggest that such support is "a mile wide and an inch deep." An average of only 23 percent of respondents in each country described their country as "completely democratic."

Both optimists and realists are correct in their assessments of what is occurring in Africa. But how can both views be valid? The answer is that each presents only one side of the story. On a continent where the record of democratization is one of partial advance in over one-half of the cases, those assessing progress toward democratization, or lack of it, tend to dwell either on what has been accomplished or on what has yet to be achieved. These divergent assessments are proverbial examples of those who view the glass as either half-full or half-empty. Optimists and realists also draw their conclusions from slightly different samples. Whereas optimists focus mainly on states that are partly free or free, realists concentrate on states that are partly free or not free.

Optimists and realists are also both right because there are several Africas rather than one. In fact, at least five Africas cut across the three broad categories of the Freedom House survey. First are the consolidated and semi-consolidated democracies—a much smaller group of countries than those classified as free. This category presently consists of only two or three cases, such as Botswana, Mauritius, and perhaps South Africa. The second group consists of approximately 15 aspiring democracies, including the remaining five classified by Freedom House as free but not yet consolidated democracies plus roughly 10 classified as partly free where the transition to democracy has not stalled. All these states have exhibited slow but continuous progress toward a more liberal and institutionalized form of democratic politics. In this group are Benin, Ghana, Madagascar, Mali, Senegal, and possibly Kenya, Malawi, Tanzania, and Zambia. Third are semi-authoritarian states, countries classified as partly free where the transition to democracy has stalled. This category consists of approximately 13 countries including Uganda, the Central African Republic, and possibly Zimbabwe. Fourth are countries that are not free, with little or no prospect for a democratic transition in the near future. About 10 countries make up this group, including Cameroon, Chad, Eritrea, Ethiopia, Rwanda, and Togo. Finally, there are the states mired in civil war, such as Angola, Congo, Liberia, and Sudan. Each of these five Africas presents a different context for the pursuit of democracy.

Inhibiting Democracy

Several conditions peculiar to the continent make Africa a difficult place to sustain democratic practice. They explain why Africa lags behind other regions in its extent of democratic advance, why political party organizations are weak, and why the ties between leaders and followers are usually based on clientelist relationships. These conditions in turn create pressures for more and more patronage, a situation that undermines electoral accountability and leads to corruption.

Africa is the poorest of the world's principal regions: per capita income averages US$490 per year. This condition does not affect the emergence of democracy but does impact its sustainability. On average, democracies with per capita incomes of less than US$1,000 last 8.5 years while those with per capita incomes of over US$6,000 endure for 99. The reasons for this are straightforward. Relatively wealthy countries are better able to allocate their resources to most or all groups making claims on the state, while poor countries are not. The result is that politics in a poor country is likely to be a zero-sum game, a reality that does not foster bargaining and compromise between competing interests or a willingness to play by democratic rules.

Almost all African countries remain agrarian societies. With few exceptions like South Africa, Gabon, and Nigeria, 65 percent to 90 percent of the national populations reside in rural areas where most people are peasant farmers.

Consequently, most Africans maintain strong attachments to their places of residence and to fellow citizens within their communities. Norms of reciprocity also shape social relations to a much greater degree than in urban industrial societies. In this context, Africans usually define their political interests—that is to say, their interests as citizens vis-à-vis the state—in terms of where they live and their affective ties to neighbors, rather than on the basis of occupation or socio-economic class.

With the exceptions of Botswana and Somalia, all African countries are multi-ethnic societies where each group inhabits a distinct territorial homeland. Africans' tendency to define their political interests in terms of where they live is thus accentuated by the fact that residents of different areas are often members of different ethnic groups or sub-groups.

Finally, African states provide much larger proportions of wage employment, particularly middle-class employment, than states in other regions do. African states have also historically been large mobilizers of capital, though to a lesser extent recently. Few countries have given rise to a middle class that does not depend on the state for its own employment and reproduction. In this context, people seek political office for the resources it confers, for their clients' benefit, and for the chance to enhance their own status. In the words of a well known Nigerian party slogan, "I chop, you chop," literally, "I eat, you eat."

POSTSCRIPT

Are Multi-Party Democratic Traditions Taking Hold in Africa?

In addition to the factors raised by Barkan as inhibiting democracy in Africa, which some would dispute, there are other conditions frequently evoked to foreground a discussion of multiparty democracy in Africa. First, prior to independence, most African nations had little to no experience with democracy (as it is conceived in the West) at the national scale. If anything, the colonial period served to reinforce undemocratic tendencies, as the main purpose of unelected colonial administrations was to extract resources. Second, most African nations inherited national borders from the colonial era that cut across ethnic boundaries. The "unnaturalness" of these boundaries has made African states more difficult to govern. Third, and a point alluded to in the introduction to this issue, external powers have meddled in African affairs, often to the detriment of more representative government. The French, in particular, have intervened on a number of occasions in their former colonies when the leadership was not supportive of French interests. Finally, the role of the military is very poorly defined in many African contexts. In the absence of strong civilian rule, there is a tendency for this institution to assert control when there are economic or political problems. Here again, the way in which the military was used in the colonial era probably has contributed to this ill-defined role. While some scholars are highly cognizant of the aforementioned factors when assessing political change in Africa, others assert that attempts to put the "blame" for Africa's nontransition to democracy on outsiders or former colonial powers is simply a convenient excuse for the corruption and mismanagement of African leaders.

For those interested in further reading, Claude Ake offers a different perspective on African democracy than Bratton and Mattes, arguing that "the democracy movement in Africa will emphasize concrete economic and social rights rather than abstract political rights." See *Democracy and Development in Africa* (The Brookings Institution, 1997). A good example of a more thoroughly pessimistic view (or realistic depending on your perspective) of the prospects for democracy in Africa is George Ayittey's *Africa in Chaos* (St. Martin's Press, 1998).

ISSUE 18

Is Foreign Assistance Useful for Fostering Democracy in Africa?

YES: Arthur A. Goldsmith, from "Donors, Dictators, and Democrats in Africa," *The Journal of Modern African Studies* (2001)

NO: Julie Hearn, from "Aiding Democracy? Donors and Civil Society in South Africa," *Third World Quarterly* (2002)

ISSUE SUMMARY

YES: Arthur A. Goldsmith, professor of management at the University of Massachusetts in Boston, examines the relationship between the amount of development assistance given to sub-Saharan African countries in the 1990s and the evolution of their political systems. He suggests that there is a positive, but small, correlation between donor assistance and democratization during this period. He views aid as insurance to prevent countries from sliding back into one-party or military rule.

NO: Julie Hearn, lecturer in development studies at the School of Oriental and African Studies, University of London, investigates democracy assistance in South Africa. She critically examines the role assigned to civil society by donors, questioning the "emancipatory potential" of the kind of democracy being promoted.

Starting in the early 1990s the international aid community began to more explicitly link donor assistance to democratic reform in Africa. This was in part due to a realization that, in the quest for constancy in strategically important African nations, the United States and other donor nations had often perpetuated the existence of corrupt and antidemocratic regimes that ultimately were unstable. Furthermore, the end of the cold war and the disintegration of the Soviet Union meant that the competing one-party, socialist ideology had lost most of its steam. Finally, there was a growing conceptual assertion that good governance was essential for economic growth and enhanced international trade. The way in which donors openly began to work to shape political structures in Africa was unprecedented. While covert attempts to influence the internal politics of African nations had always existed, it was once considered an infringement of national sovereignty for one nation to work actively and openly at shaping the political discourse in another.

The link between foreign assistance and democratic reform has taken two major forms to date. The first is "political conditionality," which links general foreign assistance (for all types of programs) to governance and democracy criteria. In other words, a country must meet certain standards of democracy as a condition for receiving any type of assistance. So, for example, the World Bank now requires its loan applicants to meet certain governance and economic criteria before it will release funds. The second type of assistance is designed specifically to promote the development and strengthening of democratic institutions and practices (often through technical assistance programs). For example, the United States Agency of International Development (USAID) works to advance democracy in Africa by funding programs that facilitate: 1) free and fair elections, 2) the rule of law, 3) a greater advocacy role for civil society, and 4) transparent, accountable, and participatory governance.

Proponents of this approach argue that it is better to rely on governance criteria than geopolitical considerations when dispensing aid. Critics suggest that the former is just a thinly veiled rationale for the latter, which will always dictate who receives foreign assistance and how much. Furthermore, they assert that democratic and economic conditionalities are now tightly woven into one package promoting a global neoliberal agenda that favors the interests of the wealthy nations over those of the poor.

In this issue, Arthur A. Goldsmith examines the relationship between the amount of development assistance given to sub-Saharan African countries in the 1990s and their levels of democratic reform. His assessment of such reform is based on a number of measures, including the staging of elections, electoral challenges to incumbent leaders, voter participation, the number of coup d'etats, and political indices (Freedom House index of political freedom and a democracy index). He states that there is a positive, but small, correlation between donor assistance and democratization during this period. Goldsmith views aid as analogous to "'maintenance therapy'—the long-term use of foreign assistance to forestall or postpone re-emergent authoritarian rule."

In contrast, Julie Hearn examines democracy assistance in South Africa and suggests that Western donors are largely concerned with establishing forms of governance that help maintain the international system. She asserts that U.S. assistance effectively has changed the debate on democracy in South Africa among civil society organizations, encouraging a focus on procedural democracy, and minimizing discourse related to economic justice. She is critical of this type of assistance because it has led to the maintenance of an "intensely exploitative economic system" in South Africa.

Arthur A. Goldsmith

 **YES**

Donors, Dictators, and Democrats in Africa

Introduction

Sub-Saharan Africa (henceforth simply Africa) is marked by weak, often dictatorial states. They are also disproportionately among the states that receive the most aid per capita in the world. Is there a connection? A growing opinion assumes that there must be. It is probably no surprise when Doug Bandow (1997) of the libertarian Cato Institute asserts that fifty years experience proves that aid usually hurts political reform. More surprising, perhaps, is a study sponsored by the Swedish Foreign Ministry that documents how large amounts of aid delivered over long periods may reduce accountability and democratic decision-making among recipient states (Bräutigam 2000). Even the World Bank (1998: 84–8) raises questions about aid's impact on governments, concluding that the Bank has often given too much money to ineffective regimes and in the process has sometimes undermined their administrative capacity.

A common theme in these criticisms is that aid is like a narcotic, fostering addictive behaviour among states that receive it. States are thought to exhibit the symptoms of dependence—a short-run benefit from aid, but increasing need for external support that does lasting damage to the country (Azam *et al.* 1999). By feeding this 'addiction', the aid donors have supposedly weakened the resolve of African states to act on behalf of their citizens. Development assistance, in other words, has had the perverse and unintended political effect of reinforcing despotic rule in the region. The implication is that countries need to 'kick the habit' of aid before they can turn to the task of building authentic democratic institutions.

This article takes a sceptical look at these convictions, asking whether it is true that heavy reliance on aid has blocked democratic development in Africa. It focuses on the period after 1990, when many donors started making democratisation an explicit object of policy. Dissenting from the prevailing critique of aid, it finds that aid dependency has not systematically set back Africa's political evolution. The evidence suggests, to the contrary, that aid may have had a small favourable effect over the past decade, encouraging more responsible self-government in some countries. . . . Getting Africa off the donor assistance 'drug' is a commendable goal, but it does not appear to be a prerequisite for democratisation in the region, as some people assume is true. Weaning countries abruptly from external financing and technical assistance might well have the opposite effect.

Foreign aid is an international transfer of resources that would not have taken place as the result of market forces. It includes grants and loans made at subsidised interest rates, provided by governments or by international financial institutions. It also includes technical assistance and debt relief. One goal of these transfers has been to encourage democracy, or pluralistic national political systems where people are reasonably free to express their political demands and to hold rulers to account. Democracy in this sense is not identical with competitive elections, held at regular intervals. The two do overlap, however, and elections are often the best tangible evidence of the extent to which a democratic system is in place. Accordingly, the discussion will concentrate on election procedures and outcomes.

The article starts by documenting the ways in which international donors have attempted to boost democracy in the region. Donors' forays into government reform have been criticised as inconsequential or counterproductive. However, analysis of the data suggests a clear, though small, correlation between the amount of aid received by a country in the 1990s, and the extent to which its political system opened up to greater accountability and competition. The issue for many African countries is how to consolidate and extend these moderate reforms on their own, with less external support.

Foreign Aid for Democracy

Africa has been using large amounts of foreign aid for years. Depending on the definition of aid, the average African country took between $600 and $1,500 (in 1995 prices) in aid per capita between the mid-1970s and mid-1990s. Starting about 1990, international donors began to link these resources expressly to democratic reform. Representative government had disappeared in most African countries following independence. Too often, African governments ignored or repressed their people. They failed to supply critical public goods—or if they did, they redirected them to the regime's narrow base of supporters. During the Cold War, the rich democracies were reluctant to say or do much about these governmental problems for fear of driving African countries into the socialist camp. The changed geopolitical scene in the late 1980s altered those calculations and freed public lenders to use their clout to extract domestic political concessions in Africa (Olson 1998).

The new official consensus on democratisation was clearly reflected in the Development Assistance Committee (DAC) policy statement in 1993. For the first time, the DAC specifically advocated that developing countries lay down procedures whereby ordinary people can help shape the policies that affect their lives. Communities and private organisations must be empowered so they can check the risk of arbitrary government, according to the DAC. Thus, client states should be encouraged to build institutions that assure the consent of the governed—and that allow the governed to withhold their consent so that their political leaders can be peacefully replaced (OECD 1993: 9–12).

To take at face value any official developed country policy statement about the importance of democratic government in a developing country would be naive. Since 1975, an amendment to the Foreign Assistance Act has made US aid conditional on respect for human rights and civil liberties. Yet, *realpolitik* often meant that human rights violations were overlooked, while national security issues dominated who got aid (Hook 1998). An oft-cited example was Zaire's Mobutu Sese Sekou, whose corrupt but reliably anti-communist regime for decades received American backing. With the 'Evil Empire' defeated, however, US foreign assistance policy turned more seriously to human rights as a basis for determining how to allocate resources (Meernik *et al.* 1998; Apodaca & Stohl 1999; Blanton 2000). The United States suspended aid to Mobutu in 1992.

There was economic logic behind making legislative and administrative reforms a centre of attention in development assistance. Greater participation and more debate in public life, the donors argued, would provide a sounder institutional grounding for economic advancement. Not only did democracy offer a better way to aggregate and employ local knowledge, the climate of open discussion would help move resources to their most productive uses. Democracy's built-in checks and balances can lead to greater public accountability, responsiveness and transparency—which can enhance the business and investment climate, and perhaps eventually make foreign assistance less important. Empirical evidence also exists that democracy helps less-developed countries make better use of the aid that they do get (Svensson 1995).

This is not the classic political defence of civil liberties and free and fair elections. The usual instrumental arguments for democratic procedures are that they give voice to majority demands and that they are a peaceful means for recruiting new leadership talent. The economic argument put forth by international financial institutions and bilateral aid agencies in the 1990s is different. Without denying the political advantages of pluralistic competition, the donors' economic argument concentrates on improving the environment for private investment and innovation.

How Aid Affects Recipient Country Politics

Foreign aid can help shape African domestic politics in four ways. The first way happens by accident. Some of the pressure for government reforms in Africa is a by-product of donor-inspired economic austerity schemes, also known as structural adjustment. Imposed with increasingly frequency since the 1980s, these market-based reforms aim to shake up moribund African economies. Yet, they also encourage political shake-ups by curbing the power of the state to pay off its partners and confidants in the private sector. Africa's hitherto 'silent majority' has been energised to begin exercising its voice—though ironically, it is sometimes to speak against the hardship associated with structural adjustment. Democratic political reforms may thus inadvertently undermine pro-market economic policies, and contradict the donors' economic aims for their African clients (Baylies 1995). Still, the political effect often is for greater mass participation in public life, especially among urban groups who have tended to see themselves as net losers under structural adjustment.

The second mode of donor political influence is through direct aid for political reform. A recent example is Chad, which held a constitutional referendum and a presidential contest in 1996. France provided the logistical support that allowed co-ordination of voting nation-wide. France also underwrote the cost of Chad's elections (May & Massey 2000: 122). The amounts the donors give directly for governance and civil service reform are limited—only 4.4 per cent of all bilateral official development assistance (ODA) commitments made world-wide in 1998 (DAC 1999, Table 19). The World Bank (1999, Appendix 12) reports that its cumulative spending for public sector management in Africa amounted to just 3.8 per cent of lending to the region as of 1999.

However, these figures grossly understate the degree of external support for political reform. They do not include additional resources that go to support non-government organisations (NGOs). NGOs are a third avenue for aid's influence on democratisation. Official donors are taking steps to strengthen these voluntary and community groups in Africa (Robinson 1995). About half the World Bank's development projects have some involvement of local and international NGOs, for example. Many bilateral donors make even greater use of NGOs. Worldwide, about $9 billion in official aid is funnelled annually through these groups, or 15 per cent of all aid, not including the private funds that NGOs raise and spend on charitable activities (Gibbs *et al.* 1999). These groups often advocate on behalf of underrepresented groups in society, which may lead to demands for changes in political institutions and procedures.

Modifying the political system is not usually the donors' immediate objective when they enter partnerships with NGOs. Sometimes the aid may have the opposite effect of politically neutralising the recipient groups by making them look like foreign agents (Robinson 1993). Critics charge that too many domestic democracy movements in Africa rely on external support for resources and validation (Ihonvbere 1996). Yet, over time, such groups may add to the domestic organisational endowment and social capital that are important foundations for democratic government. Case studies in Ghana, Uganda and South Africa find that aid-supported groups can become key players in setting the course of national political reform (Hearne 1999).

The fourth, and perhaps most important, way in which aid can affect a recipient country's politics is through specific political conditionalities. Donors put political strings on loans for seemingly unrelated projects and programmes, using the financial resources as a reward and punishment to induce political reform. Often working in collaboration, donors make plain to the client state that it must shape up politically or lose access to credit. Such pressure is common. After Niger's President Ibrahim Maïnassara was killed in a coup in April 1999, for example, France immediately cut off development assistance, pending a return to civilian rule. France, which is Niger's largest international patron, restored its support seven months later, after successful completion of the presidential election. When a recent study looked at World Bank and International Monetary Fund (IMF) programmes in Africa from 1997 to 1999, it found that nearly three fourths of the loan conditions (or about 80 per country) pertained to governance (Kapur & Webb 2000).

Direct political aid, NGO support, and political conditionalities are often used in concert to try to bring about a favourable result, as illustrated by Zimbabwe. The IMF and the World Bank suspended aid to Zimbabwe in 1998. There were several reasons, the most serious being the sending of troops to the Democratic Republic of Congo (DRC), which compromised efforts to control Zimbabwe's budget deficit. The IMF and World Bank offered to restore the money only if Zimbabwe showed a serious effort to curb its military and to undertake other administrative and political reforms.

President Robert Mugabe's subsequent land policies hardened the donors' position. In an effort to buttress political support in the run-up to legislative elections in 2000, Mugabe openly sanctioned illegal seizures of commercial farms. His government also drew up a list of properties to be taken officially without compensation. Opposition groups charged that these farms were earmarked for government cronies instead of people who really needed land. Britain signalled its willingness to release funds for land reforms and to organise a conference to solicit support from other Western countries to redistribute farms to the truly landless. However, first all unauthorised land invasions had to stop and legal procedures be restored.

The US Agency for International Development (AID) forged links in Zimbabwe with private groups on grounds that government institutions were too politicised to be effective. For the 2000 national elections, the US government funded the training of 10,000 domestic election monitors and helped the semi-independent Electoral Supervisory Commission to cope with the administrative demands of election logistics. In Washington, the United States Congress considered the Zimbabwe Democracy Bill. It would impose an embargo on the country pending a certified return to the rule of law, respect for ownership of property, and freedom of speech and association.

Doubts and Controversies

How effective have such activities been? Official statements usually reflect self-assurance. In recent testimony before Congress, for example, AID's assistant administrator for Africa, Vivian Lowery Derryck (1999) expressed the following position:

> As a whole, Africa has made major progress towards expanding and consolidating democracy during the past ten years. The widespread increases in freedom of speech and the media, freedom of assembly and association, competitive national and local elections, and the growth of civil society have givern more Africans greater freedom and stability in their lives than at any time in the recent past . . . USAID has been involved in many of these transitions, and we are proud of the achievements of these new democracies.

When policymakers speak for themselves, however, they typically sound less confident about the political outcome of their work. Former chief economist at the World Bank, Joseph Stiglitz (1999) has said that policy conditionalities actually tend to undermine participatory processes in developing countries, which should be building their own capacity to make democratic decisions. Ex-deputy administrator for AID Carol Lancaster (1999: 66) goes farther, suggesting that foreign aid in Africa has encouraged poor governance, by giving dishonest and incompetent regimes a sense of security.

It is not hard to find anecdotal evidence to support scepticism about foreign aid's beneficial political impact. Consider the Zimbabwe case, just cited. Despite the pressure, President Mugabe remained defiant. He charged that the opposition Movement for Democratic Change was the surrogate of the British and American governments. Foreign observers were turned away from the 2000 elections. Challenger candidates and their supporters were intimidated. Zimbabwe defaulted on its foreign debt that year. The government refused to obey court rulings against the land seizures, and it kept its troops engaged in the DRC. World Bank and IMF credits remained on hold. Political conditionalities appear to work best when they are in response to a specific event or tip the balance towards domestic opposition groups (Crawford 1997)—neither of which were evidently the case in Zimbabwe.

Nor have donors been completely consistent about whom they help. They have pushed harder for civil rights with some clients than others. In June 1990, French President François Mitterrand warned African heads of state that French aid would henceforth be conditional on progress in the direction of democracy. France pointedly did not intervene to protect the dictator Hissène Habré of Chad in a coup six months later, despite having troops in his country. Yet France ignored human rights abuses by Habré's successor, Idriss Déby.

The problem with negative anecdotes like these is that one can find others that cast foreign pressure and support for democratisation in a more positive light. Though it never received much direct financial assistance, South Africa is perhaps the best example of how the international community has helped give birth to a liberal constitution and multiparty political system in Africa. To assess aid's political impact systematically, we cannot rely just on stories from selected countries; we need to look specifically for two things: (a) comprehensive indications of the degree of democratisation across Africa; and (b), if democratisation can be substantiated, for signs that aid contributed to it. . . .

Evidence of Democratisation

Political scientists who specialise on Africa tend to question the authenticity and depth of democracy in the region. The social context of electoral rules obviously differs from what is found in Europe or North America, and superficially similar institutions and procedures may hold very different meanings south of the Sahara. With most African political reforms so new, no one can make confident predictions about where they are heading. Nonetheless, three types of evidence can be cited in support of a bona fide democratic tendency. One indication is that across the sub-continent more liberal rules of political engagement are being adopted. Second, is the growing number and competitiveness of elections in recent years. Third, are overall indexes of political freedom, which, because they capture the previous two types of evidence, have been on the rise in Africa. . . .

This article has looked into the relationship between foreign aid and steps towards democracy in African countries over the past decade and earlier. Among specialists in international relations, it is often considered axiomatic that aid has an addictive effect on government that subtly reinforces repression in countries with single party rule, even if the opposite result is intended. Many area experts also doubt the ability and good will of outsiders to sponsor human rights and authentic multiparty competition in Africa. The result is seen as 'donor democracy' without significant domestic bloodlines. Civil society critics of the international economic regime, such as the Fifty Years is Enough Coalition or the Jubilee 2000 movement, are apt to agree that aid (at least in its current forms) is anti-democratic in character. Even the donors sometimes concede the point that their efforts in Africa have been politically counterproductive.

This article's analysis suggests a somewhat more favourable outlook. First, the data indicates that democratisation is more than a false front put up for donors. In some countries, at least, meaningful though tenuous changes in governing style and substance look as if they are taking place. Second, while we must be cautious about attributing causation, foreign assistance appears to have a positive association with these welcome political trends. This conclusion is supported by statistical analysis of quantitative data. Donors, it seems, can work with some client states or with civil society groups to obtain somewhat greater democratisation in Africa. There is no evidence here for the prevalent view that the 'compulsive use' of aid has net perverse political effect, or that 'going cold turkey' with less aid would speed the pace of democratic reform. Rather than the metaphor of addiction, a closer analogy may be 'maintenance therapy'—the long-term use of foreign assistance to forestall or postpone re-emergent authoritarian rule.

The difficulty for development policymakers is that aid's political benefit appears to be quite small and perhaps transient. Local opposition to arbitrary leadership and public corruption seems far more important. With many other factors pulling against democratic reform, any gains won with the help of development funds can easily be undone. At issue for many African countries is whether they can find the means to use official assistance to greater political effect, consolidating and extending the modest progress made since 1990. Given pressure from the developed world to break the cycle of aid dependence in Africa, these societies probably will soon have far less external help for their efforts to head for more responsive and open systems of government.

Aiding Democracy? Donors and Civil Society in South Africa

Democracy Assistance: Promoting Stability

During the 1980s a new branch of the aid industry was born, democracy assistance. Although 'democracy' often entered the foreign-policy-making vocabulary of the North in the postwar period, it was not the dominant form in which the North related to the South. The principal form was the development of strategic alliances with authoritarian regimes. The new industry arose out of a major reconsideration of Western foreign policy towards the South, particularly within the USA. With the US defeat in Vietnam, the Nicaraguan revolution and other nationalist victories in the South, US foreign policy towards the Third World had reached crisis-point by the late 1970s. It had failed to stop popular anti-US regimes taking power in South East Asia, Central America and Southern Africa and its capacity to shape events abroad appeared severely curtailed. By the early 1980s, a new consensus began to emerge among policy-makers around the strategy of 'democracy promotion'. This involved two key elements.

First was the recognition that coercive political arrangements had failed to deal with the social movements that had challenged authoritarian rule and that formal liberal democracies were better able to absorb social dissent and conflict. It is important to understand that the rationale for turning to liberal democracy was that it was perceived to be a better guarantor of stability. The goal remained the same: social stability, it was simply that the means to achieve the end had changed. This becomes clear when we examine the kind of democracy being promoted in the Third World. It is about creating political structures that most effectively maintain the international system. It has no more to do with radical change than its predecessor, authoritarianism does. As Samuel Huntington, one of the most influential proponents of formal democracy, clearly states: 'The maintenance of democratic politics and the reconstruction of the social order are fundamentally incompatible.' At its core, the contemporary political and industry is about effective system maintenance.

The second point is that, where earlier foreign policy had focused almost exclusively on the strength of the client state and its governmental apparatus, the new democracy strategy began to recognise the important role of civil society. It was from within civil society that opposition to authoritarian rule had emerged and therefore it was imperative 'to penetrate civil society and *from therein* assure control over popular mobilization' (emphasis in original). [William I.] Robinson continues:

> The composition and balance of power in civil society in a given Third World country is now just as important to US and transnational interests as who controls the governments of those countries. This is a shift from social control 'from above' to social control 'from below ...'

This is an important part of democracy assistance. Aid is targeted at a country's most influential, modern, advocacy-orientated civil society organs which include: women's organisations, human rights groups, national or sectoral NGO [nongovernmental organization] for business associations, private policy institutes, youth and student organisations, and professional media associations. As commentators, including those who direct the new political aid, have pointed out, this is not very different from what the CIA used to do, particularly within the context of counter-insurgency and 'low-intensity conflict'. However, former CIA director William Colby makes a key point: 'Many of the programs which . . . were conducted as covert operations [can now be] conducted quite openly, and consequently, without controversy'.

As Robinson points out: 'Transferring political intervention from the covert to the overt realm does not change its character, but it does make it easier for policymakers to build domestic and international support for this intervention.' This is the trump-card of democracy promotion, it diffuses opposition to Northern intervention. Advisor to the State Department and academic, [Howard] Wiarda, clearly sums up:

> A US stance in favor of democracy helps get the Congress, the bureaucracy, the media, the public, and elite opinion to back US policy. It helps ameliorate the domestic debate, disarms critics (who could be against democracy?) . . . The democracy agenda enables us, additionally, to merge and fudge over some issues that would otherwise be troublesome. It helps bridge the gap between our fundamental geopolitical and strategic interests . . . and our need to clothe those security concerns in moralistic language . . . The democracy agenda, in short, is a kind of legitimacy cover for our more basic strategic objectives.

Since its inception in the early 1980s, democracy assistance has continued its take-off. In the 1990s this was fuelled by three important developments: the academic and donor preoccupation with 'governance' as the root of underdevelopment, the practice of political conditionality, that is, making aid conditional on political reforms, and the changing balance of power in North–South relations. During the 1980s an orthodoxy developed that Africa's development crisis was precipitated by a failure of the state and that 'governance' had to be reconstructed, from the bottom up. Shaping civil society became the road to reforming the state. Making aid dependent on such changes has been the site of sharp confrontations between the governments of many sub-Saharan African countries and donors. In South Africa, no such crude coercion was needed. As international opponents of apartheid, including Western states, united with domestic combatants, a broad consensus was forged over the form that a new liberal democracy would take. With the demise of nationalist and socialist ideologies, such foreign, overtly political, interference was no longer viewed with the same levels of distrust. The latter has allowed the North to intervene in the (civil) societies of the South with an unprecedented degree of perceived legitimacy.

South Africa has had a long history of Western support to civil society. The highly conflictual politics of apartheid, particularly of the late 1970s and 1980s, generated a 'vast array of more or less popular, more or less institutionalised organisations and initiatives in broad opposition to the apartheid state'. These included trade unions, community organisations, sectorally mobilised movements of youth, students and women, as well as business, lawyers and religious associations. It is these kinds of civil society organisations (CSOs) (although they were hardly ever referred to as such) that donors funded. This support began with Denmark in the mid-1960s and was followed by Norway and Sweden in the 1970s. It culminated in the mid-1980s with the imposition of sanctions by Western governments and the international isolation of the apartheid regime. The Nordic countries were joined at this time by the European Union and the USA, who each provided an unprecedented $340 million over a nine year period to CSOs, before the end of apartheid and the 1994 elections. Despite such a significant involvement, a comprehensive account of foreign assistance to civil society in this period is still to be written, not least because of the covert nature of that support. With the election of an internationally recognised government, foreign donors began to provide aid to the new South African state as well as continuing some funding to civil society, though on a smaller scale. Although this loss of finance has had a considerable impact, it has not been fully analysed and thus there is substantial dispute as to how much funding was withdrawn and how significant this was.

This article examines foreign assistance to civil society in South Africa since 1994. The premise of this research is that political aid is 'political', that it is about consciously influencing the 'rules of the game'. . . .

The Importance of Democracy Assistance

Unlike in other African countries, democracy assistance forms a major part of foreign aid to South Africa. The principal objective of aid programmes to the country is to influence the political transition and to focus on democratic consolidation. This kind of aid is in stark contrast to other donor programmes on the continent, which primarily supplement the meagre national budget of African governments so that they can provide basic services in the areas of welfare, agriculture, energy and infrastructure. These are the predominant categories of aid in most African countries, where democracy assistance on average accounts for less than 5% of total aid. South Africa, with an average per capita GNP [gross national product] of about $3200, is categorised as an upper middle income country, along with Brazil and Chile. As such it would normally be dis-qualified as an aid recipient. Danish aid, for example, is restricted to countries with an average GNP per capita of less than $2000. Aid to the country is seen as a temporary measure, because, as USAID/South Africa points out: 'South Africa has substantial resources to address its problems over the long term.' Donor programmes are termed 'transitional', mirroring South Africa's own transition, and were to end soon after the 1999 elections, commonly perceived as the for-mal end-point of the passage from one political system to another. In South Africa, nation-wide poverty is not the motivating force behind donor activity. The whole thrust of aid involvement is about deepening the political changes that have taken place since 1994 and ensuring that, as far as possible, 'a point of irreversibility' is reached before the ending of external assistance. Foreign aid to South Africa is very much a case of democracy assistance, and if we wished to study contemporary foreign intervention in a country's democratisation process, South Africa could not be a better case study. . . .

Such a strong emphasis on political aid in South Africa is hardly surpris-ing. Since the beginning of the century, South Africa has been important to the West, both politically and economically. It has been described as 'a reliable, if junior, member of the Northern club'. The West's relationship with Africa was largely predicated on a long-term strategic alliance with the apartheid regime in South Africa. This continued throughout the 1970s and into the 1980s. At the same time South African capital, some of which was in the same league as that of any Northern-based corporation, became thoroughly intermeshed with capi-tal originating in Europe, the USA and Japan. As globalisation advanced, South Africa became a key outpost of international capital. However, with the political unrest that began with the 1976 Soweto uprising, capital outside South Africa began to push its South African counterparts to search for a political solution that would involve a transition from racial to non-racial capitalism. . . . [P]rior-ity [was] given to democracy building in South Africa by the West after 1994. . . .

In the situation proffered by 1994, civil society was not a priority. There are a number of reasons for this. In other African countries, aid to civil society is about establishing something that is not there. It is about creating a modern, advocacy-orientated civil society. In South Africa this already exists. Using the lowest common denominator of pluralist theory, which says the more civil so-ciety groups, the better, South Africa scores highly. It has a dense, long-standing associational life. Second, this associational life is modern, a key characteristic of the donor model. Third, it is strongly advocacy-orientated and has proven its lobbying abilities on the world stage. Fourth, in South Africa, in 1994 at least, donors did not need to convince the new government that civil society had a le-gitimate role to play and that the government must give it the political space in which to operate. That was already what the government believed, since many new government officials came from the NGO sector, and as a result there were unusually close and sympathetic relationships between civil society and the government.

However, perhaps most importantly, many donors had established close links with civil society throughout the 1980s and early 1990s. For example, a large number supported the work of IDASA [Institute for a Democratic South

Africa] in the late 1980s and early 1990s. IDASA is arguably one of the most pro-fessional, effective and high profile CSOs in South Africa today, and thoroughly steeped in democratic liberalism. A 1994 annual report states that the 'personal credibility with international donors' of cofounder, Boraine, 'was pivotal in se-curing a generous funding base for the organisation, allowing it to expand country-wide and make interventions with national impact'. After the 1994 elections, IDASA was renamed the Institute for Democracy in South Africa. It now runs some 12 national programmes and projects and has a staff of 140. This close association with donors continues today, possibly making it the most donor-associated democracy NGO in South Africa. We have indicated how the German foundations developed strategic relationships with sections of civil so-ciety, how the Nordic countries were stalwart supporters of an underground civil society, and the enormous resources that the USA spent on cultivating links with CSOs. Already from the mid-1980s the USA had grasped the pivotal role that civil society would play in shaping the new South Africa. Subse-quently, it began to attempt to influence it, checking the growing radicalism among the black population by developing counterweight forces conducive to the establishment of a liberal order. This was achieved by supporting an emer-gent black middle class of professionals who could be incorporated into a post-apartheid order; developing a network of grassroots community leaders who could compete with more radical leadership; and cultivating a black business class that would have a stake in stable South African capitalism. From the stand-point of 1994, donors had already made an impact in civil society. The next sec-tion suggests one important role that donors are financing civil society to play in the new South Africa.

What Role for Civil Society?

A report on the democratic outlook for South Africa by an influential South African think-tank concludes that the chief threat to democratic consolidation is:

> the limited capacity of the state to govern—and, more particularly, to cement a 'social contract' with society in which government protects the rights of citi-zens who, in turn, meet their obligations to democratic government.

The US National Democratic Institute (NDI) describes the situation facing South Africa in the following terms: 'the twin challenges of rebuilding a new united South Africa while simultaneously developing the institutions that will conduct the daily business of government'. We have shown how foreign aid has prioritised the institutional development of government, thereby attempting to meet the need to build the capacity of the state to govern. However, as both analysts note, the other side of the equation is to link society with this new in-stitutional framework, to cement the social contract and to build, in the words of the then Deputy President, Thabo Mbeki, 'a democratized political culture'. This is where the role of civil society is so important to the architects, both do-mestic and external, of the new South Africa. It is a particular kind of civil soci-ety that can help to legitimise the new state in the eyes of the South African citizenry and help to build a culture where citizens meet their obligations to liberal democratic government. First we show why this is needed in South Africa if liberal democracy is to succeed. Then we examine which civil society institutions are available to fulfil this function and how donor organisations are supporting this very role for civil society.

The same South African report writes about 'widespread citizen non-compliance—reflected in, among other indicators, crime and widespread non-payment for public services'. It continues: 'A variety of factors produce outcomes in which citizen dissatisfaction is expressed in withdrawal from the public arena and in attempts to evade the reach of government'. It concludes: 'This threatens democracy as much as overt resistance to democratic order . . . ' An annual review of political developments in South Africa in 1997, produced for the quarterly journal of the South African Coalition of NGOs and Interfund, a consortium of Northern NGOs, mainly Nordic, provides the following commentary:

> Whether currently higher or merely consistent with historical trends, crime rates in South Africa are unusually high . . . Public insecurity aside, the effect of crime is to reduce citizens' confidence in public institutions at a point when the success of the democratic transition requires enhanced trust and participation.

And a report written for the consortium of non-governmental donors, Interfund, writes:

> A key challenge for actors in civil society is to capture and steer rising (and probably inevitable) frustration into constructive, non-violent forms of conflict, namely political pressure channeled through social movements.

The purpose of the transition was to provide the political settlement that would allow the passage from racial to non-racial capitalism in South Africa. As we have noted, aid programmes have maintained this focus on political stability rather than on socioeconomic transformation, as a result of which South Africa continues to be a highly unequal society with areas of extreme deprivation. Indeed [Robert] Mattes and [Hermann] Thiel assert that the inequality is worsening. [Hein] Marais points out how the kind of instability facing South Africa, described by the above commentators, is 'symptomatic of the extreme inequalities that scar the society'. In a situation of such contradiction between economic inequality and political stability, political stability or 'democracy' will be simply about managing that tension. Civil society is the lynchpin in holding together that tension by fostering support among citizens for a government that maintains the inequality that undermines their lives. This, in effect, is recognised by the above report where it identifies the need for civil society to capture and steer the rising frustration. Obviously, social concessions will be made from time to time to maintain stability, but the primary function of political society, that is the state and civil society, will be to manage the tension between the economic and political spheres. Constituents of civil society must understand that this is their function in the political economy of South Africa in the eyes of those who wish to maintain capitalism, and ask themselves is this the role that they want to play?

Civil society plays this role in two ways. First it is the key mediator between the new state in South Africa and its citizens. From the following ratings we can see how urgently this is needed. . . . [A]pproval of the government's performance dropped between 1995 and 1997. The national government's apart-ing fell from 57% to 47%; parliament's went from 53% to 46%; and overall approval of all the provincial governments declined from 42% to 36%. In 1997 the approval rating for the new local governments introduced at the beginning of 1996 stood at a meagre 30%. No wonder then that the Konrad Adenauer foundation asserts that 'if democracy is to be understood and accepted by South Africans it must be seen to work well, particularly at the local level'. Second, the CPS [Centre for Policy Studies] report points out that democratic consolidation can only occur when there is widespread agreement among political elites and citizens on institutional rules. They cite political development theorists [Juan] Linz and [Alfred] Stepan who see democracy as consolidated when democratic rules become 'the only game in town'. Civil society has a key role to play in creating among the population an adherence to the values of liberal democracy and an acceptance of the rules of the game. Mattes and Thiel argue that, as well as structural features, an attitudinal commitment to democracy is essential. They write: 'The level of elite and citizen commitment to democratic processes is the single direct determinant of the probability of democratic endurance or consolidation.' In their analysis of public opinion polls taken on 'democracy' they comment: 'The results raise important questions about South Africans' understanding of democracy.' While only 27% rated as 'essential' such key procedural elements of democracy as regular elections, 48% said that equal access to houses, jobs and a decent income was 'essential' to democracy. They go on to explain why this might be so:

> While 'one man, one vote' was always the goal, the key liberation movements subscribed to and spread to their poverty-stricken followers an economic, as opposed to a procedural view of democracy.

This is disturbing to the authors and, of course, to those, within and outside the country, who wish to maintain the distinction between the political and economic sphere and maintain the tension between gross inequality and political stability. This is the wrong kind of democracy. The authors advise the following re-education:

> Thus one might urge South Africa's educational system, civil society, and political parties to shift their emphasis . . . to the . . . task of teaching people to value democratic institutions and processes more for their own sake than for what they may deliver in terms of immediate and tangible benefits.

What is interesting in South Africa, compared with other African countries, is the number and calibre of CSOs geared towards doing precisely that, encouraging a popular commitment to procedural democracy. What is more, these kinds of CSOs feature predominantly in donor political aid programmes. In our research we asked over a dozen different foreign donors what kind of civil society organisations they funded through their democracy assistance. There were five main categories: democracy organisations, concerned with the overall relationship between states and citizens; human rights and legal aid groups; conflict resolution agencies; organisations servicing or representing the nongovernmental sector; and think-tanks. Of these categories, democracy organisations were the largest. Not only were they the most numerous, but they also received the largest amounts of aid and were supported by the broadest cross-section of donors.

This is not to suggest that donor focus on the other categories, particularly in the human rights and legal aid field, is unimportant. For example, the National Institute for Public Interest Law & Research (NIPLAR), a consortium, received $3.25 million from USAID between 1996 and 1998 to create 18 human rights and democracy centres nation-wide. Along with the Legal Resource Centre (LRC) and the Black Sash Trust, NIPLAR receives donor funds to provide free legal advice. Between 1996 and 1999 USAID funded the Independent Mediation Service of South Africa (IMSSA) with $3 million as an umbrella grantee providing subgrants to conflict resolution NGOs working with local government. Another significant grant from the USA in 1997 went to the Free Market Foundation, a core partner of the German liberal Friedrich Naumann Foundation. This was to fund a programme promoting market-orientated economic policies in the South African parliament and administration. Many of these internationally funded projects form part of a broad liberal democratic discourse; however, it is crucial to investigate what the specific category of democracy organisations do in civil society.

The most prominent is IDASA. It is fully committed to procedural democracy, for example, the manager of the public opinion service at IDASA co-wrote the above article. As we mentioned, it is now an organisation with a staff of 140. It is probably also the most donor-funded CSO in South Africa. The other organisations are the Institute for Multiparty Democracy, whose name could not be more indicative of procedural democracy, and Khululekani Institute for Democracy, aimed at bringing parliament closer to the people. A fourth, the Electoral Institute of South Africa, deals with that key aspect of procedural democracy. A fifth, the Helen Suzman foundation, undertakes similar democracy surveys to those of IDASA, and has a map of Southern Africa on the back of its quarterly enblazoned with 'promoting liberal democracy'. A sixth, the South African Institute of Race Relations has a foreign donor-funded Free Society project which aims to monitor South Africa's democratic development and to promote the rule of law, ethics, justice, the concept of limited government and economic freedom. As a 1995 project description, written by one of its foreign funders, the US National Endowment for Democracy, explains: the programme will inform key government and non-government officials on activities that hinder the development of a free society. The project seeks to achieve these aims through three principal means: (1) publishing *'Frontiers of Freedom'*, a quarterly newsletter; (2) sponsoring specific research projects; and (3) hosting and attending special events, including briefings and lectures, frequently in conjunction with other non-profit institutions. It is interesting to note how some of the language has changed in the 1997 project description, which introduces the relatively new term 'civil society organisations', talks explicitly about researching 'public policy alternatives', and replaces the ideologically unambiguous 'limited government' with the much more ambiguous and widely accepted term, 'good governance'.

It is not altogether surprising to find out that these civil society organisations at the forefront of promoting procedural democracy are very much part of the South African liberal landscape. The South African Institute of Race Relations is one of the oldest liberal institutions in the country. The Helen Suzman foundation is named after arguably the most prominent South African liberal politician. IDASA was started in 1987 by van Zyl Slabbert, former leader of the opposition and Alex Boraine, former Progressive Federal Party MP. What these CSOs have done is to put procedural democracy high up on the agenda for civil society, and for the nation, and to establish the terms of the debate. This is not surprising given the resources allocated to them by the international donor community. IDASA has received grants not simply of tens of thousands of dollars, but of $1 million. In 1996 it received $1.165 million from the Ford Foundation. This is an exceptionally large grant by the Foundation's standards, which normally provides grants from $200 000 to $50 000 to CSOs in Africa, and is by far the largest grant to any grantee in South Africa. At the same time IDASA received a $1 million grant from USAID for a two-year period. The South African Institute of Race Relations and the Institute for Multi-Party Democracy received similar grants from USAID over the same period.

Conclusion

As we have seen, since 1994 South Africa has received an unprecedented amount of international political aid aimed at consolidating its liberal democracy. Even without the benefit of an in-depth, detailed study, one can safely conclude that there are few, if any, aspects of the new South African political system that have not been shaped by donor input. The external involvement in the construction of the new South African state raises important questions. What is the nature of the state of a middle income country that has been so extensively 'advised' by a myriad of international players? How do we understand such a state within existing theoretical frameworks? Does a state that is so permeable and malleable to external shapers exhibit the same autonomy as the states of advanced economies? Does it change the nature of the state? Is it a case of autonomy compromised? What impact does it have on the state's foreign policy and, perhaps even more pertinently, on its domestic policy?

And what of civil society? This article suggests that political aid to civil society has had two major consequences. First it has changed the debate on democracy. During the past five years, it is possible to see a process in which democracy has been redefined. Although half of South Africans still believe that access to housing, jobs and a decent income are essential components of a democratic society, this residual belief in social democracy is being eroded and replaced by the norms and practice of procedural democracy. It is our argument that the North has played its role in this process by funding the liberal proponents of procedural democracy in civil society, and that, subsequently, political aid has successfully 'influenced the rules of the game'. The second consequence is that this has facilitated a newly legitimatised South African state to preside over the same intensely exploitative economic system, but this time unchallenged. External and domestic support for procedural democracy has successfully removed all challenges to the system. It has ensured that democracy in the new South Africa is not about reconstructing the social order but about effective system maintenance.

POSTSCRIPT

Is Foreign Assistance Useful for Fostering Democracy in Africa?

One aspect to be aware of when interpreting these articles is that Goldsmith's research was funded by USAID, the foreign assistance arm of the U.S. government that funds a number of democracy programs in Africa. The reader should also note that some of the differences in assessment boil down to fundamentally different views on the way democracy should be conceptualized.

Many scholars are critical of the type of democracy and governance assistance being offered by USAID in Africa. A major concern is that the United States has one particular view of what democracy is and how it should function. Many believe that it is culturally presumptuous for Western nations and international financial institutions to seek the universal imposition of one type of political system.

In many African countries, district- and provincial-level officials are appointed by a central government. As such these are, relatively speaking, highly centralized political and administrative systems. While this level of centralization is not an uncommon situation in some European countries, e.g., France, this is different than the federal system in the United States where local and state officials are directly elected. This type of reform is being pushed by donors (mainly the U.S. and international financial institutions) in Africa because of a belief that decentralized government is more efficient and more responsive to the people. Critics of these policies argue that it is less about transferring power to the people than an opportunity for central government to shed its responsibilities.

For those interested in further reading, Mark Robinson has written a paper concerning donor assistance to civil society organizations entitled "Strengthening Civil Society in Africa: The Role of Foreign Political Aid," *IDS Bulletin* (no. 26, 1995). Kent Glenzer also has a very interesting book chapter, under the title "State, Donor NGO Configurations in Malian Development 1960–1999," in which he looks at the decentralization question in Mali. See B. I. Logan, ed., *Globalization, the Third World State, and Poverty-Alleviation in the Twenty-First Century* (Ashgate 2002). Finally, Kempe Hope examines a new African initiative, the New Partnership for African Development (NEPAD), designed to foster good governance and democracy in Africa in an article in *African Affairs* (2002) entitled "From Crisis to Renewal: Towards a Successful Implementation of the New Partnership for Africa's Development." Gorm Rye Olsen describes how the European Union has backed away from direct support for the promotion of democracy in Africa in a 2002 article in *International Politics* entitled "Promoting Democracy, Preventing Conflict: The European Union and Africa."

ISSUE 19

Are African Governments Inherently Disposed to Corruption?

YES: Robert I. Rotberg, from "Africa's Mess, Mugabe's Mayhem,"
Foreign Affairs (September/October 2000)

NO: Arthur A. Goldsmith, from "Risk, Rule, and Reason: Leadership in Africa," *Public Administration and Development* (2001)

ISSUE SUMMARY

YES: Robert I. Rotberg, director of the Program on Intrastate Conflict and Conflict Resolution at Harvard University's John F. Kennedy School of Government, holds African leaders responsible for the plight of their continent. While he admits that Africa's failure to develop in the postcolonial period has many causes, he suggests that "the visible hand of individual leaders can also be discerned." Rotberg concludes that a large part of the problem is that absolute power corrupts, and that there are limited checks and balances to curb this tendency in Africa. In this regard, he states that "Mugabe's mayhem" was aided and abetted by an underdeveloped civil society and the fact that the rest of the world has failed to judge Zimbabwe's president more harshly.

NO: Arthur A. Goldsmith, professor of management at the University of Massachusetts in Boston, suggests that African leaders are not innately corrupt, but are responding rationally to incentives created by their environment. He argues that high levels of risk encourage leaders to pursue short-term, economically destructive policies. In countries where leaders face less risk, there is less perceived political corruption.

Rightly or wrongly, corruption is perceived to be a major obstacle to development in Africa. Transparency International's annual corruptions perceptions index (a survey of surveys on this issue) ranked Nigeria, Madagascar, Angola, and Kenya among those countries perceived to be the most corrupt in the world in 2002. Nigeria was the second worst in the world of the 102 countries surveyed. Among those countries perceived to be the least corrupt in Africa were Botswana (ahead of France and Portugal in the listing), Namibia (ahead of Taiwan and Italy), and South Africa (ahead of Costa Rica and South Korea).

Arguments about the genesis of corruption in Africa typically break into those that emphasize internal factors and those that stress structural or political/economic conditions (often termed externalist explanations). The first type of argument typically looks to aspects of African culture, society, and tradition to explain the presence and persistence of corruption. For example, regional ties, ethnic allegiances, obligations to the extended family, patron-client relationships, and (occasionally) moralistic explanations are used to elucidate why corruption takes place in Africa. In contrast, structural arguments begin with the premise that a proclivity for corruption is universal, but that history and macro-level political economic conditions have created or reinforced opportunities for corruption in Africa. Examples of such factors include a colonial experience that reinforces antidemocratic traditions, cash-strapped governments that cannot afford to pay civil servants a living wage, or predatory international corporations that seek to access markets and resources through bribery rather than normal government channels. The two points of view presented in this issue represent the two sides of this debate, yet they approach a middle ground of explanation that is slightly more nuanced than the typical arguments heard from either side.

In this issue, Robert I. Rotberg holds Africa's kleptocratic, patrimonial leaders, like Robert Mugabe of Zimbabwe, responsible for giving Africa a bad name, creating poverty and despair, and inciting civil wars and ethnic conflict. While admitting that Zimbabwe suffers from a number of structural difficulties (it is a landlocked country that suffered 90 years of white settler occupation and maldistribution of land), he suggests that this country's problems are largely linked to "the self-aggrandizement of the very man who led the nation out of white supremacist rule." Rotberg points to the axiom that "absolute power corrupts absolutely" to explain what went wrong with Mugabe and Zimbabwe. He then goes on to assert that, given this maxim, the real cause of "Mugabe's mayhem" is a weak civil society and an overly forgiving international community, two forces that could counterbalance the power of African leaders.

Arthur A. Goldsmith maintains that Africans leaders are not innately corrupt but are responding rationally to incentives created by their environment. In finding that there is a correlation between political risk and corruption, and between low political risk and liberal economic reform, he argues that high levels of risk encourage leaders to pursue short-term, economically destructive policies. He states that the risks of governing may be reduced by the spread of multiparty democracy, a form of governance that will make transitions in power more orderly and will reduce the chances of execution or imprisonment for leaders upon departure from office.

Robert I. Rotberg  **YES**

Africa's Mess, Mugabe's Mayhem

A Crisis of Leadership

Venal leaders are the curse of Africa. If sub-Saharan Africa is "in a mess," to quote Julius Nyerere, Tanzania's founding president, it is a mess made by its leaders. To be sure, Africa has its geographical constraints, a cascade of tropical medical ills, and a complex colonial legacy. But where visionary leadership lifted Asia up out of poverty since the 1960s, too many African heads of state in the same period presided over massive declines in African standards of living while carefully enriching themselves and their cronies.

Some of Africa's current and recent leaders are capable, honest, and effective. But kleptocratic, patrimonial leaders—like President Robert Gabriel Mugabe of Zimbabwe—give Africa a bad name, plunge its peoples into poverty and despair, and incite civil wars and bitter ethnic conflict. They are the ones largely responsible for declining GDP [gross domestic product] levels, food scarcities, rising infant-mortality rates, soaring budget deficits, human rights abuses, breaches of the rule of law, and prolonged serfdom for millions—even in Africa's nominal democracies.

These authoritarians, many of whom win or manipulate elections and thereby claim a democratic façade, have proved hard to control and harder to oust; witness the Kenyan failure to vote out President Daniel arap Moi and this summer's [2000] only partially successful attempt to reduce Mugabe's dominance of Zimbabwe's parliament.

The elected autocrats, sometimes termed illiberal or quasi-democrats, have built-in advantages that are hard for even popular opposition movements to overcome: incumbency; state financing for official political parties; state control of television, radio, and newspapers; friendly security forces; crackdowns on opposition rallies; control over the voter rolls; and such tricks as gerrymandering [dividing a voting district unfairly], stuffing ballot boxes, and fiddling with the election count itself. Most of all, ruling parties know how to intimidate voters, particularly semiliterate rural voters acquainted with only one ruling party since independence.

Mugabe's efforts during the weeks before Zimbabwe's June parliamentary elections (he is not up for reelection until 2002) were a depressing case in point.

The Western media noticed when Mugabe sent supposed war veterans onto thousands of white-owned farms to cow rural workers and white supporters of the new opposition Movement for Democratic Change (MDC), which is headed by the union leader Morgan Tsvangirai. Hundreds of MDC supporters, black and white alike, were beaten, and several dozen were killed. Government orders that rural clinics and hospitals refuse treatment to MDC backers got less attention, as did having teachers suspected of supporting the MDC hauled from their classrooms and beaten.

The vote itself was a shambles, with doctored voter rolls and Mugabe-designed constituency boundaries finalized a mere three weeks before the election and released to the opposition only by court order. The United Nations, which led an election-monitoring effort, learned that 10 to 25 percent of registered voters were in fact dead and threw up its hands two weeks before the election. Other outside monitors had no more luck, leaving Mugabe cronies from his party, the Zimbabwe African National Union–Patriotic Front (ZANU–PF), to oversee a campaign in which MDC candidates were beaten up and had their homes firebombed. A week before the election, the head of the Zimbabwe Human Rights Forum reported a "complete subversion of the democratic electoral process."

In a fair contest, Mugabe's record would have given the MDC plenty of political ammunition. Mugabe single-handedly destroyed his country's economic equilibrium twice in three years. In 1998, after sending Zimbabwean troops into the Democratic Republic of the Congo to protect President Laurent Kabila's predatory regime, Mugabe began using antiwhite rhetoric and confiscating white-owned farms while granting lavish pensions to veterans of the country's 1971–79 war. Zimbabwe's business-confidence index plunged, foreign direct investment dried up, exchange rates plummeted, and fuel and commodity shortages became common—while the president continued to line his pockets.

A referendum last February [2000] was intended to give Mugabe the constitutional authority to take white farms without compensation. But the voters, emboldened by the fledgling efforts of the MDC and a fed-up urban civil society, decisively rebuffed the president. In response, Mugabe incited the war veterans further and refused to let the police enforce High Court injunctions against the farm invasions. This refusal to abide by the inconvenient rule of law was hardly new; the torture of two journalists in early 1999 and the kidnapping in mid-2000 of two Cuban physicians trying to defect were only two authoritarian outrages among many. Even former associates of Mugabe have been known to suffer fatal "accidents" while driving. Within State House, the shadowy Central Intelligence Organization (CIO) enforces the president's dictates and represses dissent by any and all means.

Zimbabwe's fall is hardly unique. Yet the Zimbabwe case, like the much earlier examples of Ghana under Kwame Nkrumah or Nigeria under a string of military presidents since the mid-1960s, is particularly vexing because Zimbabwe has always been comparatively wealthy. Unlike Zambia and Botswana, which rely on one primary export (copper and diamonds, respectively), Zimbabwe has almost always been able to feed itself and export tobacco, maize, gold, and minerals. It has a healthy entrepreneurial tradition of manufacturing for export and, with the Victoria Falls and many well-managed game reserves, an attractive tourist industry.

Africa's failure to thrive at the end of the last century has many causes, not least of which is mismanagement. One can pin the blame on shifts in world commodity prices, misguided World Bank and International Monetary Fund (IMF) policies, civil wars, climatic disasters, unchecked population bulges, and so on. But the visible hand of individual rulers can also be discerned. And no single exemplar of failed leadership surpasses Mugabe, who has been prime minister or president of Zimbabwe since 1980. In the annals of human-made disasters in Africa, his has not yet equaled the inspired debacles of Mobutu Sese Seko in Zaire, Idi Amin in Uganda, and Jean Fidel Bokassa in the Central African Empire. "I'm no Idi Amin," Mugabe resentfully told London's *Sunday Times* in June. Yet those three were poorly educated potentates pursuing peculiar personal visions and vendettas in countries less robust and prosperous than Zimbabwe. By contrast, the gifted Mugabe inherited a well-run, well-off territory. He robbed Zimbabwe of its potential.

Things Fall Apart

Mugabe's assessment of his own abilities is not wrong; he is no mere thug like Amin. Rather, he is well-trained and capable of matching wits with many of the world's most sophisticated leaders. Mugabe was trained by Jesuit missionaries in rural Zimbabwe, became a teacher, and studied for a B.A. in English and history at the University College of Fort Hare in South Africa when it was one of the five best institutions of higher learning in sub-Saharan Africa. After graduating from Fort Hare in 1951, Mugabe taught first in Zimbabwe and then in Zambia at a teacher-training institution. Along the way, he obtained a diploma and a bachelor's degree in education from the University of South Africa and another bachelor's degree in economics from the University of London, all by correspondence. Subsequently, Mugabe taught in a teacher-training school in Ghana, from 1956 to 1960, where he met Sally Hayfron, his first wife.

Because Mugabe is genuinely talented and because Zimbabwe is a modern state that could have one of Africa's better economies and has, per capita, one of its best-educated populations, his madcap abuse has been that much more tragic. He has been fully in charge and fully aware, lording it over his party's authoritative politburo and central committee, the cabinet, and parliament.

Zimbabwe has been on an economic slide since 1995 but went into free fall in 1997. The comparatively (for Africa) wealthy country's per capita real GDP slumped in the late 1990s, from $645 in 1995 to $437 in 1999. Its real GDP growth rates tumbled from 7.3 percent in 1996 to −1 percent in 1999, with even worse shrinkage (perhaps almost −10 percent) expected for 2000. Consumer-price inflation has shot up from 22 percent in 1995 to 58.5 percent in 1999 and 80 percent in mid-2000. Zimbabwe's foreign-currency reserves were essentially exhausted in mid-2000. Its currency exchange rate against the U.S. dollar collapsed from 8 Zimbabwean dollars (Z$ hereafter) to $1 in 1995 to Z$23 to $1 in 1998, and then plunged to an artificially controlled level of Z$38 to $1 since mid-1999. Just after the 2000 elections, the unofficial trading rate was Z$60 to

$1, and without controls, the overvalued local currency would trade at about Z$80 to $1. Domestic interest rates on treasury bills have risen to 70 percent, up from 20 percent in 1997. These stratospheric levels make nongovernmental borrowing—and thus routine commerce—increasingly difficult.

The people of Zimbabwe are one-third poorer than they were at independence. With the onset of AIDs, their life expectancies have declined considerably, and infant mortality rates are rising rapidly. The modern look of Zimbabwe's cities belies their new poverty and the depths to which the living standards of most Zimbabweans have tumbled.

Countries cannot easily be bankrupted, but since Zimbabwe has been unable to pay its petroleum or energy bills since 1999, it is coming close. This spring, Zimbabwe could not cover the running costs and salaries of its overseas diplomatic missions. Thanks to charitable loans from Libya and Kuwait and the benevolence of South Africa, which understands the devastation that a total Zimbabwean collapse would wreak on the economy of southern Africa, Zimbabwe somehow managed in the first half of 2000 to import a dribble of fuel for cars, trucks, and tractors.

For Zimbabwean consumers, life since December 1999 has been a succession of mile-long lines: for gasoline for their cars, diesel fuel for their tractors, kerosene for heating, cooking oil, and bread. More than 60 percent of urban adults are unemployed. In an agricultural land where national self-sufficiency has been assumed except in years of dramatic drought, wheat and other staples (including the white maize that most Africans eat) have been periodically unavailable. Even local vegetables are scarce and expensive. Consumers, both urban and rural, have borne the brunt of the country's financial decay. Further shortages, fuel rationing, and damaging price controls are predicted for the last quarter of 2000.

Harare, Zimbabwe's seemingly prosperous capital, had the air in July of a modern Potemkin village. Hotels were empty, firms were going bankrupt, and tourists were absent. The sprawling city was a lifeless shell.

Ordinary commerce had started to grind to a halt in the months before the June elections, particularly with the threat of farm confiscations. In May and June, when cured Virginia leaf tobacco is usually sold, nearly all of it was sitting unprocessed in barns because so many white producers had been forced off their farms.

The root of this financial meltdown is the government's failure to control its fiscal deficit, which has risen alarmingly from 8 percent of GDP in 1998 to 12 percent in 1999 and probably to as much as 20 percent in 2000. Despite promises to the IMF throughout the 1990s, Zimbabwe failed to trim its official wage bill by reducing government workers. Excess civil-service and military employment kept fiscal deficits high, but the precipitating cause of Zimbabwe's economic troubles was Mugabe's personal decision (without prior consultation with parliament, the cabinet, or his party's central committee) to send 6,000 Zimbabwean soldiers into the Democratic Republic of the Congo in 1998 to bolster Kabila's government against rebels supported by Rwanda and Uganda.

Mugabe says that helping Kabila was a simple matter of aiding a fellow southern African leader in a time of need, and the 6,000 Zimbabwean troops have since swelled to 13,000. But Mugabe also wanted to show President Yoweri Museveni of Uganda and President Nelson Mandela of South Africa that he still counted. Equally important, Kabila offered Mugabe, his close political cronies, and some Zimbabwean generals a chance to line their pockets with the Congo's cobalt and diamond wealth. Although Zimbabwe paid the troops, purchased the ammunition, and obtained the fuel that let Mugabe's troops help contain the rebels in eastern Congo, a few individuals are widely suspected of having received rewards from Congolese mineral concessions in exchange. Zimbabwe's distraught finance minister, nominally in charge of the government's treasury, certainly never knew how the military incursion into the Congo was to be afforded. For years, Zimbabwean finances have been run from State House, and Mugabe regularly refused to let unhappy finance ministers resign.

If sending troops to the Congo were not damaging enough, Mugabe compounded Zimbabwe's misery this year by hiring rent-a-thug "veterans" to invade 1,600 white-owned farms—more than a third of the total number of such farms—and by threatening loudly to confiscate every last one of the remaining farms without compensation. Whatever moral justification there was for evicting whites from farms that had been in nonindigenous hands for 50 to 100 years after being "purchased" from Africans during the European occupation, Mugabe's bullying immediately jeopardized the employment and wages of 400,000 African farm laborers and their families, inhibited reinvestment by farmers, and chilled domestic trade and banking, which has massive loans outstanding to white farmers. Two South Africa–based construction companies scaled back their operations, and a huge sugar firm suspended fresh investments, as did the Nissan automobile company. Bata Shoes reported that its retail trade had halted. Two of Zimbabwe's largest firms warned that profits in 2000 would fall by 60 percent or more. Since the white farms produce more than half of Zimbabwe's exports and foreign-exchange earnings and since the unrest vitiated tourist earnings (which are about eight percent of annual GDP), Mugabe's actions drove a stake into the already weakened heart of his national economy.

To be clear, Africans were heartlessly pushed off the best lands in Zimbabwe by the more powerful whites. This began in the 1890s but continued well into the 1920s. These inequities, which were confirmed by a 1930 Zimbabwean commission chaired by Sir Morris Carter, relegated Africans to the less well watered, stonier, and less loamy lands of the nation. After independence, that disparity was meant to be rectified by official purchases of white-owned farms and the resettlement of truly landless African peasants. Some of that occurred, gradually, but much of the transferred farmland—especially the very best bits—somehow found its way into the hands of Mugabe's associates. The recovered lands were used as patronage spoils, and they still are. Moreover, in those cases of genuine peasant resettlement, no resources were provided to maintain irrigation facilities or other equipment. Once broken up into peasant holdings, large farms can prove inefficient and uneconomical, whatever the talents of the small farmer—especially in situations like Zimbabwe's, where the small farmers are not given title to the new landholdings. To compound the problem, the government agricultural-extension services collapsed for want of funds, cutting off a source of advice. Mugabe sowed disaster.

Absolute Power . . .

Had the farm invasions not been just the sort of electoral ploy that Mugabe had often used before, and had the president and his cronies not been suspected by most Zimbabweans of having profited since the mid-1990s from every conceivable form of graft, the high-handedness of the past two years would not have caused so much consternation among a hitherto accepting population. Mugabe, after all, was the country's founder. But when citizens suffer and suffer, leaders can forfeit their credibility. This happened to Mugabe because of the Congo campaign and a cascade of corruption.

Tracing corruption anywhere, and particularly in the developing world, is never easy, and rumors are not fact. Still, there is no doubt that Mobutu salted billions of ill-gotten dollars in banks and properties in Switzerland and France, that Hastings Banda of impoverished Malawi built 13 large palaces throughout his nation during his 30-year reign—and that Mugabe has purchased or constructed six "mansions" in Zimbabwe in recent years. It is also a fact that Mugabe has often commandeered jets from the national airline (thus inconveniencing scheduled passengers) for state "shopping" visits abroad. He has reportedly purchased property in Europe, and he has bank accounts in all the usual overseas places. A complicated web of companies registered in the Isle of Man and the British Virgin Islands may control many of his investments.

Last October, Transparency International rated Zimbabwe the world's 45th most corrupt country, below South Africa (34th) and Poland but well above Jamaica, Turkey, Belarus, Bulgaria, Russia, Ghana (63rd), Nigeria (98th), and Cameroon (99th). According to the *Africa Competitiveness Report* for 2000, Zimbabwe is one of the very least competitive African nations, pulled down significantly in the rankings by its reputation for corruption. Zimbabwe is also listed in the same report as the African country "most needing improvement," with the most volatile exchange rate, poorest export position, and least growth in all of Africa.

No wonder. Large construction contracts, including those for Harare's new international airport, have found their way to consortia controlled by Leo Mugabe, the president's nephew. In 2000, friends of the president created a diamond-harvesting scheme in the Congo and attempted to float stock on the London market. In 1997, Mugabe lost three court battles trying to prevent a local black entrepreneur from starting a successful cellular-telephone franchise; Mugabe wanted his nephew to dominate that business. (The entrepreneur became an MDC supporter and found it prudent to remain outside Zimbabwe after June's election results were announced.) Whatever the totality and truth of this circumstantial evidence, ordinary Zimbabweans in 1999 and 2000 simply rolled their eyes when questions arose about the president's corruption. He and his administration were seen as thoroughly rotten. Indeed, part of the job of the CIO is to keep tabs on the corruption of subordinates so that Mugabe always has leverage over those who have profited from their positions of public trust. Few have not.

Graft has, of course, helped block economic reform. Throughout the 1990s, advisers from the World Bank and the IMF, as well as a team from Harvard

University, recommended that Zimbabwe follow Zambia's and South Africa's leads (and the Asian "tiger" model) and sell off its money-losing state-owned monopolies in such capital-intensive areas as communications and energy. Doing so would dramatically reduce government deficits; provide better, modernized, and cheaper services; create new jobs and entrepreneurial opportunities; provide infusions of cash for social services; and not tie up scarce official capital. These arguments were well received in almost all sectors of society and even within Mugabe's cabinet and central committee. Outsiders personally discussed these ideas with Mugabe. But the sale of significant state-owned enterprises proceeded glacially and without presidential enthusiasm, for three obvious reasons: state-owned concerns were sources of patronage, which Mugabe directed; the resources of state enterprises could be diverted (like the national airline) for his own use; and state-owned firms could offer contracts that Mugabe and his associates could control and from which they could (and did) profit.

In 1996, I visited the cabinet minister responsible for privatization and beseeched him to accelerate the sale and restructuring of state enterprises. "You don't need to try to persuade me," he smiled, holding up his hands. "We're all in favor of privatizing." So why wasn't it happening? The minister rolled his eyes and gestured upward. "The boss won't let us," he said. With deliberate naiveté, I asked if the cabinet were in favor. "Of course," my tutor explained. And how would the cabinet vote on the question? "It depends," he said, "on whether the ballot were open or secret."

The Autumn of the Patriarch

Like most of Africa, Zimbabwe has structural and geographical weaknesses that go beyond poor leadership. It is landlocked, for one, and its 12 million people are burdened with one of the highest adult HIV infection rates in the world (27 percent). It also endured 90 years of white settler occupation and racist misallocation of resources, as well as a bitter civil war. But Zimbabwe's glum failure to realize its potential has much to do with the self-aggrandizement of the very man who led the nation out of white supremacist rule. In the 1960s, many Africa specialists—myself included—extolled Mugabe's clear-eyed intelligence, character, and Jesuit-instilled ascetic integrity. Mugabe's disdain for his citizens is bold, freshly arrogant, and jarring in one previously so politically acute. What went wrong? What changed?

After Ian Smith attempted to create a settler-ruled Rhodesia from 1965 to 1979, Mugabe and other Zimbabwean nationalists were severely tested. They were jailed and exiled, and their edges were honed by the determined aggression of the white settlers. Mugabe spent ten years, from 1964 to 1974, in detention. When Zimbabweans, backed by China, Cuba, Libya, Algeria, and Marxist Ethiopia, took to the bush and began an eight-year guerrilla war against Rhodesia in 1971, Mugabe emerged ascendant when his rivals struck the guerrillas as insufficiently tough. Mugabe fled Zimbabwe in 1975 and rapidly consolidated his embryonic claim to national leadership by obtaining the funds and weapons necessary to pursue the war and make combat costly for Smith's whites. Mugabe was never the field strategist, but he fueled the efforts of the guerrillas and refined their political message.

By late 1979, white Rhodesia was on the ropes, and its apartheid-minded South African friends had thrown in the towel. A British-negotiated settlement created the modern Zimbabwe. At the talks in London, Mugabe was the unquestioned leader of the team. In the subsequent 1980 election, Mugabe skillfully deployed his political troops, campaigned imaginatively throughout what was soon to become Zimbabwe, and claimed the premiership of the new country.

Zimbabwe, as an African-run, crypto-Marxist enterprise, was Mugabe's creation. Members of his ZANU-PF party were officially called "comrade." The ruling party was more powerful than the government. The nation's foreign policy was "nonaligned" and pro-Soviet. Mugabe presided over the nationalization of newspapers, the national airline, and the marketing of minerals. Hindered for seven years by the British-imposed constitution, he waited until 1987 to scrap the country's inherited parliamentary system and become executive president and unquestioned leader. (Britain said that it did not then view the inauguration of single-party rule as "an erosion of democracy.")

In the early 1990s, Mugabe's diktats became less and less subject to criticism. As Mugabe consolidated his grip on power at home and became a recognized figure abroad, he became more autocratic and avaricious. Those close to Mugabe, now 76, attribute his increasing arrogance and vanity to his age, his anxiety about being eclipsed as an African leader by Mandela after South Africa held its first democratic elections in 1994, and the waning influence of the president's then-wife, Sally. Around 1990, the president and Grace Marufu, his young secretary, became increasingly intimate, and she subsequently gave birth to his three children. (His only child with Sally died young.) After Sally herself died in 1992, Mugabe married Marufu, and the upsurge in Mugabe's caprice and greed has closely tracked her rise. Senior ZANU-PF cabinet members called her "grasping Grace"—behind her back.

The MDC's Tsvangirai has proposed a truth commission to provide a national accounting for past crimes (including the appalling Matabeleland massacres in the 1980s, in which over 20,000 opposition supporters were murdered), human rights abuses, and corruption. After the June elections, he urged Mugabe's impeachment. Mugabe's leadership decisively lost its legitimacy in the late 1990s. The unanticipated loss of the February referendum was a precursor to his party's narrow and much-questioned parliamentary victory in June, which gave 62 predominantly rural seats to ZANU-PF, 57 to the MDC, and 1 to a small opposition party.

Having received nearly half the votes cast, the MDC controls enough seats to prevent Mugabe from amending the constitution. The MDC also emerged as a new kind of African political party: it has an urban power base but also controls many of the country's rural areas. Moreover, black voters elected four MDC whites, enabling the MDC to claim that it is a truly nonracial, pan-ethnic party.

By comparison, ZANU-PF kept its seat majority by winning—sometimes in a questionable manner—a succession of entirely rural Shona-speaking constituencies. The leader of the European Union monitoring team termed the election "hardly free and fair." Even within ZANU-PF, there were strident calls immediately after the election for Mugabe's resignation.

The long reach of the CIO still makes that unthinkable. Mugabe is nimble, clever enough to reform, and aware of how energized urban African publics can be. But easing toward retirement would mean moderating his cronies' greed and his own. Following the rule of law would hardly work to Mugabe's benefit.

Zimbabwe could, under revamped leadership, recover its promise with much greater ease than neighboring Zambia and Malawi after their recent traumas of corrupt, autocratic rule. The immediate challenges are formidable but less daunting than they would be in African countries with poorer human resources: bringing the troops home from the Congo, restoring law and order, stabilizing the national finances, fighting corruption, improving health and education, strengthening infrastructure, privatizing state monopolies, wooing foreign direct investment, and starting to practice good governance. But what about the fundamental leadership problems of the rest of Africa? Here, the lesson of Mugabe is richly instructive.

Destructive Engagement

[Historian (1834-1902)] Lord Acton's dictum that "absolute power corrupts absolutely" is suggestive in the Mugabe case, but hardly conclusive. In Zimbabwe, as in so many other places in Africa since 1960, a weak civil society provided openings for autocracy; the Cold War, with its tolerance for strongmen, offered others. Zimbabwe, like many other African states, lacked an independent media to hold politicians accountable. Mugabe originally also created a party-dominated socialist state. Given the comparative prosperity of the 1980s, the absence of any real victimization of whites, and the repression and then co-optation of the main opposition party, anti-Mugabe voices were muted. Moreover, ZANU-PF was ethnically and politically formidable. Those politicians who did oppose Mugabe and the ZANU-PF steamroller either lacked national legitimacy or were harassed and arrested. The CIO kept internal ZANU-PF dissidents in line, as did Mugabe's patronage machine. By the early 1990s, Mugabe was unassailable, and the resultant sense of entitlement engendered excesses.

The forgiving attitude of the international community also contributed to Mugabe's sense of invincibility. No one in Africa ever publicly called Mugabe to account, including its giants. Nyerere was greatly displeased when Mugabe rebuffed his plea not to send troops to the Congo. Mandela regarded Mugabe as an insufferable autocrat, and South Africa's current president, Thabo Mbeki, probably feels the same way. But no one said so publicly.

Outside of Africa, the international lending institutions always thought it better to woo Mugabe than to criticize him. Indeed, the IMF rewarded him with tranche [a bond series issued for sale in a foreign country] after tranche of support. Even after Mugabe and his finance ministers refused over and over again to fulfill their bargains, even after he sent troops into the Congo, the IMF clung to engagement. The World Bank and other donors also never flagged in their support.

One recipe to curb the rise of future Mugabes is to hold strictly to conditions for help. To curb autocratic excesses, international lending institutions should use tough love: an absolute refusal to lend and donate in the absence of the rule of law, good governance, and sensible economic policy. Likewise, bilateral donors should cease supporting those who self-aggrandize or abuse human rights. Better yet, donors and international lending agencies should shun all dealings with those who breach their own national norms. There should be no state visits for the Mugabes of the world or their foreign ministers or finance ministers. Ostracism can be a powerful weapon, especially if the refusal to pursue business as usual with dictators and illiberal democrats becomes widespread.

Continuing big-power relations with the Mugabes of the world is usually excused by the term "constructive engagement." Better to retain some influence, however limited, with despots—thus runs the usual rationale. But it almost never works. Mugabe grew more and more insufferable because he could thumb his nose at the international lending institutions, the Commonwealth, and the big powers. No one has yet called his bluff. In May and June 2000, even Mbeki—heir to Mandela, the continent's shining light—consciously refrained from publicly criticizing Mugabe.

Good leadership in Africa should be rewarded, participatory leadership supported, and sensible economic management backed—but not bad leadership and bad policy. Mugabe's growth as an unlimited autocrat just might have been checked by enough international cold shoulders. If he had been made persona non grata abroad, especially in Europe and the United States, Zimbabwean civil society might have taken heart. So might his critics in and out of government. At the very least, the IMF and the World Bank should have abided by the letter and spirit of their own conditions. Depriving Mugabe's Zimbabwe of foreign aid because of his proclivities might have made a difference. All of those snubs were certainly worth trying. Constructive engagement, in other words, ought to be employed in Africa only sparingly and surgically.

Such prescriptions are useful but difficult and negating. One could also be proactive about Africa's leadership problem. It is now more necessary than ever to find ways to help elected African politicians themselves lead—not by working in society at large (although that is not harmful) and not by working at the grassroots (which is also useful but not sufficient). The models of modern Botswana, Mauritius, and South Africa need to be offered to emerging African elected leaders through carefully tailored courses and seminars. Doing so would not be a conclusive remedy for African leadership weaknesses, but it should help limit the rise of future Mugabes and their ilk. The old ways have hardly worked.

Arthur A. Goldsmith  **NO**

Risk, Rule, and Reason: Leadership in Africa

Introduction

Sub-Saharan Africa is poorly led. The region has far too many tyrants and 'tropical gangsters', far too few statesmen, let alone merely competent officeholders. Too often, these leaders reject sound policy advice and refuse to take a long and broad view of their job. They persecute suspected political rivals and bleed their economies for personal benefit. With a handful of exceptions, notably South Africa under Nobel laureate Nelson Mandela, countries in the sub-Saharan area are set back by a personalist, neopatrimonial style of national leadership (Aka, 1997).

Better leadership is not the cure-all for Africa's lack of development, but it would be an important step in the right direction. A few years back some observers saw hope in a new generation of supposedly benevolent dictators, such as Isaias Afwerki in Eritrea, Meles Zenawi in Ethiopia, or Yoweri Museveni in Uganda (Madavo and Sarbib, 1997; Connell and Smyth, 1998). Subsequent events (war between Eritrea and Ethiopia, invasion of the Congo Republic by Uganda) chilled the optimism (McPherson and Goldsmith, 1998; Barkan and Gordon, 1998; Ottaway, 1998). In most countries, it seems progressive leadership soon reverts to the more familiar form of autocratic one-man rule.

There is no shortage of macro-level explanations for this pattern. Authoritarian political traditions, lack of national identity, underdeveloped middle classes and widespread economic distress are among the sweeping, impersonal forces cited as factors that produce poor leader after poor leader. Foreign aid may have enabled some of these leaders to hang on longer than they would have otherwise, especially during the Cold War. This article instead takes a micro-level view of leadership. Without denying that macro-level social and economic factors bear on leaders' behaviour, I find it also useful to look at these people as individuals and to speculate about the incentives created by their environment.

In the tradition of political economy, we can begin with the assumption that African leaders are usually trying to do what they think is best for themselves. We can posit that they choose actions that appear to them to produce the greatest benefit at least cost, after making allowances for the degree of risk involved. Such a leader also is capable of learning, and takes cues from what is happening to other leaders in neighbouring countries. He can improve his behaviour if he has to.

While no African leader fully exemplifies this rational actor model, all these individuals' behaviour can be illuminated by it. After all, even the best leaders have mixed, sometimes egoistic motives. To the extent that it represents reality, the rational actor model also may suggest how changing the political incentive system might induce African leaders to behave less autocratically.

I start this article by speculating about how these leaders might react to perceived levels of risk in their political environment. Next, I investigate the actual level of risk, guided by a new inventory that covers every major leadership transition in Africa since 1960. Then, I assess how risk appears to have distorted the way African leaders act in office. Finally, I consider the ways in which democratization may be changing political incentives for the better.

Leadership and Individual Motivation

Perhaps the most troubling thing about African leaders is their tendency to reject (or simply not follow through on) conventional economic advice (Scott, 1998). Africa is the graveyard of many well-intended reforms. The vacillating public attitude of Kenya's President Daniel Moi is emblematic. In March 1993, he rejected an International Monetary Fund (IMF) plan for being cruel and unrealistic. One month later, he reversed himself, and agreed to the plan. In June 1997, the IMF cut off lending to Kenya after Moi refused to take aggressive steps to combat corruption. Again, his initial reaction was defiance, swiftly followed by a more accommodating line.

Why are African leaders apt to resist advice to carry out market-friendly reforms that could boost national rates of economic growth? If one accepts the premise that, with sufficient time, open market policies will work in Africa, such a choice can look senseless. Certainly, no African leader would prefer to perpetuate mass poverty and economic stagnation in his country, which can only make governing more difficult. More to the point, perhaps, cooperating with the international financial institutions is the best way to assure continued diplomatic support and financial credit. Yet, many African leaders apparently see political rationality in choosing policies that are economically damaging or irrational. Miles Kahler (1990) refers to this as the 'orthodox paradox'.

Political economy offers a theory of micro-level behaviour that may explain the paradox. Mancur Olson (1993) argues that time is the key. According to this theory, the predicament facing any individual national leader is that the pay-offs to most economic reforms lie in the future, but he also has to hold on to power now. An insecure power base is likely to encourage either reckless gambling for immediate returns or highly cautious strategies to preserve political capital; it is unlikely to promote measured actions to obtain long-range returns. Whether a leader acts for the short or the long term, therefore, is influenced by his sense of the level of threat to his career.

A more technical way to understand a leader's intertemporal choices is to think of a 'political discount rate'. One of political economy's core ideas is that future events have a present value, which one can calculate by using a rate of discount. That rate of discount rises with risk and uncertainty. When an outcome is doubtful over time, it makes sense to mark down its present value. The more doubtful the outcome, the more valuable are alternative activities that yield immediate dividends, even if the expected return of those activities is low. Thus, under conditions of political uncertainty, the narrowly 'rational' leader will systematically forgo promising political 'investments'—ventures whose benefits he may not survive to reap. Whenever he is given a choice, according to this argument, such a leader will usually prefer current political 'consumption'. It follows that free-market reforms look like a poor bargain, requiring immediate political pain in exchange for distant (and therefore questionable) gain.

High political discount rates are also a possible explanation for the extensive and destructive political corruption seen in Africa. The Democratic Republic of Congo's Mobutu Sese Sekou is the archetype. The late dictator erased the line between public and private property, accumulating a vast personal fortune and bankrupting his country. His is an extreme case, yet every national leader has opportunities to profit individually from his office. According to the premises of political economy, it is the leader with the least certainty about his fate who has the strongest incentives to take his rewards now—and to take as much as possible. A more self-assured leader may calculate that it is safe to defer most personal financial gain until after he has left office. Some of the misuse of public office also may be due to the need to buy support from friends and extended family members. Olson (1993), for example, postulates that leaders with an insecure grip on power have an incentive to take steps to patronize favoured ethnic groups, often at the expense of national economic health. This sort of pork-barrelling is well known in Africa.

Political economy thus presents a cogent theory for why African rulers act they way they do. Short-term policy making and political corruption are 'rational' ways of trying to manage the risks associated with governing in an unsettled political system, as we typically find in Africa. According to this thesis, overly cautious or corrupt leaders may simply be attempting to maximize utility under conditions of personal and political uncertainty. Their assessment of risk is affected by their personal experiences and by their perceptions about larger trends in their country and region. Unfortunately for the social welfare, their effort to protect their individual interests has spillovers that hurt everyone else.

The issue for this article is whether the facts support this theory. First, is it true that African leaders face a high degree of risk? We can reasonably assume these people are tolerant of risk, or they would not have chosen political careers. Thus, we need to look for evidence of extraordinary occupational hazards for leaders. The second question is whether political risk in Africa is associated with 'bad' (anti-market) economic policy choices or with corruption. As we will see below, the answers to both questions seem to be affirmative: there is significant physical risk for leaders, and that risk correlates with anti-market policies and with corruption. Those two findings, in turn, suggest scope for enhancing the area's national leadership by reducing the risks of governing, a goal that may be abetted by democratization. . . .

Risk and Leaders' Behaviour

There is little doubt . . . that holding high office in Africa poses acute risks. To what extent do those risks affect leaders' behaviour, specifically their behaviour in the areas of economic reform and corruption? . . . That question is difficult to answer fully without detailed case studies of the individuals involved. In the absence of such information, however, we can look for approximate answers in national indicators of economic policy and corruption. To the extent we believe that country leaders control public policy or set the tone for public honesty, aggregate data may give us clues about how these leaders conduct themselves.

To represent a country's commitment to free market economics, I use the Heritage Foundation's Index of Economic Freedom (Johnson *et al.*, 1999). The index is calculated by aggregating country scores on 10 policy indicators and measures of the business climate. Depending on their scores, countries are categorized as free (none in Africa), mostly free, mostly unfree, or repressed. While I do not see eye to eye with the Heritage Foundation on many subjects, I suspect that these categories offer a good approximation for how fully countries comply with IMF-style structural adjustment programs. My grouping of countries is based on the average economic freedom rating for 1995–1999.

I hypothesized earlier in this article that low-risk environments would tend to produce more reform-minded leaders, or at least leaders who would be more willing to go along with economic reform in exchange for financial credit. . . . [A] correlation exists between the hazards of leadership and the degree of 'economic freedom'. Leaders in the so-called mostly free countries were the least likely to be overthrown, killed, arrested or exiled. Leaders in the mostly unfree and repressed countries, by contrast, experienced a greater number of negative outcomes. Low political risk and liberal economic programmes seem to go together in Africa.

Correlation does not prove causation, especially in making inferences about micro-level behaviour based on macro-level data. We cannot say whether a safer political environment encourages leaders to opt for the market, or conversely, whether leaders who opt for the market make their political environment safer (though the latter possibility seems less likely, at least in the short run). In either case, however, the results are consistent with political economy theory.

What is the relationship between political risk and corruption? For a measure of the latter, I use Transparency International's Corruption Perception Index for 1999. Transparency International is a watchdog organization formed to help raise ethical standards of government around the world. It compiles an annual index that assesses the degree to which public officials and politicians are believed to accept bribes, take illicit payment in public procurement, embezzle public funds, and otherwise use public positions for private gains. The index is based on several international business surveys, using different sampling frames and varying methodologies (Transparency International, 1999). While

Transparency International is careful to point out that the rankings only reflect perceptions about corruption, I find it reasonable to assume that they correspond roughly to reality.

I have conjectured that leaders in the riskier African countries would have the greatest propensity to use their public offices for personal ends. Once more, the data lend support to my hypothesis. . . . The pattern is striking. There has never been a successful coup in the less corrupt group of countries. None of their ex-leaders has been arrested or exiled, and only one was killed while in office (South Africa's Verwoerd). The more corrupt countries, by contrast, have many coups and many leaders who suffered personally upon losing power.

As with the economic freedom index, these correlations do not prove that a hazardous political environment encourages leaders to become corrupt. The opposite is also plausible: corrupt rulers seem likely to invite coups and to bring personal suffering on themselves. To the extent that risk and corruption are related, the relationship between the two probably is mutually reinforcing. The important point for this article is that the observed association of risk and corruption conforms to what you would expect, based on the assumption of 'rational' behaviour among national rulers. Without overstating the case, the correlation lends support to a political economy account of poor leadership in Africa.

Democratization and Improved Leadership

Political economy also suggests that one solution to poor leadership is to make the political environment less hazardous. A safer environment would reduce the incentives to engage in political misbehaviour and, in principle, encourage more responsible and forward-looking activity. In this context, Africa's recent moves toward more pluralistic national political systems, where people can express their political opinions and take part in public decisions, are reasons for hope. It is fashionable—and correct—to observe that democracy has shallow roots in most African countries (Joseph, 1997; van de Walle, 1999). Much of the impetus for reform comes from abroad, from the region's creditors. Yet, when we observe the patterns of leadership transitions, it is hard to deny that genuine changes are taking place.

No sitting African leader ever lost an election until 1982, when Sir Seewoosagur Ramgoolam of Mauritius was voted out. Since then, 12 more incumbents have been turned out of office by voters—accounting for about one-sixth of the leadership transitions in the 1990s. The threat of losing an election also may account for the increasing rate of leader retirements—nine in the 1990s versus only eight in the previous three decades.

Democratization appears to be altering the outcomes of the many coups that still occur. In the past, the new heads of military juntas often declared themselves permanent leaders (sometimes after doffing their uniforms and becoming 'civilians'). Now, it is becoming the norm for coup leaders quickly to organize internationally acceptable elections—and, more importantly, to honour the results afterwards (Anene, 1995). Recent examples include Niger and Guinea-Bissau. . . . The fact we see more transitions of this type in the 1990s is an indirect reflection of the region's growing democratization.

[There are] additional reasons to think that contemporary presidential elections are not simply façades in many countries. The entire sub-Saharan region had only 126 elections for top national office in the 30 years through 1989. Most of those were show elections, with an average winner's share of close to 90%. Conditions have changed significantly in the 1990s. There were 73 leadership elections during that decade, or more than half as many as in the three prior decades. All but five of sub-Saharan countries were involved. Equally important, the winner's share dropped to an average of about two-thirds of the votes cast. Such results would be considered landslide victories in the developed world. No president in the history of the United States has ever reached two-thirds of the popular vote. Still, in African terms, the tendency clearly is toward greater competitiveness at the ballot box.

The classic liberal defences of free and fair elections are that they give voice to majority demands and that they are a means for recruiting new leadership talent. Political economy and African experience suggest three additional benefits, all associated with reducing the hazards leaders face.

First, elections have the virtue of softening the penalties of losing political office. The defeated candidate in an election campaign, as opposed to the victim of a coup plot, is far less likely to be executed, jailed or exiled by his successor. By providing a low-risk avenue of exit, elections thus reduce the stakes in political competition. If the arguments in this article are correct, that would free African leaders to take a more purposeful, pragmatic view of their jobs.

A second benefit occurs if elections become institutionalized, and take place according to a schedule. Countries that hold regular elections reduce speculation about when (and how) the next political transition is likely. Again, the probable impact in Africa would be to change the political calculations made by the region's chief power holders, to allow them to worry less about how to hold onto power and to think more about the long term. Predictable political transitions might also reduce anxiety among private investors, and thus mitigate the harmful political business cycle that exists in some countries.

The third benefit stems from the more rapid turnover among national rulers that results when elections become a regular part of political experience. As leaders come to see their jobs less as an entitlement and more as a phase in their careers, that actually may liberate them to 'do the right thing', and not always feel forced to do what is politically expedient. Merilee Grindle and Francisco Thoumi (1993) have remarked on this phenomenon among lame-duck presidents in Latin America. Knowledge that their positions are transitory can, somewhat ironically, concentrate the incumbents' attention on how best to leave a lasting legacy. Similar results are possible in Africa.

Concluding Observations

Before multi-party competition and elections can have these positive effects on leaders, Africa's competitive political systems must become institutionalized. This has yet to happen in most countries, according to the results of Samuel Huntington's (1968) 'two-turnover test'. Huntington notes that institutionalized democracies prove themselves by repeatedly carrying out peaceful transfers of power through the ballot box. The first time an opposition leader replaces an incumbent power holder does not necessarily establish a tradition of peaceful political change. It is only after the new incumbent is defeated and leaves office that one can begin to be confident that constitutional procedures have taken root.

Second turnovers are almost unheard of in Africa. Botswana has not had one. The same party has ruled that country since independence. Mauritius has had two election-based leadership turnovers, but many observers question whether that island nation properly deserves classification in the region. Bénin is the only other African country where incumbent power holders have twice lost elections. The dictator Mathieu Kérékou fell to Nicéphore Soglo in 1991, but he regained the presidency by defeating Soglo in the election 5 years later. Kérékou's continued role raises some doubt whether Bénin's second transition indicates much other than the persistence of narrow, personalistic politics in that country.

Nonetheless, the last decade does offer hope that some African societies will be able to establish more orderly systems of political competition. That could change the incentives for African leaders, and encourage them to act more responsibly and even-handedly. As a means of redressing decades of oppression and economic stagnation, that cannot happen soon enough.

POSTSCRIPT

Are African Governments Inherently Disposed to Corruption?

A less-discussed element of corruption is the extent to which international corporations may contribute to the problem in their dealings with African governments. In some instances, it is standard business practice for companies to provide financial incentives (or bribes) to government officials in order to win contracts, obtain permits, or gain certain rights. Interestingly, it is the government officials who are often accused of corruption while little is heard of the companies' role in this process. A recent exception to this involves a Canadian company that was convicted by the Lesotho high court for bribing a senior Lesotho government official in order to win contracts on that country's $8 billion Lesotho Highlands Water Project (a joint project of the governments of Lesotho and South Africa). Commenting on the case, a South African official said that it is often assumed that corruption is "a peculiarly African problem. This case shows that such a perception is wrong. It takes two to tango." See "Government Cracks Down on Western Corruption," *New African* (December 2002).

For those interested in further reading on this topic, good examples of the structural perspective on corruption include an article by M. M. Munyae and M. M. Mulinge in a 1999 issue of the *Journal of Social Development in Africa* entitled "The Centrality of a Historical Perspective to the Analysis of Modern Social Problems in Sub-Saharan Africa: A Tale From Two Case Studies" and an article by N. I. Nwosu in a 1997 issue of the *Scandinavian Journal of Development Alternatives* entitled "Multinational Corporations and the Economy of Third World States."

There are roughly two types of particularlistic or internalist explanations regarding corruption in Africa, one is negative and the other is positive. The first, essentially a negative perspective, views internal African characteristics as flaws that contribute to a universally accepted problem known as corruption. An example of this perspective is an article by M. Szeftel in a 2000 issue of the *Review of African Political Economy* entitled "Clientelism, Corruption and Catastrophe." The second, more positive or postmodern interpretation, often views "corruption" as a relative concept that is culturally defined. In other words, what is viewed as corruption in one culture is not necessarily corruption in another. Examples of this perspective include a volume by Patrick Chabel and Jean-Pascal Daloz entitled *Africa Works: Disorder as Political Instrument* (Indiana University Press, 1999) and a text by Jean-Francois Bayart under the title *The State in Africa: The Politics of the Belly* (Longman, 1993).

ISSUE 20

Are International Peacekeeping Missions Critical to Resolving Ethnic Conflicts in African Countries?

YES: John Stremlau, from "Ending Africa's Wars," *Foreign Affairs* (July/August 2000)

NO: William Reno, from "The Failure of Peacekeeping in Sierra Leone," *Current History* (May 2001)

ISSUE SUMMARY

YES: John Stremlau, professor and head of the Department of International Relations at the University of Witwatersrand, South Africa, argues that far too little is being done to check conflict in Africa and that the United States needs to do more in this regard. According to Stremlau, "[w]hile preventing conflict in Africa is primarily a task for Africans . . . the 1990s showed that outside help is needed."

NO: William Reno, associate professor of political science at Northwestern University, contends that no peacekeeping is better than bad peacekeeping. In his discussion of the failed Lomé Peace Accords, a settlement negotiated between warring parties in Sierra Leone, he notes that "[m]any Sierra Leoneans regarded positions taken by the UN and foreign diplomats who stressed reconciliation as offensive." As opposed to the more bureaucratic peacekeeping approaches taken by the United States and the UN, he lauds the hands-on tactics of the British.

T wo watershed events for American peacekeeping efforts in Africa were the failed military intervention in Somalia in 1993 and the lack of intervention during the Rwandan genocide in 1994. In the twilight of the presidential administration of George Bush in 1992, the United States intervened militarily in Somalia to dispense food aid that was not being effectively distributed due to the presence of armed militias. The American force was then reduced under the new presidential administration of Bill Clinton and made part of a UN operation whose mandate was to protect the delivery of humanitarian assistance and disarm the warring factions. Following the deaths of 26 Pakistani soldiers, the U.S. sent in army rangers and delta force commandos to try to capture a particularly problematic warlord named Mohammed Aideed. Then in October of 1993, 18 U.S. soldiers were killed and 50 wounded in a street war that also cost the U.S. military two Black Hawk helicopters. With the sight of a dead U.S. army ranger being dragged through the streets of Mogadishu on international television, then-President Clinton ordered the withdrawal of all U.S. troops. This debacle is now commonly referred to as "crossing the Mogadishu line." It resulted in a new U.S. Presidential Directorate (#25), stating that the United States should not become involved in a war unless there is a clear national interest, and the conflict could be won.

The events in Somalia had a chilling effect on U.S. interest in addressing the humanitarian crisis in Rwanda in 1994. Following the death of Rwandan President Juvenal Habyarimana in a mysterious plane crash in April of 1994, the systematic elimination of Tutsis and moderate Hutus was begun by hard-line elements in the military and the Habyarimana government. With most expatriates having left the country, and the meager UN force ordered to protect itself, it is estimated that over one million Tutsis and moderate Hutus were murdered while hundreds of thousands fled the country as refugees. Following these atrocities, the minority Tutsis amassed their forces and began a process of retribution. While the situation in Rwanda eventually was quelled, the inaction of the United States and other foreign powers has been roundly condemned. The specter of Somalia and Rwanda continue to haunt U.S. peacekeeping efforts in Africa today, with a fear of failure lingering from the first case and the regret of inaction from the latter.

In this issue, John Stremlau argues for a preventive approach to conflict in Africa. He suggests that external support for human rights groups and non-governmental organizations is beneficial because it helps put pressure on troubled regimes to reform and to resolve factional differences. Given that Africa experiences more conflicts than any other continent, Stremlau contends that the United States needs to expand its efforts in the realm of conflict prevention and that the Republic of South Africa would be the logical partner in this continent-wide initiative.

William Reno counters that no peacekeeping is better than bad peacekeeping. In his selection, he describes the conflict between the government of Sierra Leone and the rebel Revolutionary United Front (RUF), as well as the connections between the RUF and Liberian leader Charles Taylor. In his discussion of the failed Lomé Peace Accords, he condemns the UN and the United States for pushing for acceptance of rebel leaders with known human rights violations in the new government. He asserts that the United States wanted to appear to have addressed an African problem while avoiding a commitment of resources and soldiers. As opposed to the bureaucratic peacekeeping approaches taken by the United States and the UN, which tend to accredit armed groups that prey on society, he lauds the more hands-on tactics of the British.

John Stremlau  **YES**

Ending Africa's Wars

Cold War, Hot Wars

Following [1999's] military interventions in defense of human rights in Kosovo and East Timor, Western leaders proclaimed a new determination to stand up to similar abuses whenever they occur. On one continent, however, warfare still rages unchecked, and far too little is being done about it. Renewed clashes in Sierra Leone and on the Ethiopian-Eritrean border are two recent examples of deadly African conflicts that have killed and displaced millions of people. Yet despite occasional bursts of aid and attention, the United States and Europe have remained largely disengaged. The reasons for their lack of involvement in · conflict prevention and peacekeeping there are fairly obvious. War in Africa seems to pose no clear and present danger to U.S. interests. Furthermore, most African conflicts are fought within, not between, states. The international norms, institutions, and political will to intervene in such hard-to-solve conflicts remain inadequate. So do Washington's defense doctrine, bureaucracy, and budgets, all of which are still dedicated to preventing or settling traditional conflicts between states.

Yet those who argue that Washington and its allies should become more involved in solving Africa's problems make a powerful case. Africa is a vast continent of 700 million people with abundant natural resources and deep historical and cultural ties to the United States. It is simply too big and too important to be neglected. The question should be not whether, but rather how, to intervene there.

Of course, preventing conflict in Africa is primarily a task for Africans. But the 1990s showed that outside help is needed. The nice-sounding nostrum of "African solutions for African problems" became an excuse for neglect, until the images of human suffering in Africa became impossible for the West to ignore—leaving humanitarian relief as the only real option.

There are alternatives, however. Preventing wars, rather than fighting them, has always appealed to American strategists, so long as no vital national interests are compromised. Washington's greatest foreign policy success—winning the Cold War peacefully through military deterrence and the building of a strong coalition of democracies—vindicated the strategy of containing con-

From John Stremlau, "Ending Africa's Wars," *Foreign Affairs,* vol. 79, no. 4 (July/August 2000). Copyright © 2000 by The Council on Foreign Relations, Inc. Reprinted by permission.

flicts before they erupt. That strategy should now be adapted for and applied to Africa, where most wars result from bad governance. Weak, authoritarian African governments lack the institutional capacity to manage factional struggles. They exclude majority or minority groups from power and suffer from poverty and gross income inequality. All of these tensions throw off sparks that can start a war.

Any strategy for preventing conflict in Africa must therefore address these fundamental flaws. In deciding how to do that, and how to do it affordably, Washington should remember the Cold War lesson that working closely with democratic partners spreads the burden and gives policies greater legitimacy. Although Africa has few ready candidates for such a partnership, the region's most politically capable and economically advanced state—South Africa—does share key interests and values with the United States. Developing a strategic partnership with Pretoria must therefore become the foundation for conflict prevention and democratic development in Africa.

Such a strategy will require serious U.S. backing for South Africa's lead. Much more generous engagement by the American government, the business community, and nongovernmental organizations is needed to help develop the economic foundations of South Africa's new democracy. Only then will South Africa be able to effectively inspire and support democracy elsewhere on the continent. A real partnership based on support and mutual respect—the kind America once created with postwar Europe and Japan—could profit the region tremendously. And it would be in Washington's interests as well. For although the world may have changed since the Cold War, America's broad goals of preventing conflict and promoting a liberal international order remain the same.

Focusing on these goals in Africa will also help the United States overcome the lingering effects of two of the worst failures of President [Bill] Clinton's foreign policy: the 1993 military debacle in Somalia and the failure to prevent the 1994 genocide in Rwanda. Both disasters had a profound impact on Washington. Although the American public and Congress would never tolerate a repeat of the Somalia intervention, unease persists about not having done more to prevent the massacres in Rwanda. Devising a new threshold for intervention, somewhere between the extremes of Somalia and Rwanda, is now necessary. But any new strategy will be easier to sell, both in Africa and in America, if it is seen as part of a broader strategy involving reliable regional partners.

Two Steps Forward, Two Steps Back

War and poverty remain dominant realities in Africa. According to the State Department, [in 1999] Africa had more major conflicts than any other continent. Wars causing at least 1,000 battle deaths per year plague Angola, both Congos, Eritrea, Ethiopia, Rwanda, Somalia, and Sudan. Meanwhile, low-intensity conflicts simmer in Burundi, Chad, Djibouti, Senegal, Sierra Leone, and Uganda. And several other countries, notably Nigeria, the region's most populous, suffer from internal instability that could erupt into greater civil strife. "Preventing such wars," wrote U.N. Secretary-General Kofi Annan in an unusually frank May

1998 report, "is no longer a matter of defending states or protecting allies [but] a matter of defending humanity itself."

Approximately 8.1 million of the world's 22 million cross-border refugees live in Africa, with many more millions having been displaced within their own countries. An August 1999 global assessment of humanitarian emergencies by the National Intelligence Council paints a very grim picture of the continent, noting that the overall "demand" for humanitarian assistance through 2000 will likely exceed the willingness of major donor countries to respond. Africa is home to 23 of the world's poorest countries, and an estimated 290 million Africans survive on less than $1 a day. External debt burdens (totaling more than $200 billion), weak governments, and widespread corruption complicate efforts to alleviate poverty. On top of this, another recent National Intelligence Council report estimates that half of the world's infectious disease deaths take place in Africa, with old and new viruses posing global threats that governments must devise collective means to combat. 11.5 million Africans have died of AIDS, and in 1998, 70 percent of the world's new AIDS infections occurred there.

Other trends point in more positive directions. Until the late 1980s Africa's only functioning multiparty democracies were Botswana, Senegal, tiny Gambia, and the island of Mauritius—and the first three had never managed to produce a change in government. But according to a recent Freedom House survey, 32 of 53 African countries are now either democratic or at least partly free, with elections of varying credibility.

In June 1999, sub-Saharan Africa's two most important countries celebrated major democratic rituals. President Thabo Mbeki succeeded Nelson Mandela in South Africa, and Nigerians ended 16 years of increasingly repressive military rule by electing Olusegun Obasanjo as president. The two new presidents then teamed up diplomatically in June, successfully persuading the Organization of African Unity to agree to sanction any African government that comes to power by military means.

Meanwhile, Senegal has moved closer to accepting human rights and democracy. In February, Senegal's High Court took the unprecedented decision to prosecute exiled former President Hissene Habre of Chad for "complicity in acts of torture" following complaints filed against him by several human rights organizations. And a month later, Senegal's long-serving President Abdou Diouf gracefully accepted defeat as voters ended 40 years of uninterrupted Socialist Party rule.

Economically, the world's poorest region is also showing some improvement. Negative rates of per capita growth in the 1980s have turned positive, with a three percent rise forecast for this year. More than 30 countries are implementing broad macroeconomic reforms, many for the first time. These include liberalizing trade and investment rules, reducing tariffs, rationalizing exchange rates, ending subsidies, stabilizing currencies, and privatizing state enterprises.

Yet setbacks are still occurring, and these highlight the need for preventive action. President Robert Mugabe's disregard for the rule of law in Zimbabwe, where he has sought to rewrite the constitution and has condoned both the intimidation of the opposition and the violent appropriation of white-owned farms, is one such example. Another is the renewed violence by warlord Foday Sankoh's Revolutionary United Front in Sierra Leone, which included the seizure of several hundred predominantly African U.N. troops in a bid to wreck an otherwise promising peacekeeping operation.

What all this means for Washington is that whichever party takes control of the White House . . . will have to take a fresh look at America's stakes in a changing Africa. To properly plan America's next move, however, requires understanding its past steps and missteps, starting with those taken by the Clinton administration.

With Friends Like These

Old Africa hands were initially surprised and delighted by the amount of attention Bill Clinton paid to Africa. His original national security adviser, Anthony Lake, was the first Africa expert ever to hold the post. Within weeks of the inauguration, an unprecedented White House Conference on Africa was held to signal a new chapter in U.S.-Africa cooperation. Since then, a steady stream of cabinet-level delegations has gone to Africa at the rate of about one every eight weeks. In March 1998, Clinton made his own unprecedented eleven-day, six-nation tour.

But neither the frequent high-level visits nor special events, such as the first-ever U.S.-Africa Ministerial Meeting, held in March 1999 for representatives from 50 countries, have resulted in major new programs. The main achievement of the ministerial, for example, was a unanimous call for the U.S. Congress to pass the stalled African Growth and Opportunity Act that grants improved U.S. market access to African textiles and other products. Yet it took more than a year for the White House and its allies in Congress to overcome pressure from U.S. textile interests (who were opposed to doubling Africa's 0.8 percent of imports) and pass this modest measure. Given the preferences already enjoyed by Asian textile producers, the reluctance to fulfill American promises to help create jobs in the world's poorest region was seen in Africa as another example of U.S. hypocrisy. . . .

Money Where Their Mouths Are

A new approach to conflict prevention and democratic development in Africa will require changes in both the substance and tone of American foreign policy. The first and easiest step will be rhetorical: the next U.S. president should designate conflict prevention as a primary goal of U.S. foreign policy. This should be followed by a concerted effort to win the support of Congress and the public. The president should call for a truly national commitment, one involving business, labor, and a broad range of civil society.

States at risk, such as Nigeria today or South Africa in the waning days of apartheid, benefit from the involvement of human rights groups and nongovernmental organizations. These groups pressure troubled regimes to reform and to resolve factional differences through political means. American businesses can have a similarly powerful impact, showing their support for peaceful change through their investment decisions, and should be encouraged and backed by Washington to do so.

Internally, the U.S. government should take a number of steps to strengthen its approach to conflict prevention. Better interdepartmental communication and cooperation is necessary, especially between the State and Defense Departments. Early warning is rarely the problem. Rather, as in the case of Rwandan genocide, it is the failure to respond in time and with sufficient force.

Current U.S. political and military programs, such as the African Crisis Response Initiative or the International Military Education and Training Program (IMET), are useful but much too small: Africa's biggest IMET program, in South Africa, amounted to only $800,000 [in 1999]. Funding for these programs must be increased to build their capacity for conflict prevention and peace enforcement while helping to ensure that African militaries remain accountable to civilian authorities. Another urgent reason to work with African militaries is to help them cope with AIDS infection rates that run as high as 50 to 60 percent among some forces—a human tragedy that threatens the viability of Africa's militaries.

On an organizational level, senior American bureaucrats and embassy personnel should be allowed greater flexibility and resources to initiate preventive diplomacy, including offering to broker domestic disputes and to provide quick support from democracy-building programs. Also necessary is a long-overdue reform of what is left of U.S. foreign assistance, bringing it into line with a foreign policy dedicated to conflict prevention. In a similar vein, the State Department's public affairs programs and the operations of the United States' foreign broadcasting services must adopt prevention as a central theme, countering local hate radio that, as in Rwanda, is often a precursor of deadly conflict.

Finally, new funds will have to be found under current U.S. budget caps. This year the United States will spend less than one percent of the federal budget on non-defense-related international affairs—about half what the Reagan administration invested in international affairs in the mid-1980s. This country cannot be a good partner, much less a leader, in conflict prevention when it has so little money to spend on it.

Apart from its own efforts, the United States should encourage the World Bank and other international donors to likewise stress conflict prevention in potentially troubled countries. The bank has already made good governance a priority, has embarked on several postconflict reconstruction efforts aimed at avoiding further fighting, and has begun cooperating with U.N. conflict-prevention efforts. But more could be done. International loans should be tied to demands for the protection of human rights, the rule of law, and transparency, while ensuring that they do not exacerbate conflict by favoring one faction over another.

Support for such moves exists in Africa itself. Subregional organizations in western and southern Africa are beginning to address abuses of power in Zimbabwe, Sierra Leone, and elsewhere. Such efforts deserve strong U.S. backing. More broadly, this new attitude was evident in May, at the first meeting of the Ministerial Conference on Security, Stability, Development, and Cooperation.

The conference arose from a 1991 initiative by Obasanjo, Nigeria's current president, when he led a group of prominent Africans to adopt a set of human rights and good governance provisions modeled on those of the Conference on Security and Cooperation in Europe [CSCE]. Whether the fledgling body will match the CSCE's achievements, however, remains to be seen.

William Reno

The Failure of Peacekeeping in Sierra Leone

[In May 2000,] Revolutionary United Front (RUF) fighters detained and disarmed a Zambian battalion of the United Nations Mission in Sierra Leone (UNAMSIL) that had been sent to break an RUF siege of Kenyan UN peacekeepers in the town of Makeni. This incident effectively ended a peace agreement between the government of Sierra Leone and the RUF that had been signed in July 1999 after more than eight years of war.

Fighting has continued since. UN Secretary General Kofi Annan recommended in August 2000 an increase in the UN mission's strength from 7,500 to 20,500 troops. Britain, Sierra Leone's former colonial power, unilaterally sent warships and a commando battalion to Sierra Leone. Diplomats in other countries began to pressure Charles Taylor, president of neighboring Liberia, for allegedly aiding RUF forces.

The breakdown of the peace agreement illustrates the difficulties facing conflict resolution when state institutions have collapsed after decades of corrupt misrule. And the RUF exemplifies the kind of insurgency that can develop that neither mobilizes mass followings nor attempts to administer "liberated zones" under the guidance of new political ideas that are an alternative to the corrupt government the insurgency fights. This development poses a significant dilemma for conventional approaches to conflict resolution that owe much more to experiences with classic civil wars in which there are clear ideological or programmatic opponents and in which negotiated settlements to share control of state institutions are stressed. Sierra Leone shows the failure of conventional diplomatic strategies while providing hints of a more radical approach by the British government that draws from British imperial experience in ruling stateless societies and carrying out counterinsurgency efforts.

Insurgency and State Collapse

Sierra Leone's war began in March 1991 when a small force of RUF fighters led by Foday Sankoh, a former Sierra Leone army corporal, crossed from Liberia

From William Reno, "The Failure of Peacekeeping in Sierra Leone," *Current History* (May 2001). Copyright © 2001 by Current History, Inc. Reprinted by permission. Notes omitted.

into Sierra Leone. Initially the RUF was lightly armed, but Sierra Leone's army was small, with fewer than 3,000 soldiers, and was unable to defeat it.

The RUF received backing from Charles Taylor, whose rebel group, the National Patriotic Front of Liberia (NPFL), had invaded Liberia in December 1989 to overthrow President Samuel Doe. Taylor used support for the RUF to expand his influence beyond Liberia; it is also alleged that he personally benefited from the RUF's control over diamonds in eastern Sierra Leone.

Taylor used the RUF in 1991 to weaken the Economic Community of West African States Monitoring Group (ECOMOG), a multilateral West African peace-keeping force that was blocking Taylor's attempt to install himself as Liberia's president. At the same time, Sierra Leone President Joseph Momoh allowed an anti-Taylor coalition of Liberian dissidents to use Sierra Leone as a base, and ECOMOG used Freetown's Lungi International Airport to launch attacks against the NPFL in Liberia.

Junior military officers overthrew Momoh in April 1992. They complained that corrupt senior officers prevented supplies from reaching front lines and that some politicians were secretly collaborating with the RUF for personal gain. The coup leaders installed themselves as the National Provisional Ruling Council (NPRC) under the leadership of Captain Valentine Strasser, who promised that the army would defeat the RUF.

By 1994 the NPRC had increased the army's strength to 14,000 soldiers, recruiting unemployed youth and members of armed gangs associated with politicians. But the strengthened army could not end the war. A former NPRC minister identified the cause: "There developed an extraordinary identity of interests between [the] NPRC and [the] RUF. This was responsible for the *sobel* phenomenon, i.e., government soldiers by day became rebels by night."

Unable to fight the RUF effectively, in 1995 the NPRC hired Executive Outcomes (EO), a South African mercenary firm. The NPRC also armed progovernment militias, known as Kamajors, which were later organized as the Civil Defense Force (CDF). The South African firm secured most of the country's towns and alluvial diamond-mining areas, which the RUF had captured in late 1994 and early 1995.

EO's intervention, while militarily effective, created serious long-term political complications. Greater security gave societal groups the chance to pressure the NPRC to hold multiparty elections promised by Momoh before he was overthrown. International organizations and foreign diplomats also pressured NPRC officials to hold elections in February 1996. But elements within the army feared that the CDF would replace them. This encouraged some soldiers to collaborate more closely with the RUF. In addition, EO's offensive against the RUF forced fighters to force closer ties with Charles Taylor for aid.

The March 1996 elections brought Ahmed Tejan Kabbah to power. President Kabbah inherited these political dilemmas. Unsure of army loyalty (and the target of at least two coup attempts between March 1996 and May 1997), Kabbah relied on CDF fighters for his security and to battle the RUF. Government reliance on CDF forces intensified when EO left Sierra Leone in January 1997 after disputes over payment of EO's fees. The company's departure and the government's response aggravated tensions between the army and Kabbah. On May 25, 1997 the army overthrew Kabbah's government. The RUF welcomed the coup and formed an alliance, the Armed Forces Revolutionary Council (AFRC), with the army's Major Johnny Paul Koroma at its head. Koroma named RUF leader Sankoh his deputy (even though Sankoh had been detained in Nigeria in March 1997, where he had traveled to discuss the peace process) and appointed RUE members to cabinet positions.

Most West African governments refused to recognize the AFRC, pitting ECOMOG (about 4,000 soldiers, most of whom were Nigerian and who remained in western parts of Freetown and at the international airport) against the AFRC regime. Nigeria and Guinea boosted ECOMOG's troop levels to about 10,000 men and finally reinstalled Kabbah in February 1998. ECOMOG became even more central to the Kabbah government's security after the restoration, underscored by Kabbah's appointment in mid-April 1998 of Nigerian Brigadier Maxwell Khobe as chief of national security.

Human rights abuses, already considerable before the 1997 coup, intensified. While in power the RUF and the AFRC singled out suspected Kabbah supporters and men of voting and fighting age for mutilation, including cutting off fingers, hands, feet, and arms. Victims also included women and children. Eyewitnesses and victims reported that amputees often were instructed to deliver messages that the AFRC and RUF would resist international pressure to relinquish power and that they wanted Sankoh released from Nigerian custody. Mutilations by the RUF and AFRC continued after Kabbah's return, reportedly to discourage citizens from giving political or military help to the restored government.

The Lomé Agreement

Kabbah again negotiated with the RUF when it became clear that ECOMOG troops would leave Sierra Leone following elections in Nigeria in 1999. This decision left Sierra Leone's government more dependent on a small UN mission in Sierra Leone for future diplomatic and military protection. The mission initially consisted of 70 military observers, but its importance lay in bringing some international attention to the issue of Liberian support for the RUF since part of the mission's mandate was to help enforce an embargo on arms supplies to Sierra Leone. This did not solve the government's security problem, however. The RUF forced ECOMOG from the Kono diamond fields in December 1998. Elements of the AFRC regime and the RUF even occupied Freetown for a few days in January 1999 and forced Kabbah to flee once again until an ECOMOG counteroffensive dislodged the invaders.

Nigeria's new civilian government recognized that it lacked resources to defeat the RUF, especially while states in the region covertly supported it. Its pragmatic solution was to hold talks. Negotiations took place under the auspices of the Economic Community of West African States, and included Liberia's and Burkina Faso's presidents, both of whom had supported the RUF. Sankoh was released from Nigerian custody on April 18, 1999 and flown to Lomé, the capital of Togo, to begin negotiations with Kabbah's government.

On July 7 the Sierra Leone government and the RUF signed the Lomé peace agreement. In October 1999 the UN expanded its observer mission to 6,000 troops to assist in implementing the agreement (2,000 Nigerian troops were incorporated into this enlarged UN force in November 1999). The agreement required the disarming of all warring groups, appointment of RUF members to cabinet positions, conversion of the RUF into a political party, and Sankoh's installation as vice president and director of the Commission for the Management of Strategic Resources, National Reconstruction, and Development, which was to regulate the diamond-mining industry. Most controversial to many Sierra Leoneans was the provision in the agreement that Sankoh be granted a pardon so that he could make his conversion from a war leader to a civilian politician. Sankoh himself mistrusted the Sierra Leone government, and did not return to Freetown until October 1999, when the UN established its expanded peacekeeping mission.

Recognizing that the RUF could prevail militarily, United States officials allegedly pressured Kabbah to negotiate with the RUF. Sierra Leone would also become a test case for how much diplomatic and military support non-African states would provide in conflicts involving so-called collapsed states.

Explanations of the United States role vary. One interpretation holds that the [Bill] Clinton administration valued the appearance of order, regardless of who ruled Sierra Leone. A United States–sponsored peace agreement would thus allow it to appear that the United States had addressed a crisis in Africa while ensuring that it would not have to make a commitment of resources and soldiers to an African problem. This would explain why the US government was interested in Sierra Leone in the first place, since the country is of no strategic interest to the United States. A more optimistic interpretation explaining American interest held that officials believed that the RUF and Kabbah's government would abide by the terms of the Lomé agreement, and that the RUF would evolve into something more like a real political party as its leaders discovered that success in a democratic coalition government required gaining public support and confidence.

The Freetown newspapers stressed American coercion, and even claimed that Kabbah had been "kidnapped" by the Reverend Jesse Jackson (who acted as a facilitator in negotiations) and other American officials and forced to sign the Lomé agreement. Some press commentators said that the agreement was largely written by officials of the United States Agency for International Development. Another interpretation, among many analysts and journalists, focuses on Liberian President Charles Taylor's relationship with the RUF and asserts that the RUF and Taylor have exploited the collapse of Sierra Leone as a state. They manipulate international agencies into believing that the RUF will transform itself into a political party. Their objective, however, is to solidify control over diamond-mining areas for their own profit and to increase Taylor's regional political influence.

Some UN and foreign officials tolerate this deception because they recognize that financial and military backing for large-scale peace enforcement is unlikely. But conflict in collapsed states is fundamentally different from wars between ideological rivals who mobilize mass followings and build "liberated zones" to practice their ideas of governance. Instead, rebel groups use the willingness of outsiders to recognize them as potential government rulers as an opportunity to acquire recognition of sovereignty in peace agreements, which they use as a cover to continue their predatory acquisition of wealth and to shield their transactions with international business partners.

Many Sierra Leoneans regarded positions taken by UN and foreign diplomats who stressed reconciliation as offensive. They pointed out that United States diplomats branded RUF leader Sankoh as a violator of human rights and had made statements that stressed rebel violations of international law. They noted that a year earlier the UN Security Council condemned "as gross violations of international humanitarian law the recent atrocities carried out against the civilian population . . . of Sierra Leone by members of the Revolutionary United Front and the deposed junta." The Lomé agreement included a provision for a Truth and Reconciliation Commission (which has never been implemented). This provision highlighted inconsistencies in international approaches to Sierra Leone's war, since it marked a shift from foreign backing for the Sierra Leone government to fight and to prosecute heads of an insurgency noted for human rights abuses, then recognized those same individuals as legitimate political leaders and included them in a coalition government to rule the country.

The Failure of Lomé

The Lomé agreement did not bring an end to war in Sierra Leone and may have contributed to its continuation. Lomé's primary shortcoming was the inability of UNAMSIL peacekeepers to enforce the terms of the agreement in the face of RUF noncompliance. UNAMSIL soldiers began to surrender their weapons to RUF fighters soon after the first contingent, a Kenyan unit, arrived on November 29, 1999. Guinea's UNAMSIL contingent was forced to turn over approximately 500 AK-47 rifles, other weapons, and several tons of ammunition in January 2000. Kenyan units were relieved of their weapons on two occasions in January 2000. Later incidents also involved weapons taken by members of the former Sierra Leone army. The largest loss of weapons occurred in conjunction with RUF attacks on UNAMSIL units, beginning in May 2000, as the last Nigerian ECOMOG troops left Sierra Leone. The RUF also took weapons from a remaining Nigerian contingent attached to UNAMSIL, and a loss occurred when the Zambian battalion sent to relieve the Kenyans under siege was detained by RUF forces.

The May 2000 crisis highlighted the role that Liberian President Taylor played in providing weapons to the RUF. This reflected a longer-term relationship in which Taylor served as a commercial channel for RUF-supplied diamonds mined in Sierra Leone. RUF attacks on and the detention of UNAMSIL peacekeepers raised the level of diplomatic attention to this connection. The resultant diplomatic pressure on Taylor, and earlier decreases in financial aid to Liberia from abroad, likely increased Taylor's reliance on this source of income, estimated at upward of $125 million annually from Sierra Leone. Taylor's failure to attract large foreign investors to the country's mining industry also likely increased his reliance on deals with the RUF to gain access to income and re-sources. As part of the bargain, Taylor harbored RUF commander Sam "Maskita" Bockarie in Liberia, where Bockarie allegedly recruited fighters for the RUF.

The RUF's relationship with Taylor underscored the importance that con-trol over diamonds plays in the RUF's overall strategy. Assistance from its Liber-ian patron is tied to the RUF's occupation of diamond-mining areas. The long-term political, implications of this reliance have been considerable for the RUF. Instead of attracting and mobilizing a popular following in Sierra Leone to overthrow the country's corrupt and inept government, RUF commanders have fought the government with guns bought with diamonds, brought from Liberia, or captured from their enemies. They do not have to rely on the good-will of local inhabitants or the contributions of their energies and wealth, and they do not have to engage in the arduous political and organizational task of building a mass movement to fight their way to power. The RUF bases its politi-cal power on control over diamonds, much as had the corrupt Sierra Leone politicians that the RUF criticized.

The RUF's failure to build grassroots support increased its reliance on dia-monds and assistance from Liberia. This meant that the RUF could never realis-tically satisfy the key provision of the Lomé agreement that it allow UNAMSIL peacekeepers to control diamond-mining areas, nor could it disarm without losing its primary basis of power. The May 2000 crisis occurred in the wake of UNAMSIL attempts to unilaterally occupy areas in the Kono and Kambia dis-tricts outside the limited territory under government control. And the disarma-ment figures themselves reflected RUF noncompliance with the terms of the Lomé agreement. By the start of hostilities on May 2, the UN reported that 24,042 former combatants (out of a rough total of 45,000) had been disarmed, but had surrendered only 10,840 weapons. (The Freetown press, suspicious of these figures, echoed widespread popular suspicions that the RUF manipulated the disarmament process to its own advantage and intimidated UNAMSIL offi-cials, and predicted that violent confrontation with UNAMSIL would occur.) Some fighters reported unilaterally to collect $300 bounties. UN officials noted that, although all factions tried to prevent their fighters from disarming and punished those who did, the RUF was the most systematic and violent. . . .

The Return to War

The RUF's attack on and seizure of 500 UNAMSIL peacekeepers in May 2000 marked the end of the Lomé agreement. RUF leader Foday Sankoh reiterated in a letter to foreign and Sierra Leone officials that the RUF had not received all the state offices provided under the conditions of the Lomé agreement. Critics specu-lated that RUF officials hoped to use these posts to sell diamonds overseas under cover of diplomatic immunity. A more apparent interest may have been an ex-pansion of participation in government as much as possible to buffer interna-tional qualms about the grant of amnesty in the Lomé agreement. This reflected Sankoh's deep suspicion about the motives of the Sierra Leone government and fears that the amnesty agreement would not protect him, given the government's reluctance to provide Sankoh with a written protection from prosecution.

Sankoh himself went into hiding on May 8 after a crowd of several thousand people attacked his Freetown residence, and was shot, seized, and paraded naked through Freetown by progovernment troops 10 days later. He remains in the custody of the Sierra Leone government and could be prosecuted under the terms of a proposed international war crimes tribunal (which remains unimplemented in early 2001). Plans for the tribunal reversed the Lomé agreement's amnesty and left open the possibility that defendants such as Sankoh who would have been exempted under the terms of the Lomé agreement would be prosecuted. This has probably reinforced RUF perceptions that the Sierra Leone government and its backers will jettison bargains when they find the resources and political will to fight, and seek agreements when resources and will are lacking.

With the breakdown of Lomé, Britain added about 650 military personnel to the 15 British military observers assigned to UNAMSIL. Operating under British command, these paratroopers and marines were operationally separate from the UNAMSIL force and aided the Sierra Leone army and progovernment militias in defending Freetown against RUF fighters. The British have also begun a training program for the new Sierra Leone army. By early 2001 an estimated 4,500 Sierra Leone soldiers had completed the program.

Meanwhile, UN Secretary General Kofi Annan's proposal to increase UNAMSIL's strength to 20,500 became wedged inside a bitter dispute within UNAMSIL. In mid-October the Indian and Jordanian UNAMSIL contingents, together numbering almost 5,000 troops, signaled their intentions to depart by February 2001 (in mid-December 2000, even before Indian and Jordanian withdrawals, UNAMSIL had 12,455 soldiers). India's decision followed the leaking of a document written by UNAMSIL's Indian commander, Major General Vijay Jetley, in which he charged that the "Nigerian Army was interested in staying in Sierra Leone due to the massive benefits [it was] getting from the illegal diamond mining" through arrangements with the RUF. He also charged that former ECOMOG commander Brigadier General Maxwell Khobe had accepted $10 million from the RUF to permit mining activities without interference.

Waging Peace

The February departure of the Indian contingent highlighted a key limitation of multilateral peacekeeping in contexts such as Sierra Leone: the inability of the UN bureaucracy in New York and UNAMSIL commander in Sierra Leone to use military force to preempt RUF attacks or to launch operations against the RUF once UN soldiers had been kidnapped. Jordanian officials announced that their contingent too would depart, reflecting similar concerns that peacekeepers were vulnerable in a context like Sierra Leone where armed groups continued to fight each other and to target foreign soldiers. The peacekeeping mandate of UNAMSIL and the militaries that contributed soldiers to UNAMSIL did not envision combat of this sort, and UNAMSIL personnel were not equipped or given logistical support to engage in sustained combat.

The British approach involved considerably greater use of violence against antigovernment forces than UNAMSIL was able to marshal militarily or diplomatically. British forces attacked West Side Boys groups in Okra Hills on August 30 and September 10 to rescue the remaining 6 of 11 British personnel who had been kidnapped in late August by the former soldiers.[1] On November 13 British marines staged military exercises around Freetown; these followed the signing three days earlier of a UNAMSIL-brokered one-month cease-fire between the RUF and Sierra Leone government officials that was to allow UNAMSIL to travel throughout Sierra Leone. The military exercise was to "remind the leadership of RUF of the need to honor that agreement," the British commander said. But the RUF did not let UNAMSIL enter areas it held (except for occasional visits of small groups), and the result was to create tensions between the British forces and UNAMSIL.

As of March 2001, UNAMSIL peacekeepers were still not deployed in RUF-held areas. This continued to generate tension between UN and British military officers. "It is as if the UN leadership has learned nothing from previous experiences," said a British officer in reference to UNAMSIL unwillingness to deploy. A senior UN officer replied that "if the British want war, they can have it and we will leave."

UNAMSIL's presence, like British support for the Sierra Leone government, helps multiply rebel factions. The RUF has failed to disarm and continues its attacks on UNAMSIL, an intransigence that Sankoh had backed with his words. This, along with international pressure that he be tried for crimes against humanity, has disqualified him from future negotiations on behalf of the RUF.

UN officials sought a new interlocutor, despite the continuing loyalty of many RUF fighters to Sankoh. Issa Sesay emerged as the RUF's putative new head as UN officials signaled that they would talk to him. The spokesman for one RUF faction, Gibril Massaquoi, stated that "90 percent were taking orders from Maurice Kallon," a commander in northern Sierra Leone loyal to Sankoh. Yet an RUF commander stated that "General Issa has betrayed them [the fighters] and they now have nothing to do with him as they will continue the struggle." Fighting later broke out among RUF factions, drawing in rebels loyal to Sam Bockarie, the RUF commander from eastern Sierra Leone with close ties to Liberian President Charles Taylor.

Taylor allegedly maintains close connections with the Bockarie and Kallon factions to pursue an offensive against Guinea, signaling a major regional expansion of this war, even after Bockarie left Liberia in early 2001. Regional disorder would keep Taylor's fighters busy and less likely to challenge him, and would give him and his associates more access to commercial opportunities connected with providing weapons and exploiting local resources. War would also destabilize Taylor's neighbors and allow him to capitalize on internal political divisions besetting the political establishments of neighboring rulers toward whom Taylor harbors personal animosity.

Beginning in October 2000, RUF attacks into Guinea intensified. RUF strategy again apparently focused on forcing the UN and the Sierra Leone government to negotiate with the RUF, but with the RUF left in control of significant territory. In the event of a postagreement election, the RUF would be in a position to intimidate citizens into voting for it, much as Taylor had done in Liberia in 1997. The international community could then consider Sierra Leone "stable."

The Future of Intervention

In trying to bring an end to conflict in Sierra Leone, both the UN and British forces have found it difficult to respond to and influence autonomous militias, whether pro- or antigovernment. These groups shift allegiances and may simultaneously fight against and profess alliance to the same organization. As British officers discovered after the August kidnapping of British personnel by the West Side Boys, when force is used against multiplying decentralized opponents there is no "army" that can surrender. Military victory against irregular forces requires physical occupation and administration. In addition, the defeat of multiple factions of the RUF, the CDF, and the West Side Boys by military means alone would also require attacks on the families and homes of fighters and the use of force at levels that are prohibited by the conventions of warfare and international agreements.

The British solution to this dilemma involves a lengthy commitment. Jonathan Riley, the British force commander in Sierra Leone, has said that "We will leave when the war is either won or resolved on favorable terms." Heir to the institutional legacy of British rule of the hinterland of Sierra Leone from 1898 to 1961, British Prime Minister Tony Blair's administration appears to have clearer ideas than the UN or the United States about political and military strategy in what has essentially become a stateless society. Indeed, former colonial officers have participated in government discussions concerning British strategies in Sierra Leone. And a former colonial district commissioner returned to Sierra Leone to engage in chieftaincy consultations to gain an understanding of the multiple grievances that lead members of communities to take up arm. The effort was also designed to build support for Kabbah's regime among local notables by showing that the government could intervene in local conflicts to their benefit.

This contrasts with the more bureaucratic approaches of the UN and United States, which are seriously out of sync with the reality of conflict in collapsed states. The Americans especially tend to search for general solutions to disorder (to the extent of trying to create computer models to predict conflicts). Some local observers complain that international agencies such as the UN Commissioner for Refugees, which draw attention to refugees in Guinea where the RUF and other groups battle one another, offer rebel groups the opportunity to use organized refugee movements as human shields to shift fighters and loot supplies.

The UN approach of engaging factions in cease-fires and peace negotiations reflects explicit recognition of the limits to the use of force. This strategy recognizes that UN peacekeepers are constrained in the use of force against local groups and that officers of foreign military contingents or their governments are unwilling to commit their troops to combat. This limited mandate constrains UNAMSIL's use of intelligence and analysis—much to the annoyance of British military officers and many Sierra Leoneans. . . .

Both approaches face serious constraints. It is uncertain whether a post-Blair administration will possess the political will to remain engaged in Sierra Leone for many years. It is not clear if British voters will countenance a long engagement. Yet the UN's preference for negotiations tends to accredit armed groups that prey on society, leading as in Liberia to the installation of a predatory warlord as head of state. This approach creates the high probability of a Sierra Leone left in the control of groups known for grave human rights abuses. Departure on these terms would humiliate UN officials and seriously undermine the credibility of future peacekeeping missions. This contradiction is likely to remain, since it does not appear that the RUF can be beaten on the battlefield. Yet negotiating with the RUF when the United States and other Western powers insist that rulers such as Yugoslav President Slobodan Milosevic face a war crimes tribunal leaves the appearance that Sierra Leone suffers from a double standard in the global application of human rights principles.

Note

1. The kidnapping episode also exposed a factional split among the former soldiers. The kidnappers were loyal to Foday Kallay, who claimed to lead the former Sierra Leone army after Johnny Paul Koroma left to take a government position as head of a Commission for Consolidation of Peace in Freetown.

POSTSCRIPT

Are International Peacekeeping Missions Critical to Resolving Ethnic Conflicts in African Countries?

Neither of these authors, not even Reno, is strictly against peacekeeping missions in Africa. The problem is that given the general marginality of Africa to U.S. security interests, peacekeeping initiatives may always be modestly funded and staffed. If U.S. efforts are so meager, then Reno questions their effectiveness. Stremlau obviously would like to see a greater investment in African peacekeeping efforts, but he might also argue that prevention is less costly.

Since the writing of the selections in this issue, the geography of the African conflict has changed slightly, but many of the questions remain the same. In early 2002 the decade-long civil war in Sierra Leone was declared over by the government and rebel leaders. This conflict resulted in the deaths of tens of thousands of people and the displacement of more than two million civilians (roughly a third of the country). The British government in particular has committed to help the country rebuild. Foday Sankoh, the rebel leader of the United Revolutionary Front, has been indicted for war crimes. He pleaded not guilty at his initial hearing at a special United Nations–run tribunal.

Liberian president Charles Taylor also has been indicted for war crimes in Sierra Leone. As civil war rages on in Liberia as of this writing, some are concerned that Taylor's indictment will prevent him from stepping down, something that President George W. Bush has called for as a precondition for U.S. involvement in the crisis. Others have suggested that it would set a terrible precedent to waive the war crime charges in order to facilitate a quicker resolution of the problems in Liberia. Taylor has been offered safe haven in Nigeria, but it is unclear, despite a statement that he would step down, that he will actually take up the offer. Given its historical ties to Liberia, a country created in the nineteenth century with U.S. support as a destination for freed American slaves, many have asserted that the United States has a responsibility to intervene militarily to restore peace in this country. While President Bush has pledged support to help resolve the Liberian conflict, he has declined to commit American troops.

For those interested in further reading on conflict, another important circumstance to understand is the war in the Democratic Republic of the Congo, a situation that some have referred to as Africa's first world war given the number of countries involved. A good article on the Congo war is one by Marina Ottaway in the May 1999 issue of *Current History* entitled "Post-Imperial Africa at War." Two good books on U.S. intervention in Somalia include *The Road to Hell* by Michael Maren (The Free Press, 1977) and *Deliver Us From Evil* by William Shawcross (Touchstone, 2001). A thoughtful article on the Rwandan genocide and the failure of outside powers to intervene is by Christopher Clapham in the March 1998 issue of the *Journal of Peace Research* entitled "Rwanda: The Perils of Peacemaking."

Contributors to This Volume

EDITOR

WILLIAM G. MOSELEY is an assistant professor of geography and coordinator of the Interdepartmental Program in African Studies at Macalester College in Saint Paul, Minnesota, where he teaches courses on Africa, environment, and development. He received a B.A. in history from Carleton College, an M.S. in environmental policy and an M.P.P in international public policy from the University of Michigan, and a Ph.D. in geography from the University of Georgia. He has also worked for the U.S. Peace Corps, the Save the Children Fund (United Kingdom), the U.S. Agency for International Development, the World Bank Environment Department, and the U.S. State Department. His research and work experiences have led to extended stays in Mali, Zimbabwe, Malawi, Niger, and Lesotho. He is the author of over 17 peer-reviewed articles and book chapters that have appeared in such outlets as *Ecological Economics*, the *Geographical Journal*, *Applied Geography*, and *Geoforum*. He has also penned editorials that have appeared in the *Christian Science Monitor* and the *Chicago Sun Times* and is coeditor of *African Environment and Development: Rhetoric, Programs, Realities* (Ashgate, 2003).

STAFF

AUTHORS

JOEL D. BARKAN is a professor of political science at the University of Iowa and senior consultant on governance at the World Bank. His research interests include democratization, macroeconomic reform in developing countries, and electoral processes.

THOMAS J. BASSETT is a professor of geography and affiliate of the Center for African Studies at the University of Illinois, Urbana-Champaign. His research interests include Third World development, African agrarian systems, political ecology, and the history of cartography. He is the author of *The Peasant Cotton Revolution in West Africa: Côte d'Ivoire, 1880–1995* (Cambridge University Press, 2001) and coeditor, with Donald Crummey, of *African Savannas: Global Narratives and Local Knowledge of Environmental Change in Africa* (James Curry and Heinemann, 2003), as well as the author of numerous journal articles and book chapters that have appeared in such publications as the *Annals of the Association of American Geographers, Africa,* and the *Review of African Political Economy*.

MICHAEL BRATTON is a professor of political science at Michigan State University. His research interests include comparative politics, public administration, and African politics. He is the author or coauthor of *The Local Politics of Rural Development: Peasant and Party-State in Zambia* (University Press of New England, 1980); *Governance and Politics in Africa* (Lynne Rienner Press, 1992); and *Democratic Experiments in Africa: Regime Transitions in Comparative Perspective* (Cambridge University Press, 1997).

GRACE BUNYI is a researcher at Kenyatta University in Nairobi, Kenya.

MIRIAM S. CHAIKEN is a professor of anthropology at Indiana University of Pennsylvania. Her research interests include colonialism, medical anthropology, applied anthropology, and rural development. She is currently focused on research concerning contemporary nutritional patterns and food security. She has had articles appear in such journals as *Human Ecology: An Interdisciplinary Journal*.

KEVIN M. CLEAVER is director of agriculture and rural development of the World Bank and heads the World Bank Board of Rural Sector Managers. His interests include environmental issues, agricultural policy and adjustment, forestry, and natural resource management. His previous publications include *A Strategy to Develop Agriculture in Sub-Saharan Africa and a Focus for the World Bank* (World Bank Publications, 1993); *An Agricultural Growth and Rural Environment Strategy for the Coastal and Central African Francophone Countries* (1992); and *Conservation of West and Coastal African Rainforests* (World Bank Publications, 1992).

MARCUS COLCHESTER is the director of the Forest Peoples Programme of the World Rainforest Movement. His primary work has involved securing the rights to land and the livelihood of indigenous peoples. He has been a fellow in the Pew Fellows Program in Conservation and the Environment, an associate editor for *The Ecologist* magazine, and honorary advisor on development in the Amazon to the Venezuelan government.

W. THOMAS CONELLY is a professor of anthropology at Indiana University of Pennsylvania. His main research interests have been ecological anthropology, agricultural systems, hunter-gatherers, rural development, and migration. He performed his dissertation research on frontier settlement and agricultural intensification on Palawan Island in the Philippines. He also has experience working as an applied anthropologist doing agricultural development research and consulting in Kenya and Indonesia. Currently, he is researching migration, agricultural history and rural development in Pennsylvania. He has been published in journals such as *Human Ecology: An Interdisciplinary Journal.*

LIZ CREEL is a population specialist and senior policy analyst at the Population Reference Bureau.

SUNDAY DARE is a Nigerian journalist. Dare's career has focused mainly on investigative reporting of political corruption, military dictatorship, and human rights violation. He is a member of the Washington-based International Consortium of Investigative Journalists, an organization composed of many leading investigative journalists from around the world.

PARTHA S. DASGUPTA is the Frank Ramsey Professor of Economics at the University of Cambridge and a fellow of St. John's College. His research interests include the economics of poverty and nutrition, environmental economics, economic measurement, and the economics of knowledge. He has authored several books, including *The Economics of Transnational Commons* (Claredon Press, 1997); *Creation and Transfer of Knowledge* (Springer, 1998); and *Social Capital: A Multifaceted Perspective* (World Bank Publications, 2000). He has many published articles published in journals such as *Scientific American, Economic Journal,* and *Ecological Economics.*

MACLEANS A. GEO-JAJA is an associate professor of economics and education at Brigham Young University. His current research concentrates on the relationship between the economy and the educational system, including human development, human capital formation, and educational reform in Africa. The primary focus of his research is evaluating rapid educational reform and its implication on human capital formation, nation building, and sustainable development. He has written articles for journals such as *International Review of Education, Education and Society Journal,* and the *Journal of Black Studies.*

ARTHUR A. GOLDSMITH is professor of management at the University of Massachusetts in Boston. He is the author of *Building Agricultural Institutions: Transferring the Land-Grant Model to India and Nigeria* (Westview Press, 1990).

ROBIN M. GRIER is an assistant professor of economics and area coordinator for Latin American studies at the University of Oklahoma. She has published numerous articles in journals such as *Economic Inquiry* and *Public*

Choice. Her areas of specialization include international finance, development, and Latin American economics.

BRIAN HALWEIL is a senior researcher for the Worldwatch Institute. He has been published in the *Christian Science Monitor,* the *New York Times,* and the *Los Angeles Times.* He is a John Gardner Public Service Fellow from Stanford University.

JULIE HEARN is a lecturer in development studies at the School of Oriental and African Studies (SOAS), University of London. She currently is working on aid and civil society in Africa. She has written articles for journals such as the *Review of African Political Economy.*

JOHN L. HOUGH is the global environment facility coordinator for biodiversity and international waters for the United Nations Development Programme—Global Environment Facility, Africa Bureau. He has worked on African conservation projects and programs for more than 20 years.

HUMAN RIGHTS WATCH is a nonprofit organization supported by contributions from private individuals and foundations worldwide. The organization is the largest of its kind based in the United States. Human Rights Watch researchers conduct fact-finding investigations into human rights abuses in all regions of the world. They then publish these findings in dozens of books and reports every year. The aim is to generate extensive coverage in local and international media that will help to embarrass abusive governments in the eyes of their citizens and the world.

AKIN JIMOH is a Knight Science Journalism Fellow at the Massachusetts Institute of Technology and program director of Development Communications, a non-governmental organization (NGO) based in Lagos, Nigeria. He holds masters degrees in both medical physiology and public health, and he has worked on development and HIV/AIDS issues for over 10 years.

GAVIN KITCHING is a professor of political science at the University of New South Wales in Sydney, Australia. He is the author of the 1980 award-winning *Class and Economic Change in Kenya: The Making of an African Petite Bourgeoisie 1905–1970* (Yale University Press).

MARÍA JULIÁ is a professor of social work at Ohio State University. She has published in journals such as the *Journal of International Development* and *International Social Work.* She is also the author of *Constructing Gender: Multicultural Perspectives in Working With Women* (Brooks/Cole, 2000) and *Multicultural Awareness in the Health Care Professions* (Pearson, Allyn & Bacon, 1996).

EZEKIEL KALIPENI is an associate professor of geography at the University of Illinois, Urbana-Champaign. His research interests include medical geography, population studies, environmental issues, and Africa. He is the coeditor of *AIDS, Health Care Systems, and Culture in Sub-Saharan Africa: Rethinking and Re-Appraisal* (Michigan State University Press, 1996). He has written numerous book chapters and published in such journals as *Geographical Review, African Rural and Urban Studies,* and *Social Science and Medicine.*

BERNARD LOGAN is a professor of geography at the University of Georgia. His specialties include Africa, development, and human-environment interactions. He has published articles in *Economic Geography, Geoforum, Applied Geography, Canadian Journal of African Studies* and the *Journal of Asian and African Studies* among others. He also is the editor of *Globalization, the Third World State, and Poverty Alleviation in the Twenty-First Century* (Ashgate 2002).

DOROTHY LOGIE is a general practitioner and active member of Medact, a health-professionals organization challenging barriers to health.

OLIVER MAPONGA is chair of the Institute of Mining Research at the University of Zimbabwe. His recent research interests have been in small-scale mining, mining investment regulations, regional mineral economics, and environmental management in the minerals industry.

GARTH MANGUM is a professor emeritus of economics at the University of Utah. His research interests include labor-management relations, anti-poverty programs, economics of education, and job training. He has authored such books as *On Being Poor in Utah* (University of Utah Press, 1998) and *Vocational Rehabilitation and Feder* (University of Michigan Press, 1967). He is also the coauthor of *The Persistence of Poverty in the United States* (Johns Hopkins University Press, 2003) and *Programs in Aid of the Poor* (Johns Hopkins University Press, 1998), as well as many others.

ROBERT MATTES is an associate professor of political studies and director of the Democracy in Africa Research Unit in the Centre for Social Science Research at the University of Cape Town. He is also cofounder and codirector of Afrobarometer (a survey of Africans' attitudes toward issues including democracy and markets) and an associate with the Institute for Democracy in South Africa. His research interests include the development of democracy/democratic political culture in Africa and the impact of race and identity on politics in South Africa. He is the author of *The Election Book: Judgment and Choice in the 1994 South African Election* (Idasa, 1996).

PHILIP MAXWELL is Metana Minerals Professor in Mineral Economics and Mine Management at the Western Australian School of Mines at Curtin University of Technology. His recent research interests include the regional economic impacts of mining and mineral commodity markets. He is the coauthor of three textbooks and has authored or coauthored more than seventy articles, book chapters, discussion papers, or monographs.

GILES MOHAN is a lecturer in development studies at the Open University. He has published or copublished over 10 articles in journals such as the *Review of African Political Economy* and *Political Geography*. He is a board member of Village Aid, a United Kingdom–based NGO working in West Africa, and he is on the editorial boards of the *Review of African Political Economy* and the *International Development Planning Review*.

MICHAEL MORTIMORE is a geographer who taught at Nigerian universities between 1962 and 1986 and subsequently was a research associate at Cambridge University and the Overseas Development Institute. He is currently with Drylands Research. He has performed research and published numerous books on the topic of environmental management by smallholders in the dry lands of Africa. He is the author of *Roots in the African Dust: Sustaining the Dry Lands* (Cambridge University Press, 1998) and the coauthor of *Working the Sahel: Environment & Society in Northern Nigeria* (Routledge, 1999) and *More People, Less Erosion: Environmental Recovery in Kenya* (J. Wiley, 1994).

SIDDHARTHA MUKHERJEE is a resident in internal medicine at Massachusetts General Hospital and a clinical fellow in medicine at Harvard Medical School. He has written articles for *The New Republic* and has been a guest on WBUR, Boston's NPR affiliate.

RODERICK P. NEUMANN is an associate professor and director of graduate studies in the Department of International Relations at Florida International University. His interests include social theory, human-environment relations, African studies, and political ecology. His work has been published in *Antipode, Society and Space,* and *Development and Change.* He is also the author of *Imposing Wilderness: Struggles Over Livelihood and Nature Preservation in Africa* (University of California Press, 1998).

WILLIAM D. NEWMARK is research curator at the Utah Museum of Natural History, University of Utah. He is involved in conservation projects in East Africa, and his research has appeared in *Science, Nature,* the *New York Times,* and the *Washington Post.* His research focuses on the patterns of extinction of vertebrate species, conservation, and development.

JOSEPH R. OPPONG is an associate professor of geography at the University of North Texas. His research interests include medical geography, development, and Africa. He is the coeditor of *AIDS, Health Care Systems, and Culture in Sub-Saharan Africa: Rethinking and Re-Appraisal* (Michigan State University Press, 1996). He has written several book chapters and published in numerous journals, including *The Professional Geographer, African Rural and Urban Studies,* and *Social Science and Medicine.*

JEFF POPKE is an assistant professor of geography at East Carolina University. His interests include urban history, political economy, and social theory. He spent two years in Durban, South Africa, where he conducted research on the history of urbanization and racial identity. He has published articles in journals such as *Political Geography* and the *Journal of Geography.*

WILLIAM RENO is an associate professor of political science at Northwestern University. He specializes in African politics and the politics of failing states. He is the author of *Warlord Politics and African States* (Lynne Renner Publishers, 1998).

ROBERT I. ROTBERG is director of the Program on Intrastate Conflict and Conflict Resolution at Harvard University's John F. Kennedy School of Government. He is also president of the World Peace Foundation and has taught political science and history at both Harvard and the Massachusetts Institute of Technology. His research focuses on political and economic issues of developing countries, especially Africa and Southeast Asia. He has authored

The Founder: Cecil Rhodes and the Pursuit of Power (Oxford University Press, 1988) and *The Rise of Nationalism in Central Africa* (Harvard University Press, 1965).

MICHAEL ROWSON is the assistant director of Medact (a health-professionals organization that challenges barriers to health) and coauthor of *Do No Harm: Assessing the Impact of Adjustment Policies on Health* (Zed Books, 2002).

WILLIAM A. RUSHING was a professor of sociology at Vanderbilt University before his death in 2001. He published numerous articles in major journals and authored or edited eight books.

GERALD SCOTT is an associate professor of economics at Florida Atlantic University in Boca Raton. His research interests include debt, development, and structural adjustment in Sub-Saharan Africa.

GÖTZ A. SCHREIBER was the principal economist in the World Bank's West Central Africa Department. More recently he has worked on Central Asian issues for the World Bank. His areas of interest include macroeconomic policy, human resources, agricultural and rural development, and natural resource management.

RICHARD A. SCHROEDER is an associate professor of geography and former chair of African studies at Rutgers University. He is the author of *Shady Practices: Agroforestry and Gender Politics in The Gambia* (University of California Press, 1999) and coeditor of *Producing Nature and Poverty in Africa* (Nordiska Afrikainstitutet, 2000). He has published numerous articles in publications such as the *Annals of the Association of American Geographers, Economic Geography,* and *Africa.*

RICHARD A. SHWEDER is professor of human development at the University of Chicago. His research focuses on ethnopsychology and cultural psychology. He has written or edited nine books, including *Engaging Cultural Differences: The Multicultural Challenge in Liberal Democracies* (Russell Sage Foundation, 2002) and *Thinking Through Cultures: Expeditions in Cultural Psychology* (Harvard University Press, 1991).

ROBERT SNYDER is an associate professor of biology at Greenville College. He has worked with environmental programs in Illinois, consulted for the Environmental Protection Agency, and spent six years in Rwanda engaged in agricultural development related to natural resources. His research interests include food production, sustainable resource use, and international development.

JOHN STREMLAU is a professor and head of the Department of International Relations at the University of Witwatersrand, South Africa. His research focuses on conflict prevention, African international relations, United States foreign policy, and multilateral organizations.

MARY TIFFEN is a historian and socioeconomist at Drylands Research. She is interested in long-term change and development, interdisciplinary research, and social and economic interactions with technology. She authored *The Environmental Impact of the 1991–92 Drought on Zambia: Report* (Iven-World Conservation Union, 1994) and *The Enterprising Peasant: Economic Development in Gombe Emirate, North Eastern State, Nigeria, 1900–1968* (HMSU, 1976). She coauthored *More People, Less Erosion: Environmental Recovery in Kenya* (J. Wiley, 1994).

VÉRONIQUE WAKERLY is a professor of modern languages at the University of Zimbabwe.

FLORENCE WAMBUGU is the CEO of Harvest Biotech Foundation International. She is working to use agricultural science to help farmers and the global community. Wambugu has successfully introduced flower and strawberry varieties into Kenya, strengthening the horticulture industry. She is the author of *Modifying Africa: How Biotechnology Can Benefit the Poor and Hungry, A Case Study From Kenya* (F. M. Wambugu, 2001) and has been published numerous times in various scientific journals.

KOLI Bi ZUÉLI is a teaching and research associate at the Institute of Tropical Geography, University of Cocody in Abidjan, Ivory Coast. His interests include West Africa, development, the environment, and cultural and political ecology.

Index

ideology of community empowerment, 69–71
ill health, contributors to, 76–77
illicit commodity trade, 15
illiteracy, correlation of, with poverty, 265
Index of Economic Freedom, 353
indigenat, 23
Institute for a Democratic South Africa (IDASA), 330–331
institutional dualism, 68
integrated conservation and development project (ICDP), 148–165; analogy to colonial conservation practices, 162; critiques of, 150–152; future directions of, 156–157; limited success of explained, 152–155; problems confronting, 155–156; rationale for the approach of, 149–150
International Center for Insect Physiology and Ecology (ICIPE), 99–101
International Center for Research in Agroforestry (ICRAF), 98, 100
International Development Association (IDA), 132–133, 177
International Finance Corporation, 132–133
International Institute for Environment and Development, 176
International Monetary Fund (IMF), 5, 15, 324; assessment of programs of, 37–40; and corruption, 39; structural adjustment policies and, 47–48

Jimoh, Akin, on whether to provide HIV/AIDS drugs free of charge, 212–215
Jubilee 2000, 82, 85, 326
Julia, Maria, on the effectiveness of NGOs at facilitating development, 54–62

Kalipeni, Ezekiel, on sexual promiscuity and the HIV/AIDS epidemic, 253–259
Kenya: biointensive farming in, 101; Hamisi Division, intensive agriculture in, 107–113; Machakos District, agriculture and increasing population density in, 109, 115–122; need for biotechnology by, 93–94; intensive farming in, 106–107
Kerewan, Gambia, 282–287; control of household cash and male authority in, 286; FAO assistance and, 283; intrahousehold conflicts on the role of women in, 284–285; number of women in agricultural production, 282–283
Kitching, Gavin, on Africa as a lost cause, 4–8
kleptocracy. *See* cleptocracy
kleptocratic leaders, 340

Lagos Plan of Action, 9
land, use of and security of title to, 121
land tenure systems, and conservation, 181
language adoption in Africa, chart, 276

language of wider communication (LWC), 273, 274, 275
languages, indigenous to Africa, 266; inadequacy of, 266–267; and national unity, 269
le politique par le bas, 13
leadership in African nations, factors that produce poor, 350
legacy of colonialism, 6; and development, 20–31
lingua francas, defined, 266
Logan, Bernard I., on "overpopulation" as a major cause of poverty, 231–242
Logie, Dorothy, on developed countries providing debt relief, 76–81
Lomé agreement, 368–370; failure of, 370–371
Luhya, 106–107

Madagascar Environmental Action Plan, 162
magendo, 60
maize, 93: drought escaping, 118; higher yielding, 107–108
Malthus, Thomas, 234
Mangum, Garth, on the effectiveness of structural adjustment policies, 44–50
Maponga, Oliver, on mineral and energy resources as a catalyst for African development, 128–138
"marching desert," 180
market prices, IMF and, 42
matrilineal societies, 247
Mattes, Robert, on multi-party democracy in Africa, 298–309
Maxwell, Philip, on mineral and energy resources as a catalyst for African development, 128–138
Merck, 216
micro-enterprises, women's, 54–62
mineral policy framework, 135–136
mineral resources: controversy over, as a catalyst for development, 128–144; re-emergence of, 132
mineral-dependent economies, in Africa, 128
mining sector reform and economic performance, 134
Mobutu, 12, 13
Mohan, Giles, on the effectiveness of NGOs at facilitating development, 63–72
Moi, Daniel arap, 9, 83, 84, 310, 340, 351; and the IMF, 351
Mortimore, Michael, on food production and population growth, 115–123
Mozambique, 9
Mugabe, Robert Gabriel, 9, 324, 362; report on leadership of, in Zimbabwe, 340–349
Mukherjee, Siddhartha, on whether to provide HIV/AIDS drugs free of charge, 216–217
Multilateral Investment Guarantee Agency (MIGA), 132–133
multi-party democracies, 362
multi-party democratic traditions, controversy